ELECTROCHEMISTRY

MORGAN

ELECTROCHEMISTRY

Principles, Methods, and Applications

CHRISTOPHER M. A. BRETT

and

ANA MARIA OLIVEIRA BRETT

*Departamento de Química,
Universidade de Coimbra,
Portugal*

Oxford   New York   Tokyo

OXFORD UNIVERSITY PRESS

1993

Oxford University Press, Walton Street, Oxford OX2 6DP
Oxford   New York   Toronto
Delhi   Bombay   Calcutta   Madras   Karachi
Kuala Lumpur   Singapore   Hong Kong   Tokyo
Nairobi   Dar es Salaam   Cape Town
Melbourne   Auckland   Madrid
and associated companies in
Berlin   Ibadan

Oxford is a trade mark of Oxford University Press

Published in the United States
by Oxford University Press Inc., New York

A catalogue record for this book is available from the British Library

Library of Congress Cataloging in Publication Data
Brett, Christopher M. A.
Electrochemistry: principles, methods, and applications/
Christopher M. A. Brett and Ana Maria Oliveira Brett.
Includes bibliographical references.
1. Electrochemistry.   I. Brett, Ana Maria Oliveira.   II. Title.
QD553.B74 1993   541.3'7–dc20   92-29087
ISBN 0-19-855389-7 (hbk.)
ISBN 0-19-855388-9 (pbk.)

Typeset by The Universities Press (Belfast) Ltd.
Printed in Great Britain
by Bookcraft (Bath) Ltd.,
Midsomer Norton, Avon

To
Cristina, Michael, and Alexandra

PREFACE

Electrochemistry has undergone significant transformations in the last few decades. It is not now the province of academics interested only in measuring thermodynamic properties of solutions or of industrialists using electrolysis or manufacturing batteries, with a huge gulf between them. It has become clear that these two, apparently distinct subjects, and others, have a common ground and they have grown towards each other, particularly as a result of research into the rates of electrochemical processes. Such an evolution is due to a number of factors, but principally the possibility of carrying out reproducible, dynamic experiments under an ever-increasing variety of conditions with reliable and sensitive instrumentation. This has enabled many studies of a fundamental and applied nature to be carried out.

The reasons for this book are twofold. First to show the all-pervasive and interdisciplinary nature of electrochemistry, and particularly of electrode reactions, through a description of modern electrochemistry. Secondly to show to the student and the non-specialist that this subject is not separated from the rest of chemistry, and how he or she can use it. Unfortunately, these necessities are, in our view, despite efforts over recent years, still very real.

The book has been organized into three parts, after Chapter 1 as general introduction. We have begun at a non-specialized, undergraduate level and progressed through to a relatively specialized level in each topic. Our objective is to transmit the essence of electrochemistry and research therein. It is intended that the chapters should be as independent of one another as possible. The sections are: Chapters 2–6 on the thermodynamics and kinetics of electrode reactions, Chapters 7–12 on experimental strategy and methods, and Chapters 13–17 on applications. Also included are several appendices to explain the mathematical basis in more detail. It is no accident that at least 80 per cent of the book deals with current–voltage relations, and not with equilibrium. The essence of any chemical process is change, and reality reflects this.

We have not filled the text with lots of details which can be found in the references given, and, where appropriate, we make ample reference to recent research literature. This is designed to kindle the enthusiasm and interest of the reader in recent, often exciting, advances in the topics described.

A major preoccupation was with notation, given the traditionally different type of language that electrochemists have used in relation to other branches of chemistry, such as exchange current which measures rate constants, and given differences in usage of symbols between different branches of electrochemistry. Differences in sign conventions are another way of confusing the unwary beginner. We have decided broadly to follow IUPAC recommendations.

Finally some words of thanks to those who have helped and influenced us throughout our life as electrochemists. First to Professor W. J. Albery FRS, who introduced us to the wonders of electrochemistry and to each other. Secondly to our many colleagues and students who, over the years, with their comments and questions, have aided us in deepening our understanding of electrochemistry and seeing it with different eyes. Thirdly to anonymous referees, who made useful comments based on a detailed outline for the book. And last, but not least, to Oxford University Press for its interest in our project and enabling us to bring it to fruition.

Coimbra   C.M.A.B.
May 1992   A.M.O.B.

ACKNOWLEDGEMENTS

Full bibliographical references to all material reproduced are to be found at the ends of the respective chapters.

Figure 3.4 is reprinted with permission from D. C. Grahame, *Chem. Rev.*, 1947, **41,** 441. Copyright 1947 American Chemical Society; Fig 7.1 is reprinted with permission from G. M. Jenkins and K. Kawamura, *Nature*, 1971, **231,** 175. Copyright 1971 Macmillan Magazines Ltd; Fig. 8.2c is reprinted by permission of the publisher, The Electrochemical Society Inc., Fig. 9.10a is reprinted with permission from R. S. Nicholson and I. Shain, *Anal. Chem.*, 1964, **36,** 706. Copyright 1964 American Chemical Society; Fig. 12.3 is reprinted by permission of John Wiley & Sons Inc. from J. D. E. Macintyre, *Advances in electrochemistry and electrochemical engineering*, 1973, Vol. 9, ed. R. H. Müller, p. 122. Copyright © 1973 by John Wiley & Sons, Inc.; Fig. 12.15a is reprinted with permission by VCH Publishers © 1991; Fig. 12.15b is reprinted with permission from R. Yang, K. Naoz, D. F. Evans, W. H. Smyrl and W. A. Hendrickson, *Langmuir,* 1991, **7,** 556. Copyright 1991 American Chemical Society; Fig. 15.9 is reproduced from J. P. Hoare and M. L. LaBoda, *Comprehensive treatise of electrochemistry*, 1981, Vol. 2, ed. J. O'M. Bockris *et al.*, p. 448, by permission of the publisher, Plenum Publishing Corporation; Fig. 16.7 is reproduced by kind permission of the copyright holder, National Association of Corrosion Engineers; Fig. 17.3 is reproduced from S. Ohki, *Comprehensive treatise of electrochemistry*, 1985, Vol. 10, ed. S. Srinivasan *et al.*, p. 94, by permission of the publisher, Plenum Publishing Corporation; Fig. 17.6 is reproduced from R. Pethig, *Modern bioelectrochemistry*, ed. F. Gutmann and H. Keyser, 1986, p. 201, by permission of the publisher, Plenum Publishing Corporation; Fig. 17.7 is reprinted with permission from M. J. Eddowes and H. A. O. Hill, *J. Am. Chem. Soc.*, 1979, **101,** 4461. Copyright 1979 American Chemical Society; Fig. 17.9 is reproduced from M. Tarasevich, *Comprehensive treatise of electrochemistry*, 1985, Vol. 10, ed. S. Srinivasan *et al.*, p. 260, by permission of the publisher, Plenum Publishing Corporation; Fig. 17.11 is reproduced with the kind permission of the Institute of Measurement and Control; Table 2.2 is reproduced by kind permission of Butterworth–Heinemann Ltd; Table 7.1 is reprinted from R. L. McCreery, *Electroanalytical chemistry*, 1991, Vol. 17, ed. A. J. Bard, p. 243, by courtesy of Marcel Dekker Inc.; Table 7.3 is reprinted by permission of John Wiley & Sons Inc. from D. T. Sawyer and J. L. Roberts, *Experimental electrochemistry for chemists*, 1974, Copyright ©

1974 by John Wiley & Sons, Inc.; Tables 9.1 and 9.2 are reprinted with permission from R. S. Nicholson and I. Shain, *Anal. Chem.*, 1964, **36,** 706. Copyright 1964 American Chemical Society; Table 9.3 is reprinted with permission from R. S. Nicholson, *Anal. Chem.*, 1965, **37,** 1351, copyright 1965 American Chemical Society, and from S. P. Perone, *Anal. Chem.*, 1966, **38,** 1158, copyright 1966 American Chemical Society; Table 15.2 is reprinted by permission of the publisher, The Electrochemical Society Inc.; Table 17.1 is reproduced from H. Berg, *Comprehensive treatise of electrochemistry*, 1985, Vol. 10, ed. S. Srinivasan *et al.*, p. 192, by permission of the publisher, Plenum Publishing Corporation; Table 17.2 is reproduced from S. Srinivasan, *Comprehensive treatise of electrochemistry*, 1985, Vol. 10, ed. S. Srinivasan *et al.*, p. 476, by permission of the publisher, Plenum Publishing Corporation.

The following are also thanked for permission to reproduce or reprint copyright material: Bioanalytical Systems Inc. for Fig. 14.8; Elsevier Science Publishers BV for Figs 8.3, 8.4, 8.6, 8.7, 11.7, Tables 8.1 and 8.2; Elsevier Sequoia SA for Figs 9.11, 9.12, 9.15, 12.4, 12.8, 12.20, and 14.3; Journal of Chemical Education for Fig. 9.13a; Kluwer Academic Publishers for Fig. 3.10; R. Kötz for Fig. 12.1; Oxford University Press for Figs 2.11, 2.12, and 17.10; Royal Society of Chemistry for Table 14.2.

CONTENTS

6   KINETICS AND TRANSPORT IN ELECTRODE REACTIONS 103

PART II   Methods

7   ELECTROCHEMICAL EXPERIMENTS 129

Appendices

Notation and Units

As far as possible without straying too far from common usage, the guidelines of IUPAC have been followed, described in *Quantities, units and symbols in physical chemistry* (Blackwell, Oxford, 1988). Other, more detailed information has been taken from the following sources in the IUPAC journal, *Pure and Applied Chemistry*:

'Electrode reaction orders, transfer coefficients and rate constants. Amplification of definitions and recommendations for publication of parameters', 1979, **52,** 233.

'Interphases in systems of conducting phases', 1986, **58,** 454.

'Electrochemical corrosion nomenclature', 1989, **61,** 19.

'Terminology in semiconductor electrochemistry and photo-electrochemical energy conversion', 1991, **63,** 569.

'Nomenclature, symbols, definitions and measurements for electrified interfaces in aqueous dispersions of solids', 1991, **63,** 896.

The units quoted are those recommended. In practice, in electrochemistry, much use is made of sub-multiples: for example, cm instead of m and μA or mA instead of A, for obvious reasons. The text tends to use the commonly employed units.

In the list of symbols, those used at only one specific point in the text are mostly omitted, in order to try and reduce the length of the list, since explanation of their meaning can be found next to the relevant equation. We have also provided a list of frequently used subscripts, a list of abbreviations, and values of important constants and relations derived from these.

Following recommended usage, $\log_e$ is written as ln and $\log_{10}$ is written as lg.

Notation: main symbols

<table>
<tr><td></td><td></td><td></td><td>Units</td></tr>
</table>

			Units
a	activity		—
a	nozzle diameter of impinging jet		m
a	radius of colloidal particle		m
A	area		m^2
A	'constant'		varies
b	Tafel slope		V^{-1}
c	concentration		$mol\ m^{-3}$
		c_0 concentration at electrode surface	
		c_∞ bulk concentration	
C	capacity		F
		C_d differential capacity of double layer	
		C_i integral capacity of double layer	
		C_s capacity in RC series combination	
		C_{sc} capacity of semiconductor space–charge layer	
D	diffusion coefficient		$m^2\ s^{-1}$
e	electron charge		C
$\mathbf{E}$	electric field strength		$V\ m^{-1}$
E	electrode potential		V
		$E^{\ominus}$ standard electrode potential	
		$E^{\ominus\prime}$ formal potential	
		E_{cell} cell potential (electromotive force)	
		E_{cor} corrosion potential	
		$E_{1/2}$ half-wave potential	
		E_j liquid junction potential	
		E_m membrane potential	
		E_p peak potential	
		E_z potential of zero charge	
		E_λ inversion potential in cyclic voltammetry	
E_c	lowest energy of semiconductor conduction band		eV
E_g	bandgap energy in semiconductor		eV
E_v	highest energy of semiconductor valence band		eV
E_F	Fermi energy		eV
E_{redox}	energy of redox couple		eV
f	frequency		Hz
f	$F(E - E^{\ominus})/RT$		—

f_{DL}	Frumkin double layer correction	—
F	force	N
g	acceleration due to gravity	$m\ s^{-2}$
g	constant in Temkin and Frumkin isotherms	—
G	Gibbs free energy	$J\ mol^{-1}$
h	height	m
H	enthalpy at constant pressure	$J\ mol^{-1}$
I	electric current	A
	I_C capacitative current	
	I_f faradaic current	
	I_L diffusion limited current	
	I_p peak current	
I	ionic strength	$mol\ m^{-3}$
j	electric current density	$A\ m^{-2}$
J	volume flux	$m^3\ s^{-1}$
k	rate constant: homogeneous first order	s^{-1}
	rate constant: heterogeneous	$m\ s^{-1}$
	k_a rate constant for oxidation at electrode	
	k_c rate constant for reduction at electrode	
	k_d mass transfer coefficient	
$k_{i,j}^{pot}$	potentiometric selectivity coefficient	—
K	equilibrium constant	—
l	length of electrode	m
m	mass	kg
m_1	mass flux of liquid	$kg\ s^{-1}$
m	molality	$kg\ m^{-3}$
n	number of electrons transferred	—
n'	number of electrons transferred in rate determining step	—
n_i	number density of species i	m^{-3}
p	$(D_0/D_R)^s$ where $s = 1/2$ (stationary electrodes and DMEs), s = 2/3 (hydrodynamic electrodes), $s = 1$ (microelectrodes)	—
p_i	partial pressure of i	Pa
P	pressure (total)	Pa
Pe	Peclet number ($Pe = vl/D$)	—
Q	electric charge	C
r	radial variable	m
	r_0 radius of (hemi-)spherical electrode	
	r_1 radius of disc electrode	
	r_2 inner radius of ring electrode	
	r_3 outer radius of ring electrode	
	r_c capillary radius	

R	resistance	Ω
	R_{ct} charge transfer resistance	
	R_s resistance in RC series combination	
	R_Ω cell solution resistance	
R	radius of tube	m
Re	Reynolds number ($Re = vl/v$)	—
S	entropy	$J\,mol^{-1}\,K^{-1}$
Sc	Schmidt number ($Sc = v/D$)	—
Sh	Sherwood number ($Sh = k_d l/D$)	—
t	time	s
t_i	transport number of species i	—
T	temperature	K
u_i	mobility of species i	$m^2\,V^{-1}\,s^{-1}$
	u_e electrophoretic mobility	
U	potential (same meaning as E, used in photo- and semiconductor electrochemistry)	V
	U_{fb} flat-band potential	
v	velocity	$m\,s^{-1}$
v	potential scan rate	$V\,s^{-1}$
V	voltage (in operational amplifiers, etc.)	V
V	volume	m^3
V_f	volume flow rate	$m^3\,s^{-1}$
W	rotation speed	Hz
x	distance	m
X	reactance	Ω
Y	admittance	S
z	ion charge	—
Z	impedance	Ω
	Z_s impedance of RC series combination	
	Z' real part of impedance	
	Z'' imaginary part of impedance	
	Z_f Faradaic impedance	
	Z_w Warburg impedance	
α	electrochemical charge transfer coefficient	—
	α_a anodic	
	α_c cathodic	
α	electrode roughness parameter	—
α	double hydrodynamic electrode geometric constant	—
β	double hydrodynamic electrode geometric constant	—
β	Esin–Markov coefficient	—
β	energetic proportionality coefficient	—

γ	activity coefficient	—
γ	surface tension	$\mathrm{N\,m^{-1}}$
γ	dimensionless concentration variable	—
Γ	surface excess concentration	$\mathrm{mol\,m^{-2}}$
δ	diffusion layer thickness	m
δ_H	hydrodynamic boundary layer thickness	m
ϵ	molar absorption coefficient	$\mathrm{m^2\,mol^{-1}}$
ϵ	permittivity	$\mathrm{F\,m^{-1}}$
ϵ_0	permittivity of vacuum	$\mathrm{F\,m^{-1}}$
ϵ_r	relative permittivity	—
ϵ	porosity of material	—
ζ	zeta (electrokinetic) potential	V
ζ	$(nF/RT)(E - E_{1/2})$	—
η	overpotential	V
η	viscosity	Pa s
θ	contact angle	
θ	fractional surface coverage	—
θ	$\exp[(nF/RT)(E - E^{\ominus})]$	—
κ	conductivity	$\mathrm{S\,m^{-1}}$
λ	value of t where sweep is inverted in cyclic voltammetry	s
Λ	molar conductivity	$\mathrm{S\,m^2\,mol^{-1}}$
μ	chemical potential	$\mathrm{J\,mol^{-1}}$
$\bar{\mu}$	electrochemical potential	$\mathrm{J\,mol^{-1}}$
v	frequency of electromagnetic radiation	$\mathrm{s^{-1}}$
v	stoichiometric number	—
v	kinematic viscosity	$\mathrm{m^2\,s^{-1}}$
ρ	resistivity	$\Omega\,\mathrm{m}$
ρ	density	$\mathrm{kg\,m^{-3}}$
σ	surface charge density	$\mathrm{C\,m^{-2}}$
σ	$v(nF/RT)$	$\mathrm{s^{-1}}$
σ	mass-transport dependent expression (Table 8.2)	
τ	characteristic time in experiment	s
ϕ	electrostatic potential	V
ϕ	inner electric potential	V
ϕ	phase angle	
χ	surface electric potential	V
ψ	outer electric potential	V
ω	angular velocity, rotation speed	$\mathrm{rad\,s^{-1}}$
ω	circular frequency	$\mathrm{rad\,s^{-1}}$

Subscripts

a	anodic	max	maximum value
c	cathodic	min	minimum value
C	capacitive	O	oxidized species
det	detector electrode	p	peak value
D	disc electrode	R	reduced species
f	faradaic	R	ring electrode
f	final value	0	at zero distance (electrode surface)
gen	generator electrode		
i	species i	∞	at infinite distance (bulk solution)
i	initial value		
L	diffusion-limited value	*	at OHP

Abbreviations

AES	Auger electron spectroscopy
AFM	atomic force microscopy
ASV	anodic stripping voltammetry
AdSV	adsorptive stripping voltammetry
BLM	bilayer lipid membrane
CDE	channel double electrode
CE	electrode process involving chemical followed by electrochemical step
C'E	catalytic electrode process involving chemical followed by electrochemical step
CV	cyclic voltammetry
DDPV	differential double pulse voltammetry
DISP	electrode process involving electrochemical followed by chemical, followed by disproportionation step to regenerate reagent
DME	dropping mercury electrode
DNPV	differential normal pulse voltammetry
DPV	differential pulse voltammetry
DSA	dimensionally stable anode
EC	electrode process involving electrochemical followed by chemical step
ECE	electrode process involving electrochemical followed by chemical, followed by electrochemical step
ECL	electrochemiluminescence
ECMS	electrochemical mass spectroscopy
EELS	electron energy loss spectroscopy
EMIRS	electrochemically modulated infrared spectroscopy
EQCM	electrochemical quartz crystal microbalance
ESR	electron spin resonance
EXAFS	extended X-ray absorption fine structure
FFT	fast Fourier transform
GC	glassy carbon
HMDE	hanging mercury drop electrode
HOPG	highly oriented pyrolytic graphite
HPLC	high-performance liquid chromatography
IHP	inner Helmholtz plane

IRRAS	infrared reflection absorption spectroscopy
ISE	ion-selective electrode
ISFET	ion-selective field effect transistor
ISM	ion-selective membrane
LEED	low-energy electron diffraction
LSV	linear sweep voltammetry
MCFC	molten carbonate fuel cell
MS	mass spectrometry
NHE	normal hydrogen electrode
NPV	normal pulse voltammetry
OA	operational amplifier
OHP	outer Helmholtz plane
OTE	optically transparent electrode
OTTLE	optically transparent thin-layer electrode
PAFC	phosphoric acid fuel cell
PAS	photoacoustic spectroscopy
PSA	potentiometric stripping analysis
QCM	quartz crystal microbalance
RDE	rotating disc electrode
RHEED	reflection high-energy electron diffraction
RRDE	rotating ring-disc electrode
SCC	stress corrosion cracking
SCE	saturated calomel electrode
SCM	surface compartment model
SECM	scanning electrochemical microscopy
SEM	scanning electron microscopy
SHG	second harmonic generation
SICM	scanning ion conductance microscopy
SIMS	secondary ion mass spectroscopy
SMDE	static mercury drop electrode
SNIFTRS	subtractively normalized interfacial Fourier transform infrared spectroscopy
SOFC	solid oxide fuel cell
STM	scanning tunnelling microscopy
SWV	square wave voltammetry
TDE	tube double electrode
TEM	transmission electron microscopy
WJRDE	wall-jet ring-disc electrode
XANES	X-ray absorption near edge structure
XPS	X-ray photoelectron spectroscopy

Fundamental physical constants

c	speed of light in vacuum	$2.99792458 \times 10^8 \text{ m s}^{-1}$
e	unit of electron charge	$1.602177 \times 10^{-19} \text{ C}$
F	Faraday constant	$9.6485 \times 10^4 \text{ C mol}^{-1}$
k_B	Boltzmann constant	$1.38066 \times 10^{-23} \text{ J K}^{-1}$
R	gas constant	$8.31451 \text{ J K}^{-1} \text{ mol}^{-1}$
h	Planck constant	$6.62608 \times 10^{-34} \text{ Js}$
N_A	Avogadro constant	$6.02214 \times 10^{23} \text{ mol}^{-1}$
ϵ_0	permittivity of vacuum	$8.85419 \times 10^{-12} \text{ J}^{-1} \text{ C}^2 \text{ m}^{-1}$
g	acceleration due to gravity	9.80665 m s^{-2}

Mathematical constants

π	3.14159265359
e	2.71828182846
ln 10	2.302585

Useful relations at 25°C (298.15 K) involving fundamental constants

RT/F	25.693 mV
$(RT/F) \ln 10$	59.160 mV
$k_B T$	25.7 meV (4.12×10^{-21} J)

1

INTRODUCTION

1.1 The scope of electrochemistry

Electrochemistry involves chemical phenomena associated with charge separation. Often this charge separation leads to charge transfer, which can occur homogeneously in solution, or heterogeneously on electrode surfaces. In reality, to assure electroneutrality, two or more charge transfer half-reactions take place, in opposing directions. Except in the case of homogeneous redox reactions, these are separated in space, usually occurring at different electrodes immersed in solution in a cell. These electrodes are linked by conducting paths both in solution (via ionic transport) and externally (via electric wires etc.) so that charge can be transported. If the cell configuration permits, the products of the two electrode reactions can be separated. When the sum of the free energy changes at both electrodes is negative the electrical energy released can be harnessed (batteries). If it is positive, external electrical energy can be supplied to oblige electrode reactions to take place and convert chemical substances (electrolysis).

In this chapter, a brief overview of electrochemistry, and particularly of electrode reactions, is given in order to show the interdisciplinary nature and versatility of electrochemistry and to introduce a few of the important fundamental concepts. Before discussing these it is worth looking briefly at the nature of electrode reactions.

1.2 The nature of electrode reactions

Electrode reactions are heterogeneous and take place in the interfacial region between electrode and solution, the region where charge distribution differs from that of the bulk phases. The electrode process is affected by the structure of this region. However, we first assume that there is no effect apart from charge separation. At each electrode, charge separation can be represented by a *capacitance* and the difficulty of charge transfer by a *resistance*. For the rest of this and the ensuing sections we consider only one of the electrodes.

The electrode can act as only a source (for reduction) or a sink (for oxidation) of electrons transferred to or from species in solution, as in

$$O + ne^- \rightarrow R$$

where O and R are the oxidized and reduced species, respectively. Alternatively, it can take part in the electrode reaction, as in dissolution of a metal M:

$$M \rightarrow M^{n+} + ne^-$$

In order for electron transfer to occur, there must be a correspondence between the energies of the electron orbitals where transfer takes place in the donor and acceptor. In the electrode this level is the highest filled orbital, which in a metal is the Fermi energy level, E_F. In soluble species it is simply the orbital of the valence electron to be given or received. Thus:

● for a reduction, there is a minimum energy that the transferable electrons from the electrode must have before transfer can occur, which corresponds to a sufficiently negative potential (in volts)

● for an oxidation, there is a maximum energy that the lowest unoccupied level in the electrode can have in order to receive electrons from species in solution, corresponding to a sufficiently positive potential (in volts).

The values of the potentials can be controlled externally. In this way we can control which way an electrode reaction occurs and to what extent.

The thermodynamics and kinetics of electrode processes are summarized in the following section. However, before this we return to the structure of the interfacial region. The change in charge distribution from the bulk in this region means that the relevant energy levels in reacting species and in the electrode are not the same as in the bulk phases, and soluble species need to adjust their conformation for electron transfer to occur. These effects should be corrected for in a treatment of kinetics of electrode processes—the thinner the interfacial region the better, and this can be achieved by addition of a large concentration of inert electrolyte.

1.3 Thermodynamics and kinetics

Electrode reactions are half-reactions and are, by convention, expressed as reductions. Each has associated with it a standard electrode potential,

$E^{\ominus}$, measured relative to the normal hydrogen electrode (NHE) with all species at unit activity ($a_i = 1$).

For half-reactions at equilibrium, the potential, E, can be related to the standard electrode potential through the *Nernst equation*

$$E = E^{\ominus} - \frac{RT}{nF} \sum v_i \ln a_i \qquad (1.1)$$

where v_i are the stoichiometric numbers, positive for products (reduced species) and negative for reagents (oxidized species). The tendency for the reduction to occur, relative to the NHE reference, is thus given by

$$\Delta G^{\ominus} = -nFE^{\ominus} \qquad (1.2)$$

under standard conditions. Thus, for example, Group IA metals which have very negative values of $E^{\ominus}$, tend to oxidize (see Appendix 4).

It is often useful to be able to employ concentrations, c_i, instead of activities, where $a_i = \gamma_i c_i$ with γ_i the activity coefficient of species i. The Nernst equation (1.1) is rewritten as

$$E = E^{\ominus\prime} - \frac{RT}{nF} \sum v_i \ln c_i \qquad (1.3)$$

in which $E^{\ominus\prime}$ is the formal potential, dependent on the medium since it includes the logarithmic activity coefficient terms as well as $E^{\ominus}$.

If the oxidized and reduced species involved in an electrode reaction are in equilibrium at the electrode surface, the Nernst equation can be applied. The electrode reaction is then known as a *reversible* reaction since it obeys the condition of thermodynamic reversibility. Clearly the applicability of the Nernst equation, and therefore reversibility, has to do with the time allowed for the electrode reaction to reach equilibrium.

The concentrations of species at the interface depend on the mass transport of these species from bulk solution, often described by the mass transfer coefficient k_d. A reversible reaction corresponds to the case where the kinetics of the electrode reaction is much faster than the transport. The kinetics is expressed by a standard rate constant, k_0, which is the rate constant when $E = E^{\ominus\prime}$. So the criterion for a reversible reaction is

$$k_0 \gg k_d$$

By contrast, an *irreversible* reaction is one where the electrode reaction cannot be reversed. A high kinetic barrier has to be overcome, which is achieved by application of an extra potential (extra energy) called the overpotential, η, and in this case

$$k_0 \ll k_d$$

Quasi-reversible reactions exhibit behaviour intermediate between reversible and irreversible reactions, the overpotential having a relatively small value, so that with this extra potential reactions can be reversed.

The potential-dependent expression for the rate constant of an electrode reaction is, for a reduction,

$$k_c = k_0 \exp\left[-\alpha_c nF(E - E^{\ominus\prime})/RT\right] \qquad (1.4)$$

and for an oxidation

$$k_a = k_0 \exp\left[\alpha_a nF(E - E^{\ominus\prime})/RT\right] \qquad (1.5)$$

In these equations α_c and α_a are the cathodic and anodic *charge transfer coefficients* and are a measure of the symmetry of the activation barrier, being close to 0.5 for a metallic electrode and a simple electron transfer process. As mentioned above, the standard rate constant is the rate constant at $E = E^{\ominus\prime}$.

An alternative way used to express the rates of electrode reactions is through the *exchange current, I_0*. This is the magnitude of the anodic or cathodic partial current at the equilibrium potential, E_{eq}. It is equivalent to measuring the standard rate constant, k_0.

Experimentally, rates of electrode reactions are measured as the current passed, to which they are directly proportional. The dependence of current, I, on potential is exponential, suggesting a linear relation between $\lg I$ and potential—this is the *Tafel relation*. However, the rate (product of rate constant and reagent concentration) cannot rise indefinitely because the supply of reactants begins to diminish and becomes transport-limited.

Whereas for reversible reactions only thermodynamic and mass-transport parameters can be determined, for quasi-reversible and irreversible reactions both kinetic and thermodynamic parameters can be measured. It should also be noted that the electrode material can affect the kinetics of electrode processes.

The rate constant of an electrode reaction does not measure the rate of electron transfer itself, as this is an adiabatic process, following the Franck–Condon principle, and occurs in approximately 10^{-16} s. What it does measure is the time needed for the species, once they have reached the interfacial region, to arrange themselves and their ionic atmospheres into position for electron transfer to be able to occur.

More complex electrode processes than those described above involve consecutive electron transfer or coupled homogeneous reactions. The theory of these reactions is also more complicated, but they correspond to a class of real, important reactions, particularly involving organic and biological compounds.

1.4 Methods for studying electrode reactions

In order to study electrode reactions, reproducible experimental conditions must be created which enable minimization of all unwanted factors that can contribute to the measurements and diminish their accuracy. Normally we wish to suppress migration effects, confine the interfacial region as close as possible to the electrode, and minimize solution resistance. These objectives are usually achieved by addition of a large quantity of inert electrolyte (around $1 \, mol \, dm^{-3}$), the electroactive species being at a concentration of 5 mM or less.

A complete study of an electrode process requires measurement of kinetic as well as thermodynamic parameters. This means that conditions in which the system is not reversible must be used. Since the standard rate constant, k_0, cannot be changed, then the mass transfer coefficient, k_d, may have to be increased until the reaction becomes at least quasi-reversible. This can be done in various ways in various types of experiment:

● *steady state methods*: hydrodynamic electrodes, increasing convection; microelectrodes, decreasing size

● *linear sweep methods*: increasing sweep rate

● *step and pulse techniques*: increasing amplitude and/or frequency

● *impedance methods*: increasing perturbation frequency, registering higher harmonics, etc.

The type of technique chosen will depend very much on the timescale of the electrode reaction.

Non-electrochemical methods can and should be used for studying electrode surfaces and the interfacial region structure, particularly *in situ* in real time where this is possible.

1.5 Applications of electrochemistry

Once electrode reactions and electrode processes are understood, this knowledge can be used for:

● tailoring electrode reactions so as to enhance required and inhibit unwanted electrode reactions, perhaps by changing electrode material or developing new electrode materials

● studying complex systems in which many electrode reactions occur simultaneously or consecutively, as in bioelectrochemistry

• measuring concentrations of electroactive species, making use of the selectivity of the potential and of the electrode material at or outside equilibrium (as in potentiometric, amperometric, voltammetric, and enzyme sensors).

Thus the range of applications is vast. Electroanalysis, potentiometric and voltammetric; industrial electrolysis, electroplating, batteries, fuel cells, electrochemical machining, and many other related applications, including minimization of corrosion; biosensors and bioelectrochemistry.

1.6 Structure of the book

This book is organized into three main sections, as its subtitle suggests.

In the first part, Chapters 2–6, some fundamentals of electrode processes and of electrochemical and charge transfer phenomena are described. Thermodynamics of electrochemical cells and ion transport through solution and through membrane phases are discussed in Chapter 2. In Chapter 3 the thermodynamics and properties of the interfacial region at electrodes are addressed, together with electrical properties of colloids. Chapters 4–6 treat the rates of electrode processes, Chapter 4 looking at fundamentals of kinetics, Chapter 5 at mass transport in solution, and Chapter 6 at their combined effect in leading to the observed rate of electrode processes.

The second part of the book discusses ways in which information concerning electrode processes can be obtained experimentally, and the analysis of these results. Chapter 7 presents some of the important requirements in setting up electrochemical experiments. In Chapters 8–11, the theory and practice of different types of technique are presented: hydrodynamic electrodes, using forced convection to increase mass transport and increase reproducibility; linear sweep, step and pulse, and impedance methods respectively. Finally in Chapter 12, we give an idea of the vast range of surface analysis techniques that can be employed to aid in investigating electrode processes, some of which can be used *in situ,* together with photochemical effects on electrode reactions— photoelectrochemistry.

In the third part of the book areas in which there are important applications of electrochemistry are described. Chapters 13 and 14 look at potentiometric and amperometric/voltammetric sensors respectively, focusing particularly on recent developments such as new electrode materials and miniaturization. Electrochemistry in industry, which produces many materials used directly or indirectly in everyday life, as well as batteries, is described in Chapter 15. The electrochemical phenomenon of corrosion, economically prejudicial, is described in Chapter 16. Finally, since many biochemical processes involve charge transfer reactions, in Chapter 17 the many possibilities that arise from their study by electrochemical methods, bioelectrochemistry, are presented.

1.7   Electrochemical literature

The electrochemical literature is very widespread. Some indication of its breadth is given below. The references at the end of each chapter complement this list.

General books

Many books on electrochemistry have been published in recent decades. Mostly the more general ones are not cited throughout the text, but this does not reflect on their quality. A list of them is given below, in chronological order.

P. Delahay, *New instrumental methods in electrochemistry*, Interscience, New York, 1954.

K. J. Vetter, *Electrochemical kinetics*, Academic Press, New York, 1967.

R. N. Adams, *Electrochemistry at solid electrodes*, Dekker, New York, 1969.

J. O'M. Bockris and A. N. Reddy, *Modern electrochemistry*, Plenum, New York, 1970.

J. Newman, *Electrochemical systems*, Prentice Hall, Englewood Cliffs, NJ, 1973.

D. T. Sawyer and J. L. Roberts, *Experimental electrochemistry for chemists*, Wiley, New York, 1974.

E. Gileadi, E. Kirowna–Eisner, and J. Penciner, *Interfacial electrochemistry. An experimental approach*, Addison-Wesley, Reading, MA, 1975.

W. J. Albery, *Electrode kinetics*, Clarendon Press, Oxford, 1975.

A. J. Bard and L. R. Faulkner, *Electrochemical methods, fundamentals and applications*, Wiley, New York, 1980.

A. M. Bond, *Modern polarographic methods in analytical chemistry*, Dekker, New York, 1980.

Southampton Electrochemistry Group, *New instrumental methods in electrochemistry*, Ellis Horwood, Chichester, 1985.

J. Goodisman, *Electrochemistry: theoretical foundations*, Wiley-Interscience, New York, 1987.

J. Koryta, *Principles of electrochemistry*, Wiley, Chichester, 1987.

P. H. Rieger, *Electrochemistry,* Prentice-Hall International, Englewood Cliffs, NJ, 1987.

D. R. Crow, *Principles and applications of electrochemistry*, 3rd edn, Chapman and Hall, London, 1988.

P. W. Atkins, *Physical chemistry*, 4th edn., Oxford University Press, 1990, Chapters 10, 25, and 30.

D. Pletcher, *A first course in electrode processes*, The Electrochemical Consultancy, Romsey, UK, 1991.

J. Koryta, *Ions, electrodes, and membranes*, Wiley, Chichester, 1991.

Series

A number of series of volumes dealing with electrochemistry have been published. Those recently issued or currently being published are listed below.

Advances in electrochemistry and electrochemical engineering, Wiley, New York. Volumes 1–9, ed. P. Delahay and C. W. Tobias; Volumes 10–13, ed. H. Gerischer and C. W. Tobias.

Advances in electrochemical science and engineering, ed. H. Gerischer and C. W. Tobias, VCH, Weinheim (continuation of *Adv. Electrochem. Electrochem. Eng.*; 1 volume until end 1991).

Comprehensive treatise of electrochemistry, ed. J. O'M. Bockris, B. E. Conway, E. Yeager *et al.,* Plenum, New York, Volumes 1–10.

Comprehensive chemical kinetics, section 10; electrode kinetics, ed. R. G. Compton *et al.,* Elsevier, Amsterdam, Volumes 26–29.

Electroanalytical chemistry: a series of advances, ed. A. J. Bard, Dekker, New York (17 volumes until end 1991).

Modern aspects of electrochemistry, ed. J. O'M. Bockris, B. E. Conway *et al.,* Plenum, New York (21 volumes until end 1991).

International journals devoted to electrochemistry

There are a number of international journals devoted primarily to electrochemistry:

Bioelectrochemistry and Bioenergetics (an independent section of *J. Electroanal. Chem.*)
Corrosion
Corrosion Science
Electroanalysis
Electrochimica Acta

Elektrokhimiya (*Soviet Electrochemistry*)
Journal of Applied Electrochemistry
Journal of Electroanalytical and Interfacial Electrochemistry
Journal of the Electrochemical Society
Selective Electrode Reviews (formerly *Ion Selective Electrode Reviews,*
until 1988)

Articles with electrochemical themes also regularly appear in a large
number of other journals.

PART I

Principles

2

ELECTROCHEMICAL CELLS: THERMODYNAMIC PROPERTIES AND ELECTRODE POTENTIALS

2.1 Introduction

An understanding of thermodynamic properties associated with electrode processes is fundamental in order to answer questions such as:

- Why is it that half-reactions in electrochemical cells proceed spontaneously in one direction and furnish current?
- What is the effect of the salt bridge?
- What is the effect of ion migration?

In this chapter we attempt to reply to these and to other related questions. To treat the topic in a concrete way, we consider two electrochemical cells:

$$Zn \mid Zn^{2+}(aq) \mid Cu^{2+}(aq) \mid Cu$$

and

$$Hg \mid Hg_2Cl_2 \mid Cl^-(aq) \parallel Zn^{2+}(aq) \mid Zn$$

where we represent only the species of interest. In these cells the symbol

| denotes a phase boundary, ¦ a junction between miscible liquids, and ‖ a salt bridge (liquid junction) whose function is to provide an electrically conducting link between two spatially separated components of the cell in the liquid phase. It should be stressed that, according to the internationally accepted IUPAC convention, the half-reactions are considered in the way the cell is depicted on paper, that is oxidation in the left half-cell (the electrode is the anode) and reduction in the right half-cell (the electrode is the cathode)[1].

2.2   The cell potential of an electrochemical cell

The cell potential of an electrochemical cell is calculated from the electrode potentials (reduction potentials) of the respective half-reactions[1]. Given that, by convention, the half-reaction on the left is considered to be an oxidation and that on the right a reduction we have

$$E_{cell} = E_{right} - E_{left} \qquad (2.1)$$

where E_{right} and E_{left} are the potentials of each half-cell, obtained from the Nernst equation.

The Nernst equation relates the activities of the species involved with the electrode potential, E, of the half-reaction and its standard electrode potential, $E^{\ominus}$, which is the value of the potential relative to the standard hydrogen electrode when the activities of all species are unity. For the generic half-reaction

$$\sum v_i O_i + ne^- \to \sum v_i R_i$$

where n is the stoichiometric number of electrons transferred for each species, the Nernst equation is

$$E = E^{\ominus} - \frac{RT}{nF} \sum v_i \ln a_i \qquad (2.2)$$

in which v_i has positive values for products (reduced species) and negative values for reagents (oxidized species). This can be written as

$$E = E^{\ominus} + \frac{RT}{nF} \ln \frac{\prod a_{O_i}^{v_i}}{\prod a_{R_i}^{v_i}} \qquad (2.3)$$

For example, for

$$MnO_2 + 4H^+ + 2e^- \to Mn^{2+} + H_2O$$

the logarithmic term is

$$\frac{RT}{2F} \ln \frac{a_{MnO_2} a_{H^+}^4}{a_{Mn^{2+}} a_{H_2O}^2}$$

a_{H_2O} is approximately constant and is neglected in the Nernst equation except in the case of a mixture with another solvent or in very concentrated solutions.

The cell potential tells us the maximum work (maximum energy) that the cell can supply[2]. This value is

$$\Delta G = -nFE_{cell} \tag{2.4}$$

It is evident that on removing energy (in the form of current or converted chemical substances) the amount of unconverted substances remaining is diminished, reflecting the changes in the concentrations of the species in the liquid phase. In the solid phase, however, there is no alteration of activity, which is normally accepted as being unity.

We now calculate the cell potential for the two cases mentioned above.

Case 1

$$Zn \,|\, Zn^{2+}(aq) \,\vdots\, Cu^{2+}(aq) \,|\, Cu$$

which means we consider the cell reaction as

$$Zn + Cu^{2+} \rightarrow Zn^{2+} + Cu$$

The half-reactions are represented by

right: $Cu^2 + 2e^- \rightarrow Cu$ $E^{\ominus} = +0.34$ V

left: $Zn^{2+} + 2e^- \rightarrow Zn$ $E^{\ominus} = -0.76$ V

If the aqueous species have unit activity, then $E^{\ominus}$ values may be used and

$$E_{cell}^{\ominus} = +0.34 - (-0.76) = +1.10 \text{ V}$$

The corresponding $\Delta G^{\ominus}$ value is

$$\Delta G^{\ominus} = -2.20F = -212 \text{ kJ mol}^{-1}$$

which is negative. This result shows that the reaction proceeds spontaneously as written.

The equivalent of the Nernst equation for the whole cell is

$$E_{cell} = E_{cell}^{\ominus} + \frac{RT}{2F} \ln \frac{a_{Cu^{2+}}}{a_{Zn^{2+}}} \tag{2.5}$$

It can be seen that if the ratio $(a_{cu^{2+}}/a_{Zn^{2+}})$ is sufficiently small, E_{cell} becomes negative and the direction of spontaneous reaction is changed.

Case 2

$$Hg \,|\, Hg_2Cl_2 \,|\, Cl^-(aq) \,\|\, Zn^{2+}(aq) \,|\, Zn$$

The stoichiometric cell reaction to consider is

$$2Hg + 2Cl^- + Zn^{2+} \rightarrow Hg_2Cl_2 + Zn$$

and the half-reactions are represented by

right: $Zn^{2+} + 2e^- \rightarrow Zn$ $E^\ominus = -0.76\ V$

left: $Hg_2Cl_2 + 2e^- \rightarrow 2Hg + 2Cl^-$ $E^\ominus = +0.27\ V$

For unit activities,

$$E^\ominus_{cell} = -0.76 - 0.27 = -1.03\ V$$

and

$$\Delta G^\ominus = +199\ kJ\ mol^{-1}$$

The negative value of $E^\ominus_{cell}$ (and positive $\Delta G^\ominus$) means that at unit activities the cell functions spontaneously in the direction opposite to that written above. Thus the spontaneous cell reaction is

$$Hg_2Cl_2 + Zn^{2+} \rightarrow 2Hg + Zn^{2+} + 2Cl^-$$

The half-cell on the left is an example of a reference electrode (Section 2.7) so called since, as Hg_2Cl_2 is a sparingly soluble salt, the activities of Hg and Hg_2Cl_2 can be taken as unity. The potential of the half-cell is altered solely by the chloride ion activity according to the expression

$$E_{cell} = E^\ominus_{cell} - \frac{RT}{F} \ln a_{Cl^-} \tag{2.6}$$

This electrode is known as the *calomel electrode*.

2.3 Calculation of cell potential: activities or concentrations?

Although the use of activities in the Nernst equation is undoubtedly correct, it is worth considering whether it is necessary and what is the difference between activities and concentrations in general.

In the context of this book, a detailed discussion of activities and concentrations is not justified. However, it is clear that in relatively concentrated solutions there will be interionic interactions that do not

occur in very dilute solutions because of the large interionic distances in
the latter. Consequently the velocity of ion migration (i.e. the momentum of each ion) will be altered, and this can reduce, or possibly increase,
ionic activity. Thus we write the relations

$$a = \gamma_m m \tag{2.7a}$$

$$a = \gamma_c c \tag{2.7b}$$

where γ_m is the activity coefficient for concentrations in relation to
molality (mol kg^{-1}), and γ_c in relation to molarity (mol dm^{-3}). Thus,
these coefficients are proportionality factors between activity and concentration, whose values vary with concentration.

It is often useful to employ concentrations instead of activities in
electrochemical experiments: for example, in preparing solutions we use
masses and volumes, that is we determine the concentration of a solution.
Thus, the Nernst equation, instead of being written as

$$E = E^\ominus + \frac{RT}{nF} \ln \frac{\prod a_{O_i}^{\nu_i}}{\prod a_{R_i}^{\nu_i}} \tag{2.3}$$

can be formulated as

$$E = E^{\ominus\prime} + \frac{RT}{nF} \ln \frac{\prod [O_i]^{\nu_i}}{\prod [R_i]^{\nu_i}} \tag{2.8}$$

In this last equation, $E^{\ominus\prime}$ is the *formal potential*. It is related to the
standard electrode potential, $E^\ominus$, by

$$E^{\ominus\prime} = E^\ominus + \frac{RT}{nF} \ln \frac{\prod \gamma_{c,O_i}^{\nu_i}}{\prod \gamma_{c,R_i}^{\nu_i}} \tag{2.9}$$

Experimentally we measure the formal potential, $E^{\ominus\prime}$, relative to a
reference electrode (Section 2.7). However, by performing measurements at different concentrations and extrapolating to zero concentration,
values of $E^\ominus$ can be obtained. Another factor that can enter into the
values of $E^{\ominus\prime}$ is perturbations caused by other reactions, normally due to
complexation.

An example of the difference between values of $E^{\ominus\prime}$ and $E^\ominus$ is the
values obtained in the potentiometric titration of Fe^{2+} with Ce^{4+} in
0.5 M H_2SO_4. These are, relative to the normal hydrogen electrode
(NHE):

	Standard electrode potential $E^\ominus/V$	Formal potential $E^{\ominus\prime}/V$
$Fe^{3+} \mid Fe^{2+}$	+0.77	+0.68
$Ce^{4+} \mid Ce^{3+}$	+1.61	+1.44

The differences reflect not only the activities of the ions involved in the half-reactions but also the fact that 0.5 M H_2SO_4 does not have pH 0 (in fact the second ionization is only partially effected).

2.4 Calculation of cell potential: electrochemical potential

Although the calculation of E_{cell} in the previous section appears satisfactory, it is not very rigorous. In this section we show how a rigorous thermodynamic argument[4] leads to the same result. For this we need the concept of the electrochemical potential $\bar{\mu}$, that obeys the same criteria at equilibrium as the chemical potential μ. Its definition for component i in phase α is

$$\bar{\mu}_i^\alpha = \mu_i^\alpha + z_i F \phi^\alpha \tag{2.10}$$

$$= \mu_i^{0,\alpha} + RT \ln a_i + z_i F \phi^\alpha \tag{2.11}$$

which is the sum of a term due to the chemical potential and another that represents the contribution from charged species described by the electrostatic potential ϕ in phase α. Since

$$\mu_i^\alpha = \left(\frac{\partial G}{\partial n_i} \right)_{T,P,n_{j\neq i}} \tag{2.12}$$

then

$$\bar{\mu}_i^\alpha = \left(\frac{\partial \bar{G}}{\partial n_i} \right)_{T,P,n_{j\neq i}} \tag{2.13}$$

where $\bar{G}$ is the *electrochemical free energy*. $\bar{G}$ is analogous to the free energy, G, but contains the electrical effects of the environment. In the case of a species without charge,

$$\bar{\mu}_i^\alpha = \mu_i^\alpha \tag{2.14}$$

Various deductions are possible from these expressions:

- For a pure phase that has unit activity,

$$\bar{\mu}_i^\alpha = \mu_i^{0,\alpha} \tag{2.15}$$

where $\mu_i^{0,\alpha}$ is the standard chemical potential in phase α.

- For a metal, activity effects can be neglected. The electrochemical potential is the electronic energy of the highest occupied level (Fermi level, E_F)

$$\bar{\mu}_e^\alpha = \mu_e^{0,\alpha} - F\phi^\alpha \tag{2.16}$$

● For species i in equilibrium between two phases α and β,

$$\bar{\mu}_i^\alpha = \bar{\mu}_i^\beta \qquad (2.17)$$

We now apply these concepts to specific cases.

As a first example, consider the sparingly soluble salt silver chloride in equilibrium with its ions:

$$AgCl(s) \rightleftarrows Ag^+(aq) + Cl^-(aq)$$

We have

$$\mu_{AgCl}^{0,AgCl} = \bar{\mu}_{Ag^+}^{AgCl} + \bar{\mu}_{Cl^-}^{AgCl} = \bar{\mu}_{Ag^+}^{aq} + \bar{\mu}_{Cl^-}^{aq} \qquad (2.18)$$

given that silver chloride is neutral and has unit activity. The standard free energy change, $\Delta G^\ominus$, for dissolution is given, using (2.11), by

$$\Delta G^\ominus = -\mu_{AgCl}^{0,AgCl} + \mu_{Ag^+}^{0,aq} + \mu_{Cl^-}^{0,aq} = -RT \ln (a_{Ag^+}^{aq} a_{Cl^-}^{aq}) \qquad (2.19)$$

thus obtaining the well-known result

$$\Delta G^\ominus = -RT \ln K_{sp} \qquad (2.20)$$

As a second example we return to the first cell of Section 2.1:

$$Cu' \,|\, Zn \,|\, Zn^{2+}(aq) \,\|\, Cu^{2+}(aq) \,|\, Cu$$

in which the full electrical circuit is now represented, with Cu' and Cu the copper conductor links to a potentiometer (or high-impedance volt-meter), and with a salt bridge—see Fig. 2.1.

The reaction is

$$Cu^{2+} + Zn + 2e^-(Cu) \rightarrow Zn^{2+} + Cu + 2e^-(Cu')$$

We identify the electrons to be transferred with the electrochemical potential of the electrode where they come from. At equilibrium

$$\bar{\mu}_{Cu^{2+}}^{aq} + \bar{\mu}_{Zn}^{Zn} + 2\bar{\mu}_e^{Cu} = \bar{\mu}_{Zn^{2+}}^{aq} + \bar{\mu}_{Cu}^{Cu} + 2\bar{\mu}_e^{Cu'} \qquad (2.21)$$

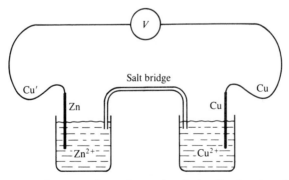

Fig. 2.1. Schematic diagram of an electrochemical cell, showing the links to a high-impedance voltmeter with copper wire.

We know that

$$2(\bar{\mu}_e^{Cu'} - \bar{\mu}_e^{Cu}) = -2F(\phi^{Cu'} - \phi^{Cu}) = 2FE_{cell} \qquad (2.22)$$

therefore (2.21) can be written

$$2FE_{cell} = \mu_{Cu^{2+}}^{0,aq} + RT \ln a_{Cu^{2+}}^{aq} + 2F\phi^{aq} + \mu_{Zn}^{0,Zn}$$
$$- \mu_{Zn^{2+}}^{0,aq} - RT \ln a_{Zn^{2+}}^{aq} - 2F\phi^{aq} - \mu_{Cu}^{0,Cu} \qquad (2.23)$$
$$= -\Delta G_{cell}^{\ominus} + RT \ln (a_{Cu^{2+}}/a_{Zn^{2+}}) \qquad (2.24)$$
$$= 2FE_{cell}^{\ominus} + RT \ln (a_{Cu^{2+}}/a_{Zn^{2+}}) \qquad (2.25)$$

which is the Nernst equation for the whole cell:

$$E_{cell} = E_{cell}^{\ominus} + \frac{RT}{2F} \ln \left(\frac{a_{Cu^{2+}}}{a_{Zn^{2+}}}\right) \qquad (2.5)$$

This type of reasoning is applicable to any cell and always leads to the corresponding Nernst equation.

2.5 Galvanic and electrolytic cells

So far the spontaneous functioning of an electrochemical cell has been described, which corresponds to the transformation of energy obtained in a chemical reaction into electron movement, that is electrical energy. This type of cell is a *galvanic cell*.

By supplying electrical energy from an external voltage source, i.e. applying a potential, we supply electrons of the corresponding energy, allowing the direction of the electrode reactions to be altered. We are able to convert electrical into chemical energy. Here we have an *electrolytic cell* which is much used in electrode reaction studies, and is used industrially in brine electrolysis, in the extraction and refining of metals, in electrosynthesis, etc.

Let us consider the charges on the electrodes in the two cases. At the anode in a galvanic cell, since the oxidation is spontaneous, there is an excess of electrons at the electrode. On the other hand, in an electrolytic cell where oxidation is forced to occur, there is a shortage of electrons and a positive charge. The two situations are:

	anode	cathode
galvanic cell	−	+
electrolytic cell	+	−

Some electrochemical cells can function as galvanic or electrolytic cells. A well-known example is the lead–acid car battery. Under discharge

(supplying current) it is a galvanic cell whose electrode reactions are
anode ($-$ve):

$$Pb + SO_4^{2-} \rightarrow PbSO_4 + 2e^-$$

cathode ($+$ve):

$$PbO_2 + 4H^+ + SO_4^{2-} + 2e^- \rightarrow 2H_2O + PbSO_4 \qquad E_{cell} = 2.05 \text{ V}$$

On recharging the battery these half-reactions are inverted, and electrical
energy has to be supplied.

2.6   Electrode classification

There are many electrode materials, with a great diversity of behaviour.
The historical classification[5,6] is a first approach, and as it is still referred
to it will be described here. Electrode materials, together with the
solutions with which they contact, are divided into four categories:

1. An electrode in contact with a solution of its ions. This can be
subdivided into two cases:
    (a) A metal in contact with its cations, e.g. $Cu \,|\, Cu^{2+}$, where

$$E = E^\ominus + \frac{RT}{F} \ln a_{M^{n+}} \qquad (2.26)$$

and the half-reaction is

$$M^{n+} + ne^- \rightarrow M$$

(b) A non-metal in contact with its ions, e.g. $H_2 \,|\, H^+$ or $Cl_2 \,|\, Cl^-$ on
the surface of an inert conducting substance such as platinum. For the
first of these,

$$E = E^\ominus + \frac{RT}{F} \ln \frac{p_{H_2}^{1/2}}{a_{H^+}} \qquad (2.27)$$

where p_{H_2} is the partial pressure of hydrogen gas.
In this type of electrode the potential arises from electron transfer
between the neutral species and the ion.

2. A metallic electrode in contact with a solution containing anions
that form a sparingly soluble salt with the metal's ions, e.g.
$Hg \,|\, Hg_2Cl_2 \,|\, Cl^-$, the calomel electrode (see Fig. 2.3). The salt activity,
being almost entirely in the solid phase, can be regarded as unity. Thus
the potential is a function only of the anion activity. For the calomel
electrode,

$$E = E^\ominus - \frac{RT}{F} \ln a_{Cl^-} \qquad (2.6)$$

These systems are much used as reference electrodes since, because of the low solubility product of the salt, the potential is very stable. Other examples are Ag | AgCl | Cl⁻ and, for alkaline solution, Hg | HgO | OH⁻.

3. This type of electrode is a source or sink of electrons, permitting electron transfer without itself entering into the reaction, as is the case for the first or second type of electrodes. For this reason they are called *redox* or *inert electrodes*. In reality the concept of an inert electrode is idealistic, given that the surface of an electrode has to exert an influence on the electrode reaction (perhaps small) and can form bonds with species in solution (formation of oxides, adsorption, etc.). Such processes give rise to non-faradaic currents (faradaic currents are due to interfacial electron transfer). This topic will be developed further in subsequent chapters.

The first 'redox' electrode materials to be used were the noble metals, namely gold and platinum, and also mercury. At present this designation includes many types of material such as glassy carbon, different types of graphite, and semiconductor oxides, so long as a zone of potential is employed where surface reactions involving the electrode material do not occur.

4. Electrodes that cannot be grouped into the above categories, e.g. modified electrodes (see Chapter 14).

This classification is useful mainly for electrodes of the first and second types. The great majority of electrodes are, however, of the third or fourth types.

2.7 Reference electrodes

Reference electrodes, as their name suggests, are used to give a value of potential to which other potentials can be referred in terms of a potential difference—potentials can only be registered as differences with respect to a chosen reference value. Thus, a good reference electrode[3,6] needs to have a potential that is stable with time and with temperature and which is not altered by small perturbations to the system—that is, by the passage of a small current. There are three types of reference electrode:

- *Type 1*: e.g. the hydrogen electrode
- *Type 2*: e.g. the calomel electrode
- *Others*: e.g. glass electrodes, Type 3 electrodes, etc.

The standard (or normal) hydrogen electrode is the most important reference electrode because it is the one used to define the standard

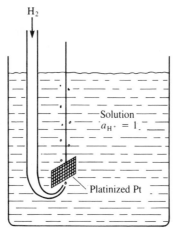

Fig. 2.2. The hydrogen electrode.

electrode potential scale. It is very reproducible, showing differences of only $10\,\mu V$ between different hydrogen electrodes. A typical design is shown schematically in Fig. 2.2.

Normally its construction consists of a platinum foil that is platinized in order to catalyse easily the reaction

$$H^+ + e^- \rightarrow \tfrac{1}{2}H_2$$

Various procedures for platinization exist, but normally involve the deposition of platinum black from a solution of 3 per cent chloroplatinic acid (H_2PtCl_6) containing a small quantity of lead acetate (0.005 per cent)

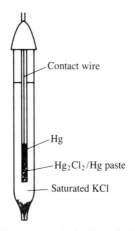

Fig. 2.3. The saturated calomel electrode.

to prolong electrode life. Hydrogen is bubbled in a solution of the electrolyte that is to be used before this is introduced into the cell.

Electrodes of Type 2 are good reference electrodes, as stated in Section 2.6. In Fig. 2.3 a schematic view of a saturated calomel electrode that can be easily introduced into any solution is shown.

Some general precautions need to be taken with reference electrodes, especially when there is a possibility of the formation of complexes involving the sparingly soluble salt. This is the case for many metallic hydroxides which have very low solubility products, suggesting their use in alkaline solution—they often form hydroxy complexes at high hydroxide concentration, which limits their use. Mercury oxide does not have this disadvantage and so is used preferentially—however, care should be taken with the anion concentration in the electrolyte.

Table 2.1 lists half-reactions for electrodes of the second type and their potential for unit activities. These electrodes have, in the majority of cases, their own electrolyte associated with them. So, to calculate the potential of a cell in relation to the standard hydrogen electrode it is necessary to take the liquid junction potential between the two electrolytes into account (Section 2.10).

These electrodes were developed principally for aqueous solution. However, they normally have a porous plug that links the electrolyte within the reference electrode to the solution in the cell (reference electrodes of this kind are rarely used nowadays to carry current, but only to control potentials). Since ion transport through the plug is very small, they can be used for short periods in non-aqueous solvents. There are reference electrodes that have been developed specifically for use in non-aqueous solvents, for example, $Li^+ \mid Li$ in dimethylsulfoxide.

Quasi-reference electrodes such as platinum or silver wires or mercury pools are sometimes used in voltammetric experiments, particularly transient experiments. The advantage is low electrical resistance, but

Table 2.1. Half-reactions for reference electrodes based on sparingly soluble salts in water solvent

	$E^{\ominus}/V$
$AgBr + e^- \rightarrow Ag + Br^-$	0.071
$AgCl + e^- \rightarrow Ag + Cl^-$	0.222
$Hg_2Cl_2 + 2e^- \rightarrow 2Hg + 2Cl^-$	0.268
$HgO + H_2O + 2e^- \rightarrow Hg + 2OH^-$	0.098
$Hg_2SO_4 + 2e^- \rightarrow 2Hg + SO_4^{2-}$	0.613
$TlCl + e^- \rightarrow Tl(Hg) + Cl^-$	-0.557

their potential may vary up to 10 or 20 mV. They are described further in Chapter 7.

2.8   The movement of ions in solution: diffusion and migration

It is important to consider the movement of ions in electrolyte solutions between anode and cathode, and thence some of the properties of electrolyte solutions in general. Solvated ions move at different velocities, according to their size and charge. Diffusion is due to a concentration gradient, and migration to electric field effects. Thus, whilst diffusion occurs for all species, migration affects only charged species (effectively, owing to the existence of dipoles, or induced dipoles in neutral species, a small electric field effect is observed).

Diffusion (Fig. 2.4) is described by Fick's first law:

$$J_i = -D_i \frac{\partial c_i}{\partial x} \tag{2.28}$$

where J_i is the flux of species i of concentration c_i in direction x, and $\partial c/\partial x$ is the concentration gradient. D_i is the proportionality factor between flux and concentration gradient, known as the diffusion coefficient. The negative sign arises because the flux of species tends to annul the concentration gradient.

In the presence of an applied electric field of strength $E = \partial \phi / \partial x$,

$$J_i = -D_i \frac{\partial c_i}{\partial x} - z_i c_i \frac{F}{RT} E \tag{2.29}$$

where the second term on the right-hand side represents migration. This term clearly shows the importance of charge of the species and of the value of $d\phi/dx$ (electric field gradient). Opposing this electric force there are three retarding forces:

- A *frictional force* that depends on the size of the solvated ion

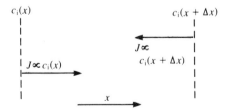

Fig. 2.4. Diffusion in one dimension. The net flux is proportional to $-c_i \, \Delta x$, due to the concentration gradient.

Fig. 2.5. The asymmetric effect on a solvated ion under the influence of an electric field: (a) no field; (b) with field.

● An *asymmetric effect*. Because of ion movement there is a tendency for greater solvation in front of the ion in the direction of its movement than behind it (Fig. 2.5).

● An *electrophoretic effect*. Ion movement causes motion of solvent molecules associated with ions of the opposite sign. The result is a net flux of solvent molecules in the direction contrary to that of the ion considered.

The combination of the attraction of the electric field and the retarding effects leads to a maximum velocity for each ion. Measurement of these velocities gives information about the structure of the solution. Different cation and anion velocities give rise to a potential difference: this is the *liquid junction potential*. It is interesting to know the magnitude of this potential, as it affects the measured potential of the whole electrochemical cell; in other words, ion conductivities need to be measured.

2.9 Conductivity and mobility

The conductivity of a solution is a result of the movement of all ions in solution under the influence of an electric field.

We consider an isolated ion. The force due to the electric field is

$$\boldsymbol{F} = ze\boldsymbol{E} \tag{2.30}$$

which is counterbalanced by a viscous force given by Stokes' equation

$$\boldsymbol{F} = 6\pi\eta r\boldsymbol{v} \tag{2.31}$$

where η is the solution viscosity, r the radius of the solvated ion and $\boldsymbol{v}$ the velocity vector. We neglect other retarding effects. The maximum velocity is, therefore

$$\boldsymbol{v} = \frac{ze\boldsymbol{E}}{6\pi\eta r} \tag{2.32}$$

$$= u\boldsymbol{E} \tag{2.33}$$

where u is the ion mobility, and is the proportionality coefficient between the velocity and electric field strength.

How are conductivity and mobility related? The flux of charge, j, is

$$j = zevcN_A \qquad (2.34)$$

where ze is the charge of each ion, v its velocity, and cN_A the numerical ion density. Writing $eN_A = F$ (one mole of electrons) and substituting

$$j = zvcF \qquad (2.35)$$

$$= zcuF\boldsymbol{E} \qquad (2.36)$$

The current, I, that passes between two parallel electrodes of area A is related to the flux of charge j, and to the potential difference between them, $\Delta\phi$, by

$$I = jA = \kappa \frac{\Delta\phi A}{l} = \kappa EA \qquad (2.37)$$

where κ is the conductivity and l the distance between the electrodes that apply the electric field of strength $\boldsymbol{E} = \Delta\phi/l$. One immediately concludes, by combining (2.36) and (2.37), that for each ion

$$\kappa_i = z_i c_i u_i F \qquad (2.38)$$

Therefore, for the solution (which contains various ions) the measured conductivity, κ, is given by

$$\kappa = F \sum_i |z_i| c_i u_i \qquad (2.39)$$

The molar conductivity of an ion, λ_i, is

$$\lambda_i = \frac{\kappa_i}{c_i} = z_i u_i F \qquad (2.40)$$

and the electrolyte molar conductivity, Λ, is

$$\Lambda = \sum_i \lambda_i = \sum \frac{\kappa_i}{c_i} \qquad (2.41)$$

As can be seen, the measurement of the conductivity of an electrolyte solution is not species selective. Individual ionic conductivities can be calculated only if the conductivity (or mobility) of one ion is known: this in the case of a simple salt solution containing one cation and one anion. If various ions are present, calculation is correspondingly more difficult. Additionally, individual ionic conductivities can vary with solution composition and concentration.

If the electric field is of high intensity (of the order of $100 \, \text{kV cm}^{-1}$) then the conductivity increases with field strength. For strong electrolytes (*first Wien effect*) this is due to the fact that the ions begin to move without their solvent sheath, since the relaxation time for the ionic atmosphere becomes too large—eventually a limiting conductivity value is reached as the field strength increases. For weak electrolytes (*second Wien effect*) the electric field interacts with the dipoles of the undissociated molecule, for example a weak acid, increasing its dissociation constant.

We next consider the relation between mobility and diffusion coefficient. This arises because a concentration gradient is also a chemical potential gradient. For a sufficiently dilute solute, i,

$$\mu_i = \mu_i^\ominus + RT \ln c_i \tag{2.42}$$

and differentiating with respect to distance

$$\left(\frac{\partial \mu_i}{\partial x}\right)_{P,T} = \frac{RT}{c_i}\left(\frac{\partial c_i}{\partial x}\right)_{P,T} \tag{2.43}$$

The diffusive force experienced by a particle i is thus

$$\boldsymbol{F} = -\left(\frac{\partial \mu_i}{\partial x}\right) \tag{2.44}$$

$$= -\frac{RT}{c_i}\left(\frac{\partial c_i}{\partial x}\right)_{P,T} \tag{2.45}$$

The number flux of ions i, J_i, is, from (2.36),

$$J_i = \frac{j_i}{z_i e} = c_i u_i \boldsymbol{E} \tag{2.46}$$

and substituting from (2.30) for the electric field intensity, $\boldsymbol{E}$,

$$J_i = \frac{c_i u_i \boldsymbol{F}}{z_i e} \tag{2.47}$$

Combining (2.45) and (2.47) leads to

$$J_i = -\frac{u_i RT}{z_i F}\left(\frac{\partial c_i}{\partial x}\right)_{P,T} \tag{2.48}$$

Comparison with Fick's first law of diffusion, (2.28), shows that

$$D_i = \frac{u_i RT}{z_i F} \tag{2.49}$$

This is the *Einstein relation,* and shows the direct proportionality between diffusion coefficient and mobility.

The relation between conductivity and diffusion coefficient, the *Nernst–Einstein relation,* is easily derived from (2.40) and (2.49):

$$\lambda_i = \frac{z_i^2 F^2 D_i}{RT} \tag{2.50}$$

This permits the estimation of diffusion coefficients from measurements of conductivity.

Another useful expression concerns the relation between the diffusion coefficient and the viscous drag. From (2.33), one can write

$$u = \frac{ze}{6\pi\eta r} \tag{2.51}$$

Substituting in the Einstein relation, (2.49), we obtain

$$D_i = \frac{k_B T}{6\pi\eta r} \tag{2.52}$$

This is known as the *Stokes–Einstein relation* and is independent of the charge of the species. Using this expression, diffusion coefficients can be estimated from viscosity measurements, so long as Stokes' Law is applicable. It is used particularly for macromolecules.

Sometimes it is useful to know what fraction of the current is transported by each ion. This is its *transport number,* and is given by

$$t_i = \frac{\lambda_i}{\Lambda} = \frac{|z_i|\, c_i u_i}{\Sigma\, |z_j|\, c_j u_j} \tag{2.53}$$

It is evident that

$$\sum t_i = 1 \tag{2.54}$$

The transport number of an ion varies with the ionic constitution of the solution, and is another way of expressing conductivities or mobilities. There are two important methods for measuring transport numbers: the Hittorf method and the moving boundary method[5].

Hittorf method (Fig. 2.6)

An electrolytic cell is divided into three compartments, and a current I is passed. After time t, It/z_+F cations have reached the cathode itself, but only $t_+(It/z_+F)$ cations have reached the cathode compartment. Thus,

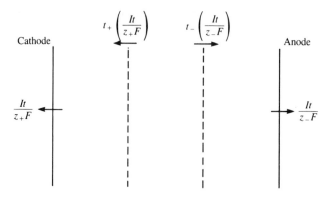

Fig. 2.6. The Hittorf method for determining transport numbers. In the diagram the passage of a current I for time t is shown. It is assumed that $t_+ + t_- = 1$. The electrolytic cell is divided into three compartments.

there is a change in the ionic concentration of the cathode compartment of

$$(t_+ - 1)(It/z_+F) = -t_-(It/z_+F) \tag{2.55}$$

In the anode compartment the change is $-t_+(It/z_-F)$. Thus, the measurement of this change in composition of the anode and cathode compartments leads directly to values of t_+ and t_-.

Moving boundary method (Fig. 2.7)

This method is used to determine the transport number of M in salt MX. A solution of higher density than MX, NX (where $u_N > u_M$), is put in a

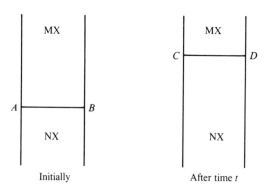

Fig. 2.7. The moving boundary method for determining transport numbers. AB and CD represent the frontiers between MX and NX at the beginning of the experiment and after time t respectively.

Table 2.2. Conductivities and mobilities of some ions in water at infinite dilution $(\lambda = u/zF)^7$

	$\lambda_i/$ S cm^2 mol^{-1}	$10^4 u/$ cm^2 s^{-1} V^{-1}		$\lambda_i/$ S cm^2 mol^{-1}	$10^4 u/$ cm^2 s^{-1} V^{-1}
H^+	349.6	36.2	OH^-	199.1	20.6
Li^+	38.7	4.0	F^-	55.4	5.7
Na^+	50.1	5.2	Cl^-	76.4	7.9
K^+	73.5	7.6	Br^-	78.1	8.1
Rb^+	77.8	8.1	I^-	76.8	8.0
Cs^+	77.2	8.0	NO_3^-	71.5	7.4
NH_4^+	73.5	7.6	ClO_4^-	67.3	7.0
Mg^{2+}	106.0	11.0	SO_4^{2-}	160.0	8.3
Ca^{2+}	119.0	6.2	CO_3^{2-}	138.6	7.5
Cu^{2+}	107.2	5.6			
Zn^{2+}	105.6	5.5			

vertical tube with MX on top. On passing a current I during time t the boundary moves upwards. All ions M from the volume V contained by the old and new boundary have to pass through the new boundary.

Thus the number of ions that pass is cVN_A, equivalent to a charge z_+cVF, and

$$t_+ = \frac{z_+cVF}{It} \tag{2.56}$$

t_- is obtained by subtraction.

Table 2.2 shows some values for mobilities in aqueous solution, extrapolated to infinite dilution. Note that the greater solvation of ions with low atomic number leads to larger values of the solvated radius. The fact that the mobilities of H^+(aq) and OH^-(aq) are so large in comparison with the other ions points to a different transport mechanism, involving the rearrangement of bonds through a long chain of water molecules. There is probably rupture of an O–H bond in H_3O^+ and the rapid formation of a new O–H bond in a neighbouring water molecule. This is the *Grotthus mechanism,* and is illustrated in Fig. 2.8.

Finally, one should note that the mobilities of K^+ and Cl^- are almost equal. It is for this reason that potassium chloride is frequently used in salt bridges in an attempt to avoid the contribution of liquid junction potentials to the cell potential.

Fig. 2.8. The Grotthus mechanism for the movement of H^+ in H_2O.

2.10   Liquid junction potentials

Liquid junction potentials are the result of different cation and anion mobilities under the influence of an electric field. The potential manifests itself in the interface between two different solutions separated by a porous separator or by a membrane. These junctions can be classified into three distinct types:

• Two solutions of the same electrolyte but with different concentrations. The typical case is a concentration cell with transport, e.g.

$$H_2, Pt \mid HCl(a_1) \mid HCl(a_2) \mid Pt, H_2$$

with the partial pressures of hydrogen equal on the two sides. There is a liquid junction between the solutions of hydrochloric acid.

• Two solutions of the same concentration of one of the ions, but the other ion differs.

• Other cases.

The total cell potential is

$$E_{cell} = E_{Nernst} + E_j$$

where E_j is the liquid junction potential. We now consider examples of the first two cases cited above for the calculation of E_j.

Case 1

Consider the liquid junction of the concentration cell given above. There are two phases α and β, which contain hydrochloric acid at different activities. The transport of ions at equilibrium is expressed by

$$t_+\bar{\mu}^{\alpha}_{H^+} + t_-\bar{\mu}^{\beta}_{Cl^-} = t_+\bar{\mu}^{\beta}_{H^+} + t_-\bar{\mu}^{\alpha}_{Cl^-} \tag{2.57}$$

or, removing the terms that are equal on the two sides,

$$t_+\left[RT \ln \frac{a^{\alpha}_{H^+}}{a^{\beta}_{H^+}} + F(\phi^{\alpha} - \phi^{\beta}) \right] = -t_-\left[RT \ln \frac{a^{\beta}_{Cl^-}}{a^{\alpha}_{Cl^-}} - F(\phi^{\beta} - \phi^{\alpha}) \right] \tag{2.58}$$

If we write

$$a^{\alpha}_{H^+} = a^{\alpha}_{Cl^-} = a^{\alpha}$$

and

$$a^{\beta}_{H^+} = a^{\beta}_{Cl^-} = a^{\beta}$$

we easily reach

$$E_j = (\phi^{\beta} - \phi^{\alpha}) = (t_+ - t_-)\frac{RT}{F} \ln \frac{a^{\alpha}}{a^{\beta}} \tag{2.59}$$

Evidently this argument is not very rigorous, in that the interfacial region is not considered. However, if we assume a linear change in concentration and invariant transport numbers in the interfacial region we arrive at the result of (2.59) corresponding to a symmetric variation of interfacial potential.

Case 2.

Here it is not very correct to assume that the concentration gradients vary linearly through the junction, especially because the concentration profiles depend on the technique of junction formation. Assuming that activities are equal to concentrations and that there is, in fact, a linear transition, we obtain the *Henderson equation*

$$E_j = \frac{\sum_i \frac{|z_i| u_i}{z_i} [c_i(\beta) - c_i(\alpha)]}{\sum_i |z_i| u_i [c_i(\beta) - c_i(\alpha)]} \frac{RT}{F} \ln \frac{\sum_i |z_i| u_i c_i(\alpha)}{\sum_i |z_i| u_i c_i(\beta)} \tag{2.60}$$

For a 1:1 electrolyte this reduces to

$$E_j = \pm \frac{RT}{F} \ln \frac{\Lambda_\beta}{\Lambda_\Lambda} \tag{2.61}$$

which is the *Lewis–Sargent relation*, the positive sign corresponding to a common cation and the negative sign to a common anion.

Minimization of the liquid junction potential is commonly carried out using a salt bridge in which the ions have almost equal mobilities. One example is potassium chloride ($t_+ = 0.49$ and $t_- = 0.51$) and another is potassium nitrate ($t_+ = 0.51$ and $t_- = 0.49$). If a large concentration of electrolyte is used in the salt bridge this dominates the ion transport through the junctions such that the two values of E_j have the same magnitude but opposing polarities. The result is that they annul each other. In this way values of E_j can be reduced to 1–2 mV.

2.11   Liquid junction potentials, ion-selective electrodes, and biomembranes

We consider again (2.59):

$$E_j = (\phi^\beta - \phi^\alpha) = (t_+ - t_-) \frac{RT}{F} \ln \frac{a^\alpha}{a^\beta} \tag{2.59}$$

If it were possible to have an interface permeable to only one ion, then the transport number of that ion would be unity and

$$E_j = \frac{RT}{F} \ln \frac{a_i^\alpha}{a_i^\beta} \tag{2.62}$$

or, in general, for an ion of charge z_i,

$$E_m = \frac{RT}{z_i F} \ln \frac{a_i^\alpha}{a_i^\beta} \tag{2.63}$$

E_m, the corresponding liquid junction potential, is called the *membrane potential* or *Donnan potential*. Ideally E_m changes in a Nernstian fashion with the activity of the ion in one of the phases, the activity in the other phase being held constant. This is the basis of the functioning of ion-selective electrodes (Chapter 13) and, to a good approximation, of biomembranes (Chapter 17).

2.12 Electrode potentials and oxidation state diagrams

In calculating the electromotive force of an electrochemical cell and to calculate the maximum possible energy produced one uses the expression

$$\Delta G = -nFE \tag{2.64}$$

In fact, each electron transfer half-reaction involves a free energy change following this formula. To reach the total variation in ΔG we sum the contributions of the two half-reactions, remembering that one is a reduction and the other an oxidation. For example, for the dissolution of silver chloride under standard conditions

$$\begin{array}{llll} \text{AgCl} + \text{e}^- \rightarrow \text{Ag} + \text{Cl}^- & E^\ominus = +0.22 \text{ V} & \Delta G_1^\ominus = -0.22 \text{ F} \\ \text{Ag}^+ + \text{e}^- \rightarrow \text{Ag} & E^\ominus = +0.80 \text{ V} & \Delta G_2^\ominus = -0.80 \text{ F} \end{array}$$

Thus, for

$$\text{AgCl} \rightarrow \text{Ag}^+ + \text{Cl}^-$$

we have $\Delta G_{\text{tot}}^\ominus = \Delta G_1^\ominus - \Delta G_2^\ominus = +0.58 \text{ F}$. From the relation

$$\Delta G^\ominus = -RT \ln K \tag{2.65}$$

we can calculate that

$$K = \exp\left(-\Delta G_{\text{tot}}^\ominus / RT\right) = \exp\left(-0.58 \text{ F}/RT\right) = 1.55 \times 10^{-10}.$$

This value of K can be identified with the solubility product—the differences that arise in relation to tabulated values of K_{sp} are due to differences in the solution conditions. In the same way we can calculate stability constants of complexes.

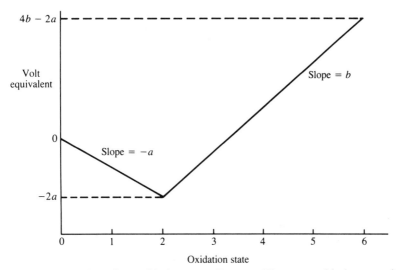

Fig. 2.9. Construction of an oxidation state diagram. The zero oxidation state has a volt-equivalent of zero (arbitrary). The slopes of the lines have values corresponding to $E^{\ominus}$.

$$M(II) + 2e^- \rightarrow M(0), \qquad E^{\ominus} = -a$$
$$M(VI) + 4e^- \rightarrow M(II), \qquad E^{\ominus} = b$$

It is very useful to be able to visualize which reactions have equilibrium constants larger than unity and which do not, without the necessity of going through these calculations. *Oxidation state diagrams*[8,9], also known as Frost diagrams, permit this visualization. These diagrams are plots of free energy on the ordinate (relative to an arbitrary zero which is generally—but not necessarily—the zero oxidation state) vs. the oxidation state on the abscissa. Since the free energy change is proportional to $nE^{\ominus}$, called volt-equivalent (VE), lines of slope $E^{\ominus}$ are drawn on the plot that join as shown in the diagram of Fig. 2.9.

Some conclusions can be drawn from these diagrams:

● For a given element, the species that is lowest on the diagram corresponds to the most stable oxidation state relative to the reference.

● The species highest on the diagram are the least stable. In practice they are those with highest oxidation states and are therefore strong oxidizing agents (great tendency to be reduced).

● Disproportionation and proportionation reactions occur when there exist situations of the type shown in Fig. 2.10. It is easy to predict these reactions from the diagrams. It should be noted, however, that these are thermodynamic predictions and kinetics can cause the reactions to

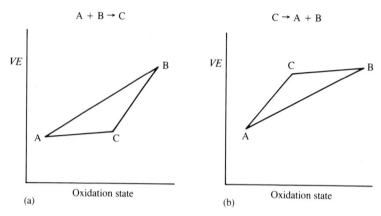

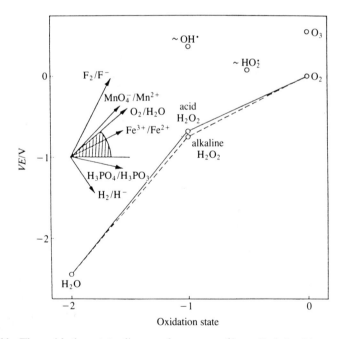

Fig. 2.10. Oxidation state diagrams for: (a) proportionation—C is below the straight line linking A and B; (b) disproportionation—C is above the straight line linking A and B.

Fig. 2.11. The oxidation state diagram for oxygen (from Ref. 8 with permission).

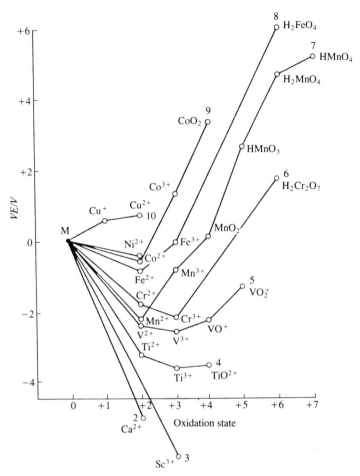

Fig. 2.12. The oxidation state diagram for transition metals of the first series (from Ref. 9 with permission).

proceed only very slowly at room temperature, as for example the disproportionation

$$3ClO^- \rightarrow 2Cl^- + ClO_3^-$$

The diagrams are normally constructed in relation to the hydrogen electrode with $a_{H^+} = 1$ as reference, which would be a horizontal straight line. However, they can be modified for the reference to be another pH or another half-reaction. If we choose as reference the reduction of oxygen to water under standard conditions, for example, which has $E^{\ominus} = +1.23$ V, the modification corresponds to a clockwise rotation of the diagram.

Examples are given in Figs. 2.11 and 2.12. Note, in Fig. 2.11, the tendency for hydrogen peroxide to disproportionate into water and oxygen and in Fig. 2.12 that the most stable species have oxidation states of 2 or 3, and that oxidation state 6 or 7 is strongly oxidizing (e.g. MnO_4^-).

More details on the use of these diagrams may be found in Refs. 8 and 9.

References

1. A. J. Bard, R. Parsons, and J. Jordan (ed.), *Standard potentials in aqueous solution,* Dekker, New York, 1985.
2. J. J. Lingane, *Electroanalytical chemistry,* Wiley-Interscience, New York, 1958.
3. D. J. G. Ives and G. J. Janz (ed.), *Reference electrodes,* Academic Press, New York, 1961.
4. J. S. Newman, *Electrochemical systems,* Prentice Hall, Englewood Cliffs, NJ, 1973.
5. D. A. MacInnes, *Principles of electrochemistry,* Dover, New York, 1961.
6. D. T. Sawyer and J. L. Roberts, *Experimental electrochemistry for chemists,* Wiley-Interscience, New York, 1974, Chapter 2.
7. R. A. Robinson and R. H. Stokes, *Electrolyte solutions,* Butterworth, London, 1959.
8. C. S. G. Philips and R. J. P. Williams, *Inorganic chemistry,* Oxford University Press, 1965.
9. D. F. Shriver, P. W. Atkins, and C. H. Langford, *Inorganic chemistry,* Oxford University Press, 1990, Chapter 8.

3

THE INTERFACIAL REGION

3.1 Introduction

In the previous chapter the physical nature of the interface where electrode reactions occur was not considered. The thermodynamic driving force and how the reactions take place depends on the structure of the interfacial region. Until recently only the solution side of the interfacial region was taken into account, and this has been the subject of many review articles, e.g. Refs. 1–5, but currently, given the increasing utilization of semiconductor electrodes, there is much interest in understanding the behaviour of the part of the interfacial region in the solid[4,6–8]. For reasons linked with the historical development of theoretical models the interfacial region in solution is known as the *electrolyte double layer region* and the interfacial region in the solid the *space–charge region* (Fig. 3.1). In metals the latter is very thin.

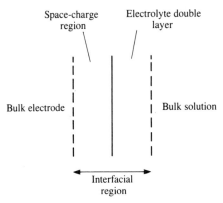

Fig. 3.1 Schematic illustration of the electrode–solution interface.

In this chapter the structure of the electrolyte double layer, and the consequences of adsorption on the electrode surface, are described. The effect of differences in structure and electronic distribution of different metals are indicated. The space–charge region in semiconductors is then discussed. Finally some properties of colloids are mentioned, given that they possess an interfacial region very similar to an electrode.

3.2   The electrolyte double layer: surface tension, charge density, and capacity

The interfacial region in solution is the region where the value of the electrostatic potential, ϕ, differs from that in bulk solution. The designation 'double layer' reflects the first models developed to describe the region, see Section 3.3. The basic concept was of an ordering of positive or negative charges at the electrode surface and ordering of the opposite charge and in equal quantity in solution to neutralize the electrode charge. The function of the electrode was only to supply electrons to, or remove electrons from, the interface: the charge at the interface depending on applied potential. More sophisticated models required accurate experimental observations.

The proportionality constant between the applied potential and the charge due to the species ordering in the solution interfacial region is the *double layer capacity*. The study of the double layer capacity at different applied potentials can be done by various methods. One much used is the *impedance technique,* which is applicable to any type of electrode, solid or liquid, and is described in Chapter 11. Another method uses *electrocapillary measurements.* It was developed for the mercury electrode, being only applicable to liquid electrodes, and is based on measurement of surface tension.

The principle of electrocapillary measurements was described more than a century ago by Lippmann[9]. It is a null-point technique that counterbalances the force of gravity and surface tension, and highly accurate results can be obtained.

The experimental system is shown in Fig. 3.2. It consists of a capillary column containing mercury up to height h regulated so that, on altering the applied potential, the mercury/solution interface stays in the same position. Under these conditions surface tension counterbalances the force of gravity, according to

$$2\pi r_c \gamma \cos \theta = \pi r_c^2 \rho_{Hg} h g \qquad (3.1)$$

where r_c is the capillary radius, θ is the contact angle (see Fig. 3.2), and γ

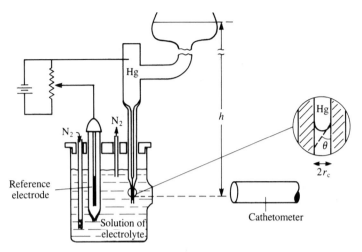

Fig. 3.2 The experimental arrangement for measurement of surface tension of mercury by Lippmann's method.

is the surface tension, ρ_{Hg} being the density of mercury. The contact angle is measured with a microscope. A plot of γ vs. E is called an *electrocapillary curve* and has the form of Fig. 3.3a.

A variation on this method consists in using the dropping mercury electrode[10] (Section 8.3). The mass flux, m_1, is

$$m_1 = \frac{\pi r_c^2 \rho_{Hg} h}{\tau} \tag{3.2}$$

where τ is the drop lifetime. Substituting in (3.1) we obtain

$$2\pi r_c \gamma = m_1 g t \tag{3.3}$$

Thus a plot of τ vs. E gives a curve of the same form as the electrocapillary curve (Fig. 3.3a).

Conversion of values of γ into capacities is done by double differentiation in relation to the electrostatic potential difference $\Delta\phi$ between its value in the metal, ϕ_M, and that in the solution, ϕ_S. The first derivative gives the charge on the interface, and is the Lippmann equation

$$\frac{\partial\gamma}{\partial\Delta\phi} = -\sigma_M = \sigma_S \tag{3.4}$$

where σ_M is the charge on the metal and σ_S the charge of the solution such that $\sigma_M + \sigma_S = 0$. If we define an arbitrary reference potential and

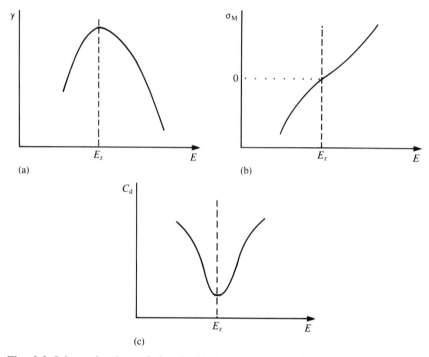

Fig. 3.3 Schematic plots of the double layer region. (a) Electrocapillary curve (surface tension, γ, vs. potential); (b) Charge density on the electrode, σ_M, vs. potential; (c) Differential capacity, C_d, vs. potential. Curve (b) is obtained by differentiating curve (a), and (c) by differentiation of (b), E_z is the point of zero charge.

the potential in relation to that reference is E_Δ then $\partial(\Delta\phi) \sim \delta(E_\Delta)$ and

$$\frac{\partial \gamma}{\partial E_\Delta} = -\sigma_M \tag{3.5}$$

The fact that the Lippmann equation is the derivative of the electrocapillary equation shows that the charge σ_M is zero when the slope of the electrocapillary curve is zero. The potential where this occurs is called the point of zero charge, E_z, and occurs at the maximum in the electrocapillary curve, see Fig. 3.3.

A second differentiation of the electrocapillary curve gives the value of the interfacial capacity. There are, however, two definitions of this:

● The *differential capacity* C_d. This is the derivative of the curve of σ_M vs. E (Fig. 3.3c), whose minimum value occurs for $E = E_z$:

$$C_d = \frac{\partial \sigma_M}{\partial E_\Delta} \tag{3.6}$$

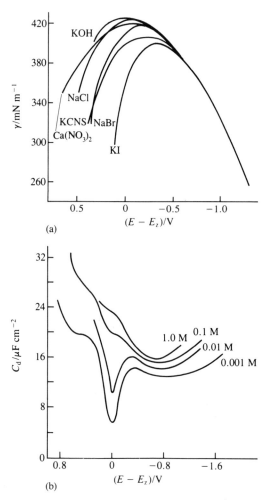

Fig. 3.4 (a) Electrocapillary curves for mercury in contact with various electrolytes (E_z is the value for sodium fluoride); (b) Variation of the differential capacity of sodium fluoride at mercury with potential. (From Ref. 11 with permission.)

● The *integral capacity*, C_i. Measuring σ_M for two reasonably different potentials, the value of the calculated capacity is the average value in that zone, assuming that C_d varies with E. This is the integral capacity that is zero at $E = E_z$, the point of zero charge

$$C_i = \frac{\sigma_M}{E - E_z} \tag{3.7}$$

$$= \frac{\displaystyle\int_{E_z}^{E} C_d \, dE}{\displaystyle\int_{E_z}^{E} dE} \tag{3.8}$$

Figure 3.4 gives examples of real electrocapillary curves and differential capacity curves. Double layer models have to explain the shape of these curves.

The impedance technique gives values of C_d directly. It consists of the application of a small sinuisoidal perturbation superimposed on a fixed applied potential. The component of the resulting current that is out of phase with the applied signal leads to calculation of the differential capacity of the interface. More details are given in Chapter 11. Thus, values of σ_M and γ are obtained by integration. Besides making the use of solid, and not only liquid, electrodes possible, another advantage is that integration tends to reduce the errors in the experimental measurements, whereas differentiation increases them.

3.3 Double layer models

Any double layer model has to explain experimental results, for example in Fig. 3.4 for sodium fluoride at a mercury electrode. Until the 1960s measurements were made almost exclusively at mercury electrodes and models were developed for this electrode. The fact that mercury is an ideally polarizable liquid in the zone negative to the hydrogen electrode means that its behaviour is often different from solid electrodes (monocrystalline and polycrystalline). These models are, therefore, of a predominantly electrostatic nature.

Nevertheless, an important application of electrostatic models is to the interface between two immiscible electrolyte solutions. This can be viewed as two electrolyte double layers arranged back to back. In reality, however, total immiscibility never occurs and the degree of miscibility increases with the presence of electrolyte, so that corrections to the models need to be introduced.

It was only after making measurements with solid electrodes that the concept of the energy associated with the electrode's electronic distribution in the interfacial region was introduced. This distribution depends on the electrode material as well as on its crystalline structure and exposed crystallographic face. However, it is interesting to see the historical evolution of the models, given that successively more factors that reflect the structure have been introduced.

The first models: Helmholtz, Gouy–Chapman, Stern and Grahame

Helmholtz Model (1879)

The first double layer model, due to Helmholtz[12], considered the
ordering of positive and negative charges in a rigid fashion on the two
sides of the interface, giving rise to the designation of double layer (or
compact layer), the interactions not stretching any further into solution.
This model of the interface is comparable to the classic problem of a
parallel-plate capacitor. One plate would be on the contact surface
metal/solution. The other, formed by the ions of opposite charge from
solution rigidly linked to the electrode, would pass through the centres of
these ions (Fig. 3.5a). So x_H would be the distance of closest approach of
the charges, i.e. ionic radius, which, for the purpose of calculation, were
treated as point charges. By analogy with a capacitor the capacity would
be

$$C_{d,H} = \frac{\epsilon_r \epsilon_0}{x_H} \qquad (3.9)$$

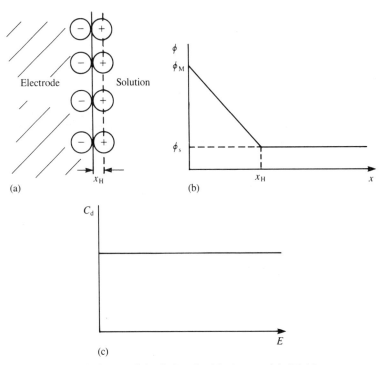

(a)

(b)

(c)

Fig. 3.5 The Helmholtz model of the double layer. (a) Rigid arrangement of
ions; (b) Variation of the electrostatic potential, ϕ, with distance x, from the
electrode; (c) Variation of C_d with applied potential.

where ϵ_r is the relative permittivity (which is assumed not to vary with distance) and ϵ_0 the permittivity of vacuum. A typical value of ϵ_r is 6–7, leading to $C_{d,H} = 10\ \mu F\ cm^{-2}$. The decay of the electrostatic potential from ϕ_M to ϕ_S is linear (Fig. 3.5b) and $C_{d,H}$ does not vary with the potential applied to the electrode (Fig. 3.5c).

The two principal defects of this model are first that it neglects interactions that occur further from the electrode than the first layer of adsorbed species, and secondly that it does not take into account any dependence on electrolyte concentration.

Gouy–Chapman Model (1910–1913)

At the beginning of this century Gouy[13] and Chapman[13] independently developed a double layer model in which they considered that the applied potential and electrolyte concentration both influenced the value of the double layer capacity. Thus, the double layer would not be compact as in Helmholtz's description but of variable thickness, the ions being free to move (Fig. 3.6a). This is called the *diffuse double layer*.

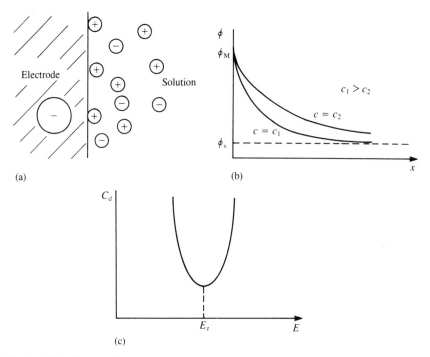

(a)

(b)

(c)

Fig. 3.6 The Gouy–Chapman model of the double layer. (a) Arrangement of the ions in a diffuse way; (b) Variation of the electrostatic potential, ϕ, with distance, x, from the electrode, showing effect of ion concentration, c. (c) Variation of C_d with potential, showing the minimum at the point of zero charge E_z.

In this model, the distribution of species with distance from the electrode obeys Boltzmann's law

$$n_i = n_i^0 \exp\left[\frac{-z_i e \phi_\Delta}{k_B T}\right] \tag{3.10}$$

where $\phi_\Delta = \phi - \phi_S$ and n_i^0 is the numerical concentration of ions i in bulk solution. Dividing the solution into slices of thickness dx, at distance x from the electrode the charge density is

$$\rho(x) = \sum_i n_i z_i e \tag{3.11}$$

$$= \sum_i n_i^0 z_i e \exp\left[\frac{-z_i e \phi_\Delta}{k_B T}\right] \tag{3.12}$$

for all ions i.

The Poisson equation relates the potential with the charge distribution

$$\frac{\partial^2 \phi_\Delta(x)}{\partial x^2} = -\frac{\rho(x)}{\epsilon_r \epsilon_0} \tag{3.13}$$

Combining (3.12) and (3.13) we obtain the Poisson–Boltzmann equation

$$\frac{\partial^2 \phi_\Delta(x)}{\partial x^2} = -\frac{e}{\epsilon_r \epsilon_0} \sum_i n_i^0 z_i \exp\left(\frac{-z_i e \phi_\Delta}{k_B T}\right) \tag{3.14}$$

This equation is precisely equal to that in the Debye–Hückel treatment of ionic interaction for dilute electrolyte solutions[14], only that the distance x refers to a central ion (point charge) and not to an electrode. In the Debye–Hückel case, since the central ion is small and ϕ_Δ small we can make the approximation $(e\phi_\Delta / k_B T)^2 \ll 1$, and use only the first term of the exponential expansion. For an electrode, which is much larger (an electrode can be thought of as a giant ion) the linear approximation is not valid.

In solving (3.14) we use the property of derivatives

$$\frac{\partial^2 \phi_\Delta(x)}{\partial x^2} = \frac{1}{2} \frac{\partial}{\partial \phi_\Delta}\left(\frac{\partial \phi_\Delta}{\partial x}\right)^2 \tag{3.15}$$

In this way the Poisson–Boltzmann equation can be rewritten as

$$\left(\frac{\partial \phi_\Delta}{\partial x}\right)^2 = -\frac{2e}{\epsilon_r \epsilon_0} \sum_i n_i^0 z_i \exp\left(\frac{-z_i c \phi_\Delta}{k_B T}\right) d\phi \tag{3.16}$$

Integrating for the following boundary conditions

$$x = 0 \qquad \phi_\Delta = \phi_{\Delta,0} \tag{3.17a}$$

$$x \to \infty \qquad \phi_\Delta \to 0 \quad (\partial / \phi_\Delta / \partial x) = 0 \tag{3.17b}$$

we get

$$\left(\frac{\partial \phi_\Delta}{\partial x}\right)^2 = \frac{2k_B T}{\epsilon_r \epsilon_0} \sum_i n_i^0 \left[\exp\left(\frac{-z_i e \phi_\Delta}{k_B T}\right) - 1 \right] \tag{3.18}$$

For a $z:z$ electrolyte

$$\frac{\partial \phi_\Delta}{\partial x} = \left(\frac{8kTn_i^0}{\epsilon_r \epsilon_0}\right)^{1/2} \sinh\left(\frac{ze\phi_\Delta}{2kT}\right) \tag{3.19}$$

This equation can be integrated, if written in the form

$$\int_{\phi_{\Delta,0}}^{\phi_\Delta} \frac{d\phi_\Delta}{\sinh\left(\dfrac{ze\phi_\Delta}{2k_B T}\right)} = -\left(\frac{8k_B Tn_i^0}{\epsilon_r \epsilon_0}\right)^{1/2} \int_0^x dx \tag{3.20}$$

The result is

$$\frac{2k_B T}{ze} \ln\left[\frac{\tanh(ze\phi_\Delta/4k_B T)}{\tanh(ze\phi_{\Delta,0}/4k_B T)}\right] = -\left(\frac{8k_B Tn_i^0}{\epsilon_r \epsilon_0}\right)^{1/2} x \tag{3.21}$$

We can write

$$\left[\frac{\tanh(ze\phi_\Delta/4k_B T)}{\tanh(ze\phi_{\Delta,0}/4k_B T)}\right] = \exp\left[-x/x_{DL}\right] \tag{3.22}$$

in which x_{DL} is a distance characteristic of the diffuse layer thickness

$$x_{DL} = \left(\frac{\varepsilon_r \varepsilon_0 k_B T}{2n_i^0 z^2 e^2}\right)^{1/2} \tag{3.23}$$

For water ($\epsilon_r = 78$) at 298 K, $x_{DL} = 3.04 \times 10^{-8} z^{-1} c^{-1/2}$ cm. Note that the drop in potential with distance is faster for higher concentrations and that $x_{DL} \propto T^{1/2}$ reflecting the thermal energy of the ions (equation (3.23)). If $c_\infty = 1.0$ M and $z = 1$ then, from (3.23), $x_{DL} = 0.3$ nm. The decay of potential is shown in Fig. 3.6b.

The value of C_d is easily obtained from (3.19). The charge density of the diffuse layer is

$$\sigma_M = \epsilon_r \epsilon_0 \left(\frac{\partial \phi_\Delta}{\partial x}\right)_{x=0} \tag{3.24}$$

$$= (8kT\epsilon_r \epsilon_0 n_i^0)^{1/2} \sinh\left(\frac{ze\phi_{\Delta,0}}{2k_B T}\right) \tag{3.25}$$

Differentiating,

$$C_{d,GC} = \frac{\partial \sigma_M}{\partial \phi_{\Delta,0}} = \left(\frac{2z^2 e^2 \epsilon_r \epsilon_0 n_i^0}{k_B T}\right)^{1/2} \cosh\left(\frac{ze\phi_{\Delta,0}}{2k_B T}\right) \tag{3.26}$$

The cosh term gives rise to the variation in capacity with potential shown in Fig. 3.6c. The minimum in the curve is identifiable with the point of zero charge, E_z, and the curve is symmetric around E_z.

For dilute aqueous solutions at 298 K,

$$C_{d,GC} = 228zc_\infty^{1/2} \cosh (19.5z\phi_{\Delta,0}) \; \mu F \; cm^{-2}$$

This model is better than a parallel-plate capacitor for simulating curves such as in Fig. 3.4b, but only close to E_z: in reality, far from E_z, C_d is, to a first approximation, independent of potential. We remember the approximation that ions are considered as point charges and that, consequently, there is no maximum concentration of ions close to the electrode surface!

Stern Model (1924)

Stern[15] combined the Helmholtz model for values of potential far from E_z with the Gouy–Chapman model for values close to E_z (Fig. 3.7a,b). He considered that the double layer was formed by a compact layer of ions

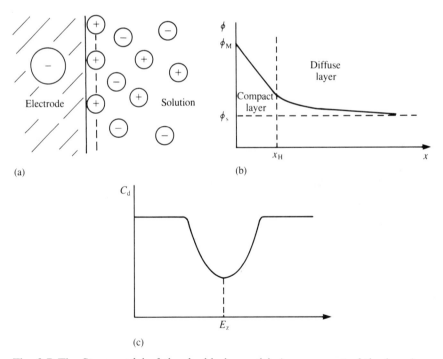

(a)   (b)   (c)

Fig. 3.7 The Stern model of the double layer. (a) Arrangement of the ions in a compact and a diffuse layer; (b) Variation of the electrostatic potential, ϕ, with distance, x, from the electrode; (c) Variation of C_d with potential.

next to the electrode followed by a diffuse layer extending into bulk solution. The physical explanation of the experimental measurements is that, far from E_z, the electrode exerts a strong attraction towards the ions that are therefore attached rigidly to the surface, all the potential drop being restricted to within the distance corresponding to the first layer of ions (compact layer). Close to E_z there is a diffuse distribution of ions (diffuse layer).

In mathematical terms this is equivalent to two capacitors in series, with capacities C_H representing the rigid compact layer and G_{GC} representing the diffuse layer. The smaller of the two capacities determines the observed behaviour:

$$\frac{1}{C_d} = \frac{1}{C_H} + \frac{1}{C_{GC}} \tag{3.27}$$

$$= \frac{x_H}{\epsilon_r \epsilon_0} + \frac{1}{(2\epsilon_r \epsilon_0 z^2 e^2 n_i^0 / k_B T)^{1/2} \cosh(ze\phi_{A,0}/2k_B T)} \tag{3.28}$$

Fig. 3.7c shows the variation of the total capacity with potential. There are two extreme cases:

- close to E_z, $C_H \gg C_{GC}$ and so $C_d \sim C_{GC}$
- far from E_z, $C_H \ll C_{GC}$ and $C_d \sim C_H$

which satisfies the assumptions of the model.

As in the Gouy–Chapman model, the more concentrated the electrolyte the less the importance of the thickness of the diffuse layer and the more rapid the potential drop. At distance x_H there is the transition from the compact to the diffuse layer. The separation plane between the two zones is called the *outer Helmholtz plane* (OHP): the origin of the inner Helmholtz plane will be discussed below.

Comparison between Figs 3.7c and 3.4b shows that this model is the best of the three so far, but does not yet explain all the facets of the curves. Indeed, as already mentioned, mercury, as a liquid, is a special case. Results with other electrolytes and with solid electrodes show a more complicated behaviour.

Grahame Model (1947)

In spite of the fact that Stern had already distinguished between ions adsorbed on the electrode surface and those in the diffuse layer, it was Grahame[11] who developed a model that is constituted by three regions (Fig. 3.8). The difference between this and the Stern model is the existence of specific adsorption (Section 3.4): a specifically adsorbed ion loses its solvation, approaching closer to the electrode surface—besides this it can have the same charge as the electrode or the opposite charge,

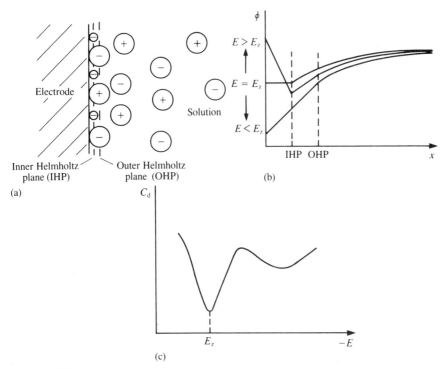

Fig. 3.8 The Grahame model of the double layer, for the mercury electrode. (a) Arrangement of ions; (b) Variation of the electrostatic potential, ϕ, with distance, x, from the electrode, according to the applied potential; (c) Variation of C_d with potential.

but the bonding is strong. The *inner Helmholtz plane* (IHP) passes through the centres of these ions. The outer Helmholtz plane (OHP) passes through the centres of the solvated and non-specifically adsorbed ions. The diffuse region is outside the OHP.

   In both the Stern and Grahame models, the potential varies linearly with distance until the OHP and then exponentially in the diffuse layer.

Bockris, Devanathan, and Müller Model (1963)

More recent models of the double layer have taken into account the physical nature of the interfacial region. In dipolar solvents, such as water, it is clear that an interaction between the electrode and the dipoles must exist. That this is important is reinforced by the fact that solvent concentration is always much higher than solute concentration. For example, pure water has a concentration of 55.5 mol dm^{-3}.

The interfacial region

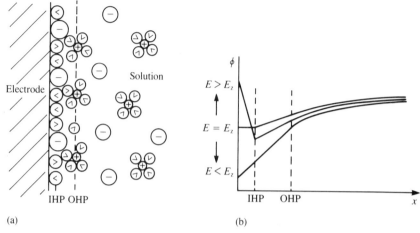

(a)   (b)

Fig. 3.9 The model of Bockris *et al.* of the double layer. (a) Arrangement of ions and solvent molecules; ⊘ represents a water molecule; (b) Variation of the electrostatic potential, ϕ, with distance, x, from the electrode.

The Bockris, Devanathan, and Müller model[16] recognizes this situation and shows the predominance of solvent molecules near the interface (Fig. 3.9). The solvent dipoles are oriented according to the electrode charge where they form a layer together with the specifically adsorbed ions.

Regarding the electrode as a giant ion, the solvent molecules form its first solvation layer; the IHP is the plane that passes through the centre of these dipoles and specifically adsorbed ions. In a similar fashion, OHP refers to adsorption of solvated ions that could be identified with a second solvation layer. Outside this comes the diffuse layer. Note that the actual profile of electrostatic potential variation with distance (Fig. 3.9*b*) is the same in qualitative terms as in the Grahame model (Fig. 3.8*b*).

These authors also defined a shear plane, not necessarily coincident with the outer Helmholtz plane, which is extremely important in electrokinetic effects (Section 3.7). The shear plane limits the zone where the rigid holding of ions owing to the electrode charge ceases to operate. The potential of this plane is called the zeta or electrokinetic potential, ζ.

'Chemical' models

The concept of double layer structure is far from being well established and evaluated. The models presented above give emphasis to electrostatic considerations. 'Chemical' models have been developed that consider the electronic distribution of the atoms in the electrode, which is related to their work function. This was only possible after experimental

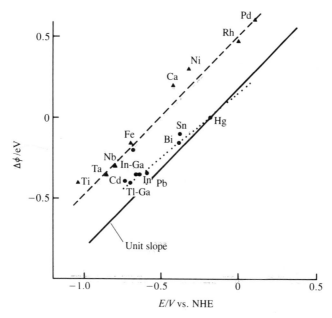

Fig. 3.10 Variation of potential of zero charge with metal work function for sp (●) and d (▲) metals; (——) line of unit slope (from Ref. 18 with permission).

measurements at a range of solid electrodes of different natures had been carried out. For example, there is a difference between sp metals and transition metals (Fig. 3.10). Since the first model of this kind proposed by Damaskin and Frumkin[17], and based on these principles, there has been a gradual evolution in the models, reviewed by Trasatti[18] and more recently by Parsons[19]. The break in the structure of the solid causes a potential difference that begins within the solid—the *surface potential* (Fig. 3.13).

The interfacial region of a metal up to the IHP has been considered as an electronic molecular capacitor, and this model has explained many experimental results with success[20]. Another important model is the jellium model[21] (Fig. 3.13*b*). From an experimental point of view, the development of *in situ* infrared and Raman spectroscopic techniques (Chapter 12) to observe the structure, and the calculation of the bond strength at the electrode surface can better elucidate the organization of the double layer. Other surface analytical techniques such as EXAFS are also valuable.

Some of these ideas are developed in Sections 3.5 and 3.6.

3.4 Specific adsorption

As explained in the description of the Grahame model for the double layer, specific adsorption is the adsorption of ions at the electrode surface after losing their solvation partially or completely. These ions can have the same charge or the opposite charge to the electrode. Bonds formed with the electrode in this way are stronger than for solvated ions.

The idea of the existence of specific adsorption appeared as an explanation for the fact that electrocapillary curves at mercury electrodes are different for different electrolytes at the same concentration (Fig. 3.4a). For sodium and potassium halides in water the differences arise at potentials positive of E_z, which suggests an interaction with the anions. As the effect is larger the smaller the ionic radius of the anion, the idea of specific adsorption with partial or total loss of hydration arose.

The degree of specific adsorption should vary with electrolyte concentration, just as there should be a change in the point of zero charge due to specific adsorption of charges. This is the *Esin–Markov effect,* expressed by the Esin–Markov coefficient, β:

$$\beta = \frac{1}{RT}\left(\frac{\partial(\delta E_z)}{\partial \ln a}\right)_{\sigma_M} = \left(\frac{\partial(\delta E_z)}{\partial \mu}\right)_{\sigma_M} \tag{3.29}$$

This derivative is equal to zero in the absence of specific adsorption. For anion adsorption, and constant charge density, the point of zero charge moves in the negative direction in order to counterbalance adsorption. For cations, E_z moves in the positive direction, assuming constant charge density. In aqueous solution, specific adsorption only occurs close to E_z; far from E_z, solvent molecules are attracted so strongly that it is difficult to push them out of the way.

Experimentally it is observed that specific adsorption occurs more with anions than with cations. This is in agreement with chemical models of the interfacial region. Since, according to the free electron model, a metallic lattice can be considered as a cation lattice in a sea of electrons in free movement, it is logical to expect a greater attraction for anions in solution.

The degree of adsorption depends on electrolyte concentration. The degree of coverage of a surface by specific adsorption of ions can be described by monolayer adsorption isotherms (Fig. 3.11). Three types of isotherm are generally considered:

• *Langmuir isotherm* (Fig. 3.11a). It is assumed that there is no interaction between adsorbed species, that the surface is smooth, and that eventually surface saturation occurs. Thus, if θ is the fraction of

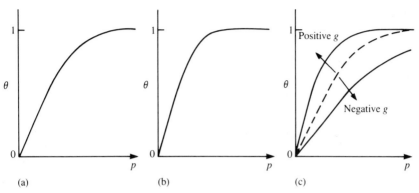

Fig. 3.11 Adsorption isotherms: (a) Langmuir; (b) Temkin; (c) Frumkin.

coverage

$$\frac{\theta}{1-\theta} = \beta_i a_{i,\infty} \tag{3.30}$$

where β_i is the energetic coefficient of proportionality and $a_{i,\infty}$ the activity of species i in bulk solution.

● *Temkin isotherm* (Fig. 3.11b). This considers that the adsorption energy is a function of the degree of coverage according to

$$\Gamma_i = \frac{RT}{2g} \ln (\beta_i a_{i,\infty}) \tag{3.31}$$

where Γ_i is the excess of species i, and g a parameter that treats the interaction energy between the adsorbed species, varying with coverage.

● *Frumkin isotherm* (Fig. 3.11c). This considers interactions in a different way:

$$\beta_i a_{i,\infty} = \frac{\Gamma_i}{\Gamma_s - \Gamma_i} \exp \frac{2g\Gamma_i}{RT} \tag{3.32a}$$

or

$$\Gamma_i = \frac{RT}{2g} \ln (\beta_i a_{i,\infty}) \ln \left(\frac{\Gamma_s - \Gamma_i}{\Gamma_i}\right) \tag{3.32b}$$

Γ_s being the maximum surface excess. A positive value of g implies attractive interaction and negative g repulsive interaction. When $g = 0$ and putting $\Gamma_i/\Gamma_s = \theta$, the Langmuir isotherm is obtained from (3.32a). Additionally, comparison of (3.31) and (3.32) shows that the Temkin isotherm is a special case of the Frumkin isotherm when $\Gamma_i/\Gamma_s = 0.5$.

Adsorption can be studied by many electrochemical methods, as can adsorption kinetics. When electroactive species are adsorbed, reagents or products of electrode reactions or both, a significant change in voltammetric response can occur. Adsorption of non-electroactive species can inhibit the electrode reaction. These processes depend on electrode material as well as on solution composition.

3.5 The solid metallic electrode: some remarks

Mercury is not a typical electrode material: it is liquid, and there is constant movement of atoms on the surface in contact with solution. A solid electrode has a well-defined structure, probably polycrystalline and in some cases monocrystalline. In a solid metallic electrode conduction is predominantly electronic owing to the free movement of valence electrons, the energy of the electrons that traverse the interface being that of the Fermi level, E_F (Section 3.6), giving rise to effects from the electronic distribution of the atoms in the metallic lattice already mentioned.

For a metal, the occupation of the electronic levels close to E_F is given more correctly by the expression[22]

$$f = 1/[1 + \exp(E - E_F)/k_B T] \tag{3.33}$$

where f is the probability of occupation of a level of energy E and k_B is the Boltzmann constant—$f = 0.50$ when $E = E_F$. The Fermi energy is the electrochemical potential of the electrons in the metal electrode. By substitution in (3.40) we see that when $E = E_F + k_B T$, $f = 0.27$ and when $E = E_F - k_B T$, $f = 0.73$. Figure 3.12 plots (3.33): at absolute zero the

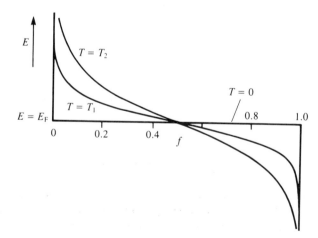

Fig. 3.12 The density of states occupied by electrons in a metal in the region of the Fermi level, E_F, at $T = 0$, T_1, and T_2, where $T_2 > T_1$.

cut-off is sharp, the probability of occupation becoming more smeared out as the temperature increases.

By convention, for a metal, it is said that only electrons with energies within $k_B T$ of E_F can be transferred. (In the case of semiconductors (3.33) cannot be applied and larger energy intervals have to be considered.)

The interfacial structure of a solid electrode depends on various factors. The interatomic distance varies with the exposed crystallographic face and with the interaction energy; between the crystallites in a polycrystalline material there are breaks in the structure and one-dimensional and two-dimensional defects, such as screw dislocations, etc. Adsorption of species can be facilitated or made more difficult, and at the macroscopic level we observe the average behaviour.

Recently the structure of the double layer associated with the interface of gold and platinum monocrystals with solution has been investigated[23]. A clear difference between crystallographic faces is noted, manifested in the values of differential capacity and in evidence of adsorption in voltammograms. Cyclic voltammograms suggest that there is a reorganization on the metal surface to give the equivalent of a surface layer of low Miller index, the identity of this face depending on the applied potential[24]. These studies are still in their early stages and concrete conclusions concerning a possible restructuring cannot yet be stated.

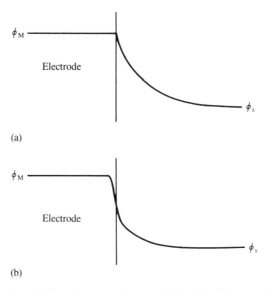

Fig. 3.13 Variation of the electrostatic potentials with distance from a metallic electrode. (a) Classical representation; (b) The 'jellium' model.

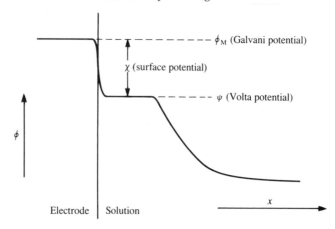

Fig. 3.14 Variation of potential with distance from a metal electrode separated from the electrolyte solution without modifying charges (not to scale), ψ is the value of the potential where the charge distributions due to the electrode and to the solution contact.

The effects of the crystallographic face and the difference between metals are evidence of the incorrectness of the classical representations of the interface with all the potential decay within the solution (Fig. 3.13a). In fact a discontinuity is physically improbable and experimental evidence mentioned above confirms that it is incorrect, the schematic representation of Fig. 3.13b being more correct. This corresponds to the 'chemical' models (Section 3.3) and reflects the fact that the electrons from the solid penetrate a tiny distance into the solution (due to wave properties of the electron). In this treatment the Galvani (or inner electric) potential, ϕ, (associated with E_F) and the Volta (or outer electric) potential, ψ, that is the potential outside the electrode's electronic distribution (approximately at the IHP, 10^{-5} cm from the surface) are distinguished from each other. The difference between these potentials is the surface potential χ (see Fig. 3.14 and Section 4.6).

3.6   The semiconductor electrode: the space–charge region

In a semiconductor electrode[6-8] the accessible electronic levels are more restricted, which has important consequences. As is well known, in a semiconductor there is a separation between the occupied valence band and the unoccupied conduction band. By convention, if the separation is greater than 3 eV the solid is called an insulator (for example diamond 5.4 eV) and if it is less it is a semiconductor. Promotion of an electron

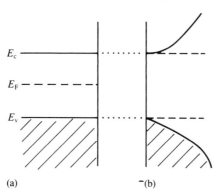

Fig. 3.15 Band model for an intrinsic semiconductor. The valence band is totally filled and the conduction band empty. Conduction occurs via promotion of electrons from E_v to E_c, the conductivity increasing with increase in temperature. (a) Definition of energy levels; (b) Variation of density of available states with energy.

from valence to conduction band leaves a hole (lack of electron), a positive charge, that can move through the crystal. For this reason it is useful to speak not only of electron movement but also of hole movement. Conduction occurs by movement of electrons in the conduction band or of holes (lack of electrons) in the valence band.

In an intrinsic semiconductor electron promotion to the conduction band occurs through thermal or photon excitation. The Fermi energy is in the middle of the bandgap, i.e. $E_F \sim (E_v + E_c)/2 = E_v + E_g/2$, given that E_F is defined by the probability of occupation of 0.5 (Fig. 3.15). In terms of the bandgap energy, E_g, substituting $E = E_c$ in (3.33), and considering the case where $E_g \gg kT$, the number of excited electrons, n, is

$$n \propto \exp\left(-E_g/2k_B T\right) \qquad (3.34)$$

Examples of important semiconductors in electrochemistry are given in Table 3.1.

Other electronic levels (surface states) can exist on the semiconductor surface due to adsorbed species or surface reorganization. These states can facilitate electron transfer between electrode and solution.

If the semiconductor is an ionic solid, then electrical conduction can be electronic and ionic, the latter being due to the existence of defects within the crystal that can undergo movement, especially Frenkel defects (an ion vacancy balanced by an interstitial ion of the same type) and Schottky defects (cation and anion vacancies with ion migration to the surface). This will be discussed further in Chapter 13, as ionic crystals are the sensing components of an important class of ion selective electrodes.

Table 3.1. Bandgap energy, E_g, and corresponding wavelength λ_{bg} (important for photo-excitation) of some semiconductors of electrochemical interest[25]. Zone of visible light ($300 \rightarrow 950$ nm)

Semiconductor	E_g/eV	λ_{bg}/nm
SnO_2	3.5	350
ZnO	3.2	390
$SrTiO_3$	3.2	390
TiO_2	3.0	410
CdS	2.4	520
GaP	2.3	540
Fe_2O_3	2.1	590
CdSe	1.7	730
CdTe	1.4	890
GaAs	1.4	890
InP	1.3	950
Si	1.1	1130

However, to simplify the discussion that follows and since at present the majority of semiconductors of electrochemical interest are not ionic crystals, we consider only electronic conduction.

Owing to difficulties in electron movement in semiconductors, when a steady state has been achieved almost all the applied potential appears within the semiconductor, creating a region of potential variation close to the surface called the space–charge region (Fig. 3.16).

In fact intrinsic semiconductors, which are necessarily pure crystals, are

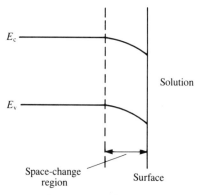

Fig. 3.16 The space–charge region in an intrinsic semiconductor.

little used. It is more common to use doped semiconductors, the doping normally being introduced externally. In an *n*-type semiconductor doping is obtained by introducing an atom into the lattice that has approximately the same size as the substrate atoms but with more electrons, e.g. silicon doped with phosphorus or arsenic, or non-stoichiometric III–V compounds; they will occupy lattice positions and supply electrons that can move through the crystal. The energy of these electrons is slightly less than E_c; electronic conduction can be by thermal excitation from the impurity band to the conduction band (Fig. 3.17a). In a *p*-type semiconductor the doped atoms have a deficiency of electrons in relation to substrate atoms leading to an unoccupied band a little above E_v; hole conduction in the valence band occurs via electron promotion from E_v to the unoccupied band (Fig. 3.17b). By doping it is possible to change a solid that under normal conditions would be an insulator, owing to the larger bandgap, into a semiconductor. However, in electrochemistry, the doping is done principally to fix E_F close to E_c (n-type) or close to E_v (p-type).

In a semiconductor electrode, almost all the potential variation in the interfacial region occurs in the space–charge region. This is due to the fact that the values for the space–charge capacity, C_{sc}, are from $0.001–1\ \mu F\ cm^{-2}$, whilst those for C_d are from $10–100\ \mu F\ cm^{-2}$, so that C_{sc} dominates. The theory of the space–charge region was developed by Schottky[26], Mott[27], Davydov[28], and more completely by Brittain and Garrett[29].

We now describe the effect of applied potential to an n-type semiconductor in its space–charge region. As there is an excess of electrons, the electron is the majority carrier; as there is also movement of a much smaller number of holes (electron deficiencies), the hole is the minority

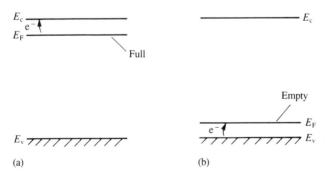

Fig. 3.17 Semiconductors: (a) n-type; (b) p-type. The mode of electron conduction is indicated by the arrows.

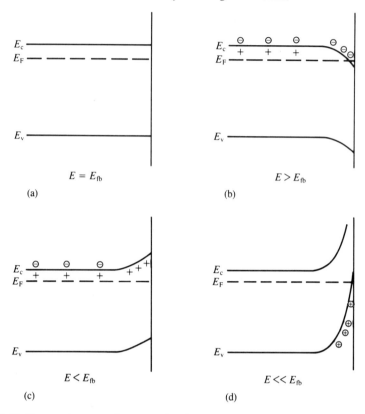

Fig. 3.18 Types of space–charge region in an n-type semiconductor, dependent on the potential applied relative to the flat band potential, U_{fb}. U represents potential (V) and $E_{c,sur}$ the electronic energy corresponding to E_c close to the surface. (a) $E_{c,sur} = E_{fb}$; no space–charge region; (b) $E_{c,sur} > E_{fb}$ $(U < U_{fb})$: formation of an accumulation layer; (c) $E_{c,sur} < E_{fb}$ $(U > U_{fb})$: formation of a depletion layer; (d) $E_{c,sur} \ll e_{fb}$ $(U \gg U_{fb})$: formation of an inversion layer.

carrier. According to the potential applied we create various types of region, as shown in Fig. 3.18.

At this point a difficulty in the nomenclature used in semiconductor electrochemistry should be noted: this is due to the symbol E being used not only for energy (J or eV) but also for potential (V). To attempt to avoid confusion, the symbol U is used within this area of electrochemistry for potential (V). The rest of the section follows this convention.

The most important situations that we should stress are:

• For a certain value of applied potential, there is equality between the number of electrons removed from and supplied to the electrode. In this situation there will be no space–charge region and the potential is called the flat-band potential, U_{fb}.

- Electrons accumulate in the space–charge layer by injection, giving rise to an accumulation layer.

- Electrons are removed from the space–charge layer, creating a depletion layer.

- The force to extract electrons from the electrode is so great that they are extracted not only from the conduction band but also from the valence band (equivalent to hole injection). An inversion layer is formed, so called because the n-type semiconductor is converted into a p-type semiconductor at the surface. Adsorbates can facilitate this process.

To have passage of current it is necessary that E_F is within the conduction or within the valence band in the space–charge region, i.e. accumulation layer in an n-type semiconductor (reduction) or hole accumulation layer in a p-type semiconductor (oxidation).

There is an analogy with the Schottky diode. When $E_{c,sur} > E_{fb}$ (negative voltage bias), and possibly for E slightly negative in relation to E_{fb}, there will be a large current flux, assuming that there are electronic levels in solution to accept the electrons from the electrode. When $E_{c,sur} < E_{fb}$ the current will be almost zero.

Another important aspect refers to adsorbates. These have their own associated energy levels, known as *surface states*, and can aid electron transfer if there is superposition of the conduction band and that corresponding to the surface state (Fig. 3.19). A better understanding of the energy distributions of the solution species in Fig. 3.19 is possible on referring to Section 4.6.

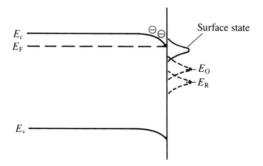

Fig. 3.19 Conjugation of an accumulation layer in an n-type semiconductor with a surface state to facilitate the reduction of O to R. In the absence of the surface state, there would be no reaction.

For a p-type semiconductor the arguments are analogous; in this case the majority carrier is the hole.

Due to the great extension of the space–charge region, almost all the potential drop occurs across it. So we can measure its capacity, C_{sc}, and calculate E_{fb} from the Mott–Schottky relation

$$C_{sc}^{-2} = \frac{2}{e\epsilon_r\epsilon_0 N_D} \left[(U - U_{fb}) - \frac{k_B T}{e} \right] \tag{3.35}$$

where N_D is the density of donor atoms. A plot of C_{sc}^{-2} vs. U is a straight line if all other voltage drops are unaffected by the applied potential—we calculate U_{fb} from the intercept and N_D from the slope. The presence of adsorbates modifies C_{sc} and is manifested in the non-linearity of the plots. Once E_{fb} is known, we can calculate E_v and E_c from the relations

$$E_v = -E_{fb} + k_B T \ln \frac{N_D}{N_v} \qquad \text{(p-type semiconductor)} \qquad (3.36a)$$

$$E_c = E_{fb} + k_B T \ln \frac{N_D}{N_c} \qquad \text{(n-type semiconductor)} \qquad (3.36b)$$

where N_v and N_c are the effective density of states in the valence and conduction bands respectively for p- and n-type semiconductors, data obtained from solid state measurements. Knowing the values of the bandgap energy, E_g (Table 3.1), we can then calculate E_c for p-type and E_v for n-type semiconductors respectively.

Semiconductors are extremely important in photoelectrochemistry, where the energy necessary to jump from the valence to the conduction band is supplied by visible light. These aspects are developed in Chapter 12.

3.7 Electrokinetic phenomena and colloids: the zeta potential

A colloidal system consists of a disperse phase suspended in a dispersion medium, which does not separate with time[30-32]. All combinations of gases, liquids, and solids are possible except for a gas dispersed in a gas. Normally when speaking of colloids one is referring to a solid suspended in a liquid, the solid particles having diameters between 10^{-7} cm and 10^{-5} cm. The solid particles are charged, which causes repulsion between the particles and gives temporal stability to the colloidal system. Recently there has been increasing interest in colloids because of their possible use as electrodes for electrolysis, each particle acting as anode and cathode at the same time. Their particular advantage is the large surface area exposed to solution in relation to their solid volume. Since the particles are charged there is an interfacial region which exhibits many of the properties of the interfacial region of a solid electrode. Therefore, the study of colloids can also lead to a better knowledge of the double layer region, especially for ionic solids and semiconductors.

A very useful type of phenomenon in the study of colloidal particles is the electrokinetic phenomenon that results from the movement of a solid phase with surface charge relative to an electrolyte-containing liquid phase. An applied electric field induces movement or, conversely, movement induces an electric field. The phenomena can be divided into two types:

● charged solid particles (colloidal particles) moving through the liquid under the influence of an electric field, *electrophoresis,* or due to gravitational force, *sedimentation.*

● liquids moving past charged solid surfaces (or possibly through membranes) under the influence of an electric field, *electroosmosis,* or an applied pressure difference, *streaming potential.* These effects are normally studied in fine capillaries in order to maximize the ratio of the solid surface area to the liquid volume.

These four manifestations of the electrokinetic effect are summarized in Table 3.2.

Table 3.2. Electrokinetic phenomena

Mobile phase	Stationary phase	Phenomenon	Force applied	Property measured
Solid	Liquid	Electrophoresis	Electric field	Electrophoretic mobility via mass transport measurement, microscope or Doppler effect
Solid	Liquid	Sedimentation potential[a]	Force of gravity	Potential difference
Liquid	Solid	Electroosmosis	Electric field	Rate of liquid movement, pressure
Liquid	Solid	Streaming potential[b]	Pressure	Potential difference

[a] Effectively the inverse of electrophoresis.
[b] Effectively the inverse of electroosmosis.

The size of the particles that is calculated from these experiments corresponds to particle dimensions plus the double layer thickness, in this case defined by the shear plane inside which the adsorbed species are rigidly held, and outside of which there is free movement. The shear plane can therefore be associated roughly with the outer Helmholtz plane, an approximation often made. The value of the electrostatic potential at the shear plane with respect to the value in bulk solution is called the *electrokinetic* or *zeta potential*, ζ[33] (see Section 3.3).

In the presence of a large quantity of inert electrolyte, all the potential drop is confined to within the compact layer and ζ is zero. By application of an appropriate potential at an electrode we can also arrange for $\zeta = 0$ – this value of the potential is called the *isoelectric point*. This is, in general, not equal to the point of zero charge, as the value of the latter is affected by the presence of specifically adsorbed species (Section 3.4).

We now consider briefly the four effects that are described in Table 3.2 and how to deduce values of the zeta potential, ζ, from experimental information.

Electrophoresis

In electrophoresis the solid moves in a liquid phase due to the application of an electric field. The forces acting on the particles are similar to those that act on solvated ions:

- Force of the electric field on the particle

- Frictional forces

- Forces due to the action of the electric field on ions of the opposite charge to that of the particle within the double layer (*relaxation effect*)

- Induction forces in the double layer caused by the electric field (*electrophoretic retardation*)

The electrophoretic mobility, u_e, is calculated by solving the Poisson equation with the appropriate boundary conditions. The final relation is of the type

$$u_e = \frac{2}{3} \frac{\epsilon \zeta E}{\eta} f(a/x_{DL}) \tag{3.37}$$

where ϵ is the permittivity, η the absolute viscosity, E the electric field strength, and $f(a/x_{DL})$ a numerical factor, where a is the particle radius and x_{DL} the double layer thickness, that varies according to which of the forces indicated above have to be considered. This depends on particle size and double layer thickness. For every small particles in dilute solution, the double layer is thick and $f(a/x_{DL}) \rightarrow 1$ (negligible relaxation effect). For large particles in concentrated solutions, where the double layer is thin, $f(a/x_{DL}) \rightarrow 1.5$ (negligible electrophoretic retardation). All other situations lead to intermediate numerical factors. Measurements of electrophoretic mobility, using (3.37) with the appropriate numerical factor, lead to values for the zeta potential.

As can be inferred, electrophoretic mobility depends on solution ionic strength since double layer thickness decreases with increasing electrolyte concentration. It also depends on the surface charge of the particles. If this charge varies in colloidal particles of similar dimensions then electrophoresis provides a basis for their separation. An example of this is in proteins, where the surface charge varies with pH in a different way according to the protein identity.

Sedimentation potential

Colloidal particles are affected by the force of gravity, either natural or through centrifugation. Sedimentation of the particles often gives rise to an electric field. This occurs because the particles move, whilst leaving some of their ionic atmosphere behind. These potentials are usually difficult to measure, and are an unwanted side effect in ultracentrifugation, where they are minimized by adding a large concentration of inert electrolyte.

Electroosmosis

In electroosmosis, the stationary and mobile phases are exchanged in relation to electrophoresis. As measurement of the rate of movement of a liquid through a capillary is difficult, the force that it exerts is measured, i.e. the electroosmotic pressure, or, alternatively, the volume of liquid transported through a capillary in a given time interval. The electroosmotic velocity, v_{eo}, is

$$v_{eo} = \frac{\epsilon \zeta E}{\eta} \tag{3.38}$$

which is of the same form as (3.37) for electrophoresis, putting $f(a/x_{DL}) = 1.5$, since the capillary radius is much larger than the double layer thickness.

The volume flow of liquid, V_f, is $v_{eo}A$, where A is the cross-sectional area of the capillary. A current will flow of magnitude $I = A\kappa E$, in which κ is the solution conductivity, and thus the electroosmotic flow, flux per unit electric current at zero pressure difference, is given by

$$\frac{V_f}{I} = \frac{v_{eo}A}{I} = \frac{\epsilon \zeta}{\kappa \eta} \tag{3.39}$$

Streaming potential

If a pressure difference, ΔP, is applied between the extremes of a capillary, then a potential difference is created, called the streaming potential:

$$\Delta \phi = \frac{e \zeta}{\kappa \eta} \Delta P \tag{3.40}$$

By comparing this with (3.39) the close relationship between streaming potential and electroosmotic flow can be seen.

Limitations in the calculation of the zeta potential

Quantitative measurements of electrokinetic phenomena permit the calculation of the zeta potential by use of the appropriate equations. However, in the deduction of the equations approximations are made: this is because in the interfacial region physical properties such as concentration, viscosity, conductivity, and dielectric constant differ from their values in bulk solution, which is not taken into account. Corrections to compensate for these approximations have been introduced, as well as consideration of non-spherical particles and particles of dimensions comparable to the diffuse layer thickness. This should be consulted in the specialized literature.

References

1. D. C. Grahame, *Ann. Rev. Phys. Chem.*, 1955, **6**, 337.
2. R. Parsons, *Modern aspects of chemistry*, Butterworths, London, Vol. 1, 1954, ed. J. O'M. Bockris and B. E. Conway, pp. 103–179.
3. R. Parsons, *Advances in Electrochemistry and Electrochemical Engineering*, ed. P. Delahay and C. W. Tobias, Wiley, New York, Vol. 1, 1961, pp. 1–64.
4. A. F. Silva ed., *Trends in interfacial electrochemistry*, Proceedings of NATO ASI (1984), Reidel, Dordrecht, 1985.
5. S. Trasatti, *Modern aspects of electrochemistry*, Plenum, New York, Vol. 13, 1979, ed. B. E. Conway and J. O'M, Bockris, pp. 81–206.
6. S. R. Morrison, *Electrochemistry at semiconductor and oxidised metal electrodes*, Plenum, New York, 1980.
7. K. Uosaki and H. Kita, *Modern aspects of electrochemistry*, Plenum, New York, Vol. 18, 1986, ed. R. E. White, J.O'M. Bockris, and B. E. Conway, pp. 1–60.
8. A. Hamnett, in *Comprehensive chemical kinetics*, ed. R. G. Compton, Elsevier, Amsterdam, Vol. 27, 1987, Chapter 2.

9. G. Lippmann, *Compt. Rend.*, 1873, **76,** 1407.
10. J. Heyrovsky, *Chem. Listy,* 1922, **16,** 246.
11. D. C. Grahame, *Chem. Rev.,* 1947, **41,** 441.
12. H. L. F. von Helmholtz, *Ann. Physik,* 1853, **89,** 211; 1879, **7,** 337.
13. G. Gouy, *Compt. Rend.,* 1910, **149,** 654; D. L. Chapman, *Phil.* Mag., 1913, **25,** 475.
14. D. MacInnes, *Principles of electrochemistry,* Dover, New York, 1962, Chapter 7.
15. O. Stern, *Z. Elektrochem.,* 1924, **30,** 508.
16. J. O'M. Bockris, M. A. Devanathan, and K. Muller, *Proc. R. Soc.,* 1963, **A274,** 55.
17. B. B. Damaskin and A. N. Frumkin, *Electrochim, Acta,* 1974, **19,** 173; B. B. Damaskin, U. V. Palm, and M. A. Salve, *Elektrokhimiya,* 1976, **12,** 232; B. B. Damaskin, *J. Electroanal. Chem.,* 1977, **75,** 359.
18. S. Trasatti, in Ref. 4, 25–48.
19. R. Parsons, *Chem. Rev.,* 1990, **90,** 813.
20. G. A. Martynov and R. R. Salem, *Electrical double layer at a metal–dilute electrolyte solution interface,* Lecture Notes in Chemistry **33,** Springer-Verlag, Berlin, 1983.
21. J. Goodisman, *Electrochemistry: theoretical foundations,* Wiley-Interscience, New York, 1987, pp. 232–239.
22. Ref. 6, p. 5.
23. A. Hamelin, *Modern aspects of electrochemistry,* Plenum, New York, Vol. 16, 1985, ed. B. E. Conway, R. E. White, and J. O'M, Bockris, pp. 1–101.
24. R. M. Cerviño, W. E. Triaca, and A. J. Arvia, *J. Electroanal. Chem.,* 1985, **182,** 51.
25. H. O. Finklea, *J. Chem. Ed.,* 1983, **60,** 325.
26. W. Schottky, *Zeit, für Physik,* 1939, **113,** 367.
27. N. F. Mott, *Proc. Roy. Soc.,* 1939, **A 171,** 27.
28. B. Davydov, *J. Physics USSR,* 1939, **1,** 169.
29. W. H. Brattain and C. G. B. Garrett, *Phys. Rev.,* 1955, **99,** 376.
30. R. J. Hunter, *Foundations of colloid science,* Oxford University Press, New York, 1987.
31. D. H. Everett, *Basic principles of colloid science,* Royal Society of Chemistry, London, 1988.
32. A. Kitahara and A. Watanabe, *Electrical phenomena at interfaces,* Dekker, New York, 1984.
33. R. J. Hunter, *Zeta potential in colloid science,* Academic Press, New York, 1981.

4

FUNDAMENTALS OF KINETICS AND MECHANISM OF ELECTRODE REACTIONS

4.1 Introduction

In this chapter the mechanisms of electrode reactions are explained for the most simple case of an electron transfer without chemical transformation, i.e. without formation or breaking chemical bonds. Other more complex cases are also referred to. Comparison with electron transfer reactions in homogeneous solution are made.

In a system involving reagents and products at equilibrium, the rates of the reactions in each direction are equal. Equilibrium can thus be seen as a limiting case, and any kinetic model must give the correct equilibrium expression. For reactions at an electrode, *half-reactions*, the equilibrium expression is the Nernst equation.

4.2 The mechanism of electron transfer at an electrode

We consider the case of an oxidation of reduction at an electrode without chemical transformation, an example being

$$Fe^{3+}(aq) + e^-(electrode) \rightleftarrows Fe^{2+}(aq)$$

The mechanism consists of the steps shown in Fig. 4.1[1,2].

- 1. Diffusion of the species to where the reaction occurs (described by a mass transfer coefficient k_d – see Chapter 5).

- 2. Rearrangement of the ionic atmosphere (10^{-8} s).

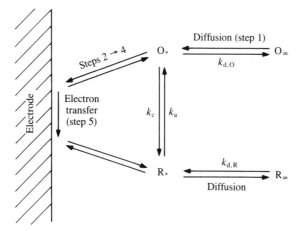

Fig. 4.1 Scheme of electron transfer at an electrode.

- 3. Reorientation of the solvent dipoles (10^{-11} s).
- 4. Alterations in the distances between the central ion and the ligands (10^{-14} s).
- 5. Electron transfer (10^{-16} s).
- 6. Relaxation in the inverse sense.

Steps 2–5 are included in the charge transfer rate constant, k_a or k_c, and include adsorption of the reagent on the electrode surface, which in the case of a soluble product will afterwards de-adsorb. Steps 2–4 can be seen as a type of pre-equilibrium before the electron transfer. During the electron transfer itself all positions of the atoms are frozen, obeying the Franck–Condon principle (adiabatic process).

4.3   The mechanism of electron transfer in homogeneous solution

The mechanism described above is also correct for electron transfer in homogeneous solution except that, instead of the reaction site being an electrode, it is the point where the two ions meet in the interior of the solution. In the equations for energy changes a factor of 2 relative to electrode reactions appears, since whole reactions rather than half-reactions are being considered. Theoretical and experimental comparisons between electrode reactions and redox reactions in solution have been made with satisfactory results[3].

4.4   An expression for the rate of electrode reactions

For any type of electrode reaction in solution, the *Arrhenius expression* relates the *activation enthalpy*, $\Delta H^{\ddagger}$, with the *rate constant*, k:

$$k = A \exp\left[-\Delta H^{\ddagger}/RT\right] \tag{4.1}$$

A being the pre-exponential factor. In an electron transfer reaction, the rearrangement of the ionic atmosphere is a fundamental step, and thus it is useful to include the *activation entropy* $\Delta S^{\ddagger}$. The reorientation and rearrangement causes the separation between the energy levels to be different in the activated complex than in the initial state. If we write the pre-exponential factor, A, as

$$A = A' \exp\left[\Delta S^{\ddagger}/R\right] \tag{4.2}$$

then

$$k = A' \exp\left[-(\Delta H^{\ddagger} - T \Delta S^{\ddagger})/RT\right] = A' \exp\left[-\Delta G^{\ddagger}/RT\right] \tag{4.3}$$

We now see how the potential applied to the electrode is reflected in the values of $\Delta G^{\ddagger}$.

Consider a half-reaction of first order occurring at an 'inert' metallic electrode;

$$O + ne^{-} \rightarrow R$$

species O and R both being soluble. The $O \mid R$ couple has an associated energy that can be related to the electrode potential (see Section 4.6). We call this energy E_{redox}. By applying a potential to the electrode, we influence the highest occupied electronic level in the electrode. This level is the Fermi level, E_{F} – electrons are always transferred to and from this level. The situation is shown schematically in Fig. 4.2, where one sees how different potentials applied to the electrode can change the direction of electron transfer. The energy level E_{redox} is fixed: by altering the applied potential, and thence E_{F}, we oblige the electrode to supply electrons to species O (reduction) or remove electrons from species R (oxidation). What is, then, the energy profile describing electron transfer?

In a similar fashion to the description of the kinetics of homogeneous reactions, in the development of a model for electron transfer parabolic energy profiles have been used for reagents and products. Nevertheless, the region where the profiles intersect is of paramount interest since this corresponds to the activated complex: in this region the energy variation is almost linear—its variation far from the intersection is not important. Figure 4.3 shows a typical profile. A change x in the free energy of O will result in a change $\alpha_c x$ in the activation energy, assuming a linear

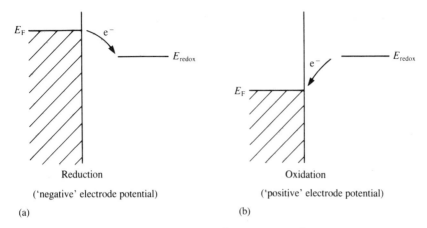

Fig. 4.2 Electron transfer at an inert metallic electrode. The potential applied to the electrode alters the highest occupied electronic energy level, E_F, facilitating (a) reduction or (b) oxidation.

intersection. So for a reduction we can write

$$\Delta G_c^{\ddagger} = \Delta G_{c,0}^{\ddagger} + \alpha_c nFE \tag{4.4a}$$

In a similar way, for an oxidation

$$\Delta G_a^{\ddagger} = \Delta G_{a,0}^{\ddagger} - \alpha_a nFE \tag{4.4b}$$

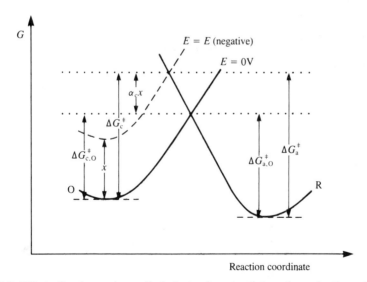

Fig. 4.3 Effect of a change in applied electrode potential on the reduction of O to R (R considered absent in bulk solution and in the electrode material).

where E is the potential applied to the electrode and α is a measure of the slope of the energy profiles in the transition state zone and, therefore, of barrier symmetry. Values of α_a and α_c can vary between 0 and 1, but for metals are around 0.5. A value of 0.5 means that the activated complex is exactly halfway between reagents and products on the reaction coordinate, its structure reflecting reagent and product equally. In this simple case of a one-step transfer of n electrons between O and R, it is easily deduced that $(\alpha_a + \alpha_c) = 1$.

We now substitute the expressions for $\Delta G^{\ddagger}$ from (4.4a) and (4.4b). We obtain for a reduction

$$k_c = A' \exp\left[-\Delta G^{\ddagger}_{c,0}/RT\right] \exp\left[-\alpha_c nFE/RT\right] \qquad (4.5a)$$

and for an oxidation

$$k_a = A' \exp\left[-\Delta G^{\ddagger}_{a,0}/RT\right] \exp\left[\alpha_a nFE/RT\right] \qquad (4.5b)$$

These equations can be rewritten in the form

$$k_c = k_{c,0} \exp\left[-\alpha_c nFE/RT\right] \qquad (4.6a)$$

and

$$k_a = k_{a,0} \exp\left[\alpha_a nFE/RT\right] \qquad (4.6b)$$

As the reaction is first order, at equilibrium

$$k_c[O]_* = k_a[R]_* \qquad (4.7)$$

where $[O]_*$ and $[R]_*$ are the concentrations of O and R next to the electrode. If $[O]_* = [R]_*$, the potential is $E^{\ominus\prime}$, the formal potential, and

$$k_c = k_a = k_0 \qquad (4.8)$$

this last constant being the standard rate constant. Substituting (4.8) in (4.6a) (4.6b),

$$k_c = k_0 \exp\left[-\alpha_c nF(E - E^{\ominus\prime})/RT\right] \qquad (4.9a)$$

$$k_a = k_0 \exp\left[\alpha_a nF(E - E^{\ominus\prime})/RT\right] \qquad (4.9b)$$

This is the formulation of electrode kinetics first derived by Butler and Volmer[4]. The observed current for kinetic control of the electrode reaction is proportional to the difference between the rate of the oxidation and reduction reactions at the electrode surface and is given by

$$I = nFA(k_a[R]_* - k_c[O]_*) \qquad (4.10)$$

where A is the electrode area.

We can draw some conclusions:

● 1. On changing the potential applied to the electrode, we influence k_a and k_c in an exponential fashion. The electrode is thus a powerful catalyst. Nevertheless, it should be noted that $k_c[O]_*$ and $k_a[R]_*$ do not grow indefinitely, being limited by the transport of species to the electrode. When all the species that reach it are oxidized or reduced the current cannot increase further. If there are no effects from migration, diffusion limits the transport of electroactive species close to the electrode; the maximum current is known as the *diffusion-limited current* (Section 5.3). Whatever the value of the standard rate constant, k_0, if the applied potential is sufficiently positive (oxidation) or sufficiently negative (reduction) the maximum current will always be reached.

● 2. As indicated, for metals the activation barrier (Fig. 4.3) is halfway between reagents and products and $\alpha \sim \frac{1}{2}$ (that is α_a or α_c). In certain less usual cases the activated complex has predominantly the structure of the oxidized or the reduced species, giving rise to values of $\alpha \approx 0$ and $\alpha \approx 1$ respectively (Fig. 4.4). These situations occur with semiconductor electrodes, since the externally applied voltage appears as a potential difference almost totally across the semiconductor space charge layer.

In many cases electrode processes involving the transfer of more than one electron take place in consecutive steps. The symmetry of the activation barrier referred to above relates to the rate-determining step. For example, a two-electron transfer involving a pre-equilibrium for the first electron transfer and the second electron transfer as the rate-determining step leads to $(\alpha n) = 1 + 0.5 = 1.5$. From this we might calculate $\alpha = 0.75$, which is not a reflection of the position of the activated complex on the reaction coordinate. Thus extreme care must be

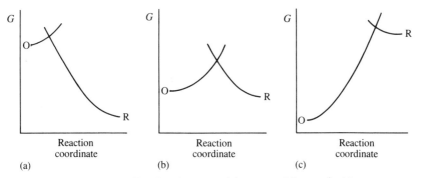

Fig. 4.4 Energy profiles for the cases (a) $\alpha_c \approx 0$; (b) $\alpha_c \approx \frac{1}{2}$; (c) $\alpha_c \approx 1$.

taken in the interpretation of experimental transfer coefficients—it is always preferable to quote values of an until the mechanism is elucidated.

Finally, since the anodic and cathodic reactions do not occur at the same potential, the mechanism for oxidation may not be the opposite of reduction. This occurs when there is multiple step electron transfer, possibly with intermediate chemical steps. If this happens then, in general, $\alpha_a + \alpha_c$ do not add up to unity.

4.5 The relation between current and reaction rate: the exchange current

As stated above, the current in a kinetically controlled electrode reaction is given by

$$I = nFA(k_a[R]_* - k_c[O]_*) \tag{4.10}$$

At equilibrium $k_c[O]_* = k_a[R]_*$ and, from (4.9)

$$[O]_* \exp[-\alpha_c nF(E_{eq} - E^{\ominus\prime})/RT] = [R]_* \exp[\alpha_a nF(E_{eq} - E^{\ominus\prime})/RT] \tag{4.11}$$

The fact that the current is zero means that there are no concentration gradients close to the electrode, so that the surface concentrations are equal to those in bulk solution, $[O]_\infty$ and $[R]_\infty$. Rewriting (4.11) we obtain

$$\exp(nF(E_{eq} - E^{\ominus\prime})/RT] = [O]_\infty/[R]_\infty \tag{4.12}$$

where we use the fact that for a simple charge transfer reaction $(\alpha_c + \alpha_a) = 1$. This is the *Nernst equation*, normally expressed as

$$E_{eq} = E^{\ominus\prime} + \frac{RT}{nF} \ln \frac{[O]_\infty}{[R]_\infty} \tag{4.13}$$

Instead of using values of k_0 directly, the exchange current, I_0, equal to one of the components $-I_c$ or I_a of the current at equilibrium, has been used:

$$I_0 = |I_c| = nFAk_0[O]_\infty \exp[-\alpha_c nF(E_{eq} - E^{\ominus\prime})/RT] \tag{4.14}$$

Multiplying the Nernst equation (4.12) by $\exp(-\alpha_c)$ and substituting in (4.14) one obtains

$$I_0 = nFAk_0[O]_\infty^{1-\alpha_c}[R]_\infty^{\alpha_c} \tag{4.15}$$

When $[O]_\infty = [R]_\infty$, and putting $[O]_\infty = c_\infty$,

$$I_0 = nFAk_0c_\infty \tag{4.16}$$

This last expression shows that I_0 and k_0 both express the rate of an electrode reaction (independent of α_c in this case).

Exactly the same result is obtained by following identical reasoning, using the anodic instead of the cathodic reaction in (4.14).

4.6 Microscopic interpretation of electron transfer

Some notions of the mechanism of electron transfer were given in Section 4.2. Any theory must be realistic and take into account the reorientation of the ionic atmosphere in mathematical terms. There have been many contributions in this area, especially by Marcus, Hush, Levich, Dognadze, and others[5-9]. The theories have been of a classical or quantum-mechanical nature, the latter being more difficult to develop but more correct. It is fundamental that the theories permit quantitative comparison between rates of electron transfer in electrodes and in homogeneous solution.

We illustrate the results obtained in the approximate model of Marcus, remembering that the activation barrier results predominantly from solvation changes. The energy profile can be represented by a parabola. Figure 4.5 shows that ΔG is the energy difference between reagents and products, ΔG_s describes the solvation change between reagents and products, and $\Delta G^\ddagger$ is the activation energy[2]. For the intersection of the two parabolas, assumed to be identical in form, one obtains, after a little algebraic manipulation,

$$\Delta G^\ddagger = \frac{(\Delta G + \Delta G_s)^2}{4\,\Delta G_s} \tag{4.17}$$

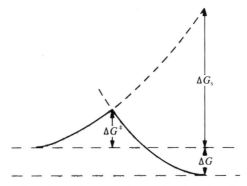

Fig. 4.5 Representation of reaction coordinate showing that the activation barrier is due principally to solvent reorganization.

So, what is the variation of the rate constant with potential? The *charge transfer coefficient*, α, anodic or cathodic, is given by

$$\alpha = \frac{RT}{F}\left|\frac{\partial \ln k}{\partial E}\right| = \frac{\partial \Delta G^{\ddagger}}{\partial E} \tag{4.18}$$

This is a form of the Tafel relation (Section 6.6) so long as the current is proportional to the rate constant. It is an example of a linear free energy relationship (a kinetic parameter, $\ln k$, varies linearly with a thermodynamic parameter, E). Substituting $\Delta G^{\ddagger}$ we get

$$\alpha = -\frac{1}{2F}\left(1 + \frac{\Delta G}{\Delta G_s}\right)\frac{\partial \Delta G}{\partial E} \tag{4.19}$$

$$= \frac{1}{2}\left(1 + \frac{\Delta G}{\Delta G_s}\right) \tag{4.20}$$

We can isolate some limiting cases:

- $\Delta G_s \gg \Delta G$: the kinetics is slow and $\alpha \approx \frac{1}{2}$ (this is the case for many reactions)

- ΔG_s small (fast reactions),

  $\Delta G \approx 0 \ (E \sim E^{\ominus\prime})$ and $\alpha \approx \frac{1}{2}$;

  $\Delta G_s \approx \Delta G$, $\alpha \to 1$ (Fig. 4.4a);

  $\Delta G_s \approx -\Delta G$, $\alpha \to 0$ (Fig. 4.4c).

So, for very fast reactions, the theory predicts a variation of α with potential. There is some evidence that this occurs, but given the multistep nature of any electrode reaction no definitive conclusions can be taken, and mechanisms can be elaborated which have constant charge transfer coefficients. Indeed the fact that the enthalpic and entropic parts of the coefficients have different temperature dependences leads to the question as to what is the real significance of the charge transfer coefficient, a topic currently under discussion[9].

Another aspect affecting electron transfer that has become more important with the increasing use of semiconductor electrodes[10-13] in, for example, solar energy conversion, but is also valid for metal electrodes, should be mentioned. Electron transfer occurs between the highest occupied energy level in the electrode (the Fermi level E_F) and the energy level of the redox pair in solution, E_{redox}. The occupation of the electronic levels close to E_F is given more correctly by the expression[14]

$$f = 1/[1 + \exp (E - E_F)/k_B T] \tag{4.21}$$

where f is the probability of occupation of a level of energy E, and k_B is

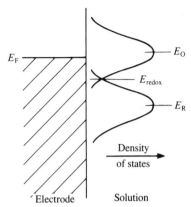

Fig. 4.6 Energy distribution of a redox couple of species O and R on the surface of a metallic electrode; E_F is altered by the applied potential, facilitating reduction of O (shown) or oxidation of R.

the Boltzmann constant $-f = 0.5$ when $E = E_F$. The Fermi energy is the electrochemical potential of the electrons in the electrode, see Chapter 3.

However, there are really two distributions of electronic energy levels associated with E_{redox}, due to the fact that O and R, having different charges, have different solvations; the energy of R is slightly lower than that of O. The density of states is shown schematically in Fig. 4.6. Overlap between E_F and the distribution for E_O shows that oxidized species can be reduced.

In order to relate E_{redox}, E_F, and electrode potentials it is important to utilize the same reference state, namely vacuum[14]. In relation to vacuum the energy of the standard hydrogen electrode is $-4.44\,eV$ (Fig. 4.7). When we measure electrode potentials, we measure the corresponding value of E_F through the relation

$$E_F = -eU \tag{4.22}$$

where e is the electronic charge and U the potential.

It therefore seems logical, when describing the mechanism of an electrode reaction, to speak of an energy associated with the redox couple corresponding to that of the electrons in the solution species that are transferred, and equal to the Fermi energy in the actual electron transfer step after solvent reorganization, etc. There is some controversy in this matter[15], but without ambiguity we can say

$$E_F = E_{redox} - e\chi \tag{4.23}$$

where χ is the surface potential of the electrode. The surface potential is

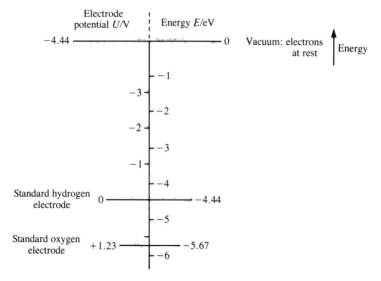

Fig. 4.7 The relation between electrode potentials, their corresponding energies, and vacuum.

the potential difference between the interface (of potential ψ, the Volta potential) and the interior of the electrode (of potential ϕ, the Galvani potential). Thus, at equilibrium and with unit activities, E_{redox} is equivalent to the energy of the Volta potential, whereas E_F is associated with the Galvani potential. χ reflects the break in the structure of the solid and consequent variations in electronic distribution (Fig. 4.8 and Chapter 3).

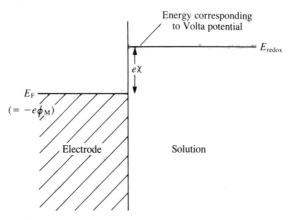

Fig. 4.8 The relation between the Galvani potential, ϕ_M, the Volta potential, ψ, and the surface potential, χ, shown schematically.

If $\chi \sim 0$ then

$$E_F = E_{redox} \tag{4.24}$$

Thus E_F is associated with the electrode potential and E_{redox} with the redox potential of the species: since in general $\chi \neq 0$, we cannot assume their equivalence. A measurement of potential gives values of electrode potentials and *never* redox potentials.

The crucial point is that the difference of potential available to effect electrode reactions and surmount activation barriers is not simply the difference between the Galvani potential (i.e. the Fermi energy) and the potential in solution. On the side of the solid it is the Volta potential and on the side of the solution it is the potential at the inner Helmholtz plane, where species have to reach to in order for electron transfer to be possible. Corrections to rate constants for the latter are commonly carried out using the Gouy–Chapman model of the electrolyte double layer and will be described in Section 6.9.

References

1. K. J. Vetter, *Electrochemical kinetics*, Academic Press, New York, 1967.
2. W. J. Albery, *Electrode kinetics*, Clarendon Press, Oxford, 1975.
3. R. A. Marcus, *J. Phys. Chem.*, 1963, **67**, 853.
4. J. A. V. Butler, *Trans. Faraday Soc.*, 1924, **19,** 729 and 734; T. Erdey–Gruz and M. Volmer, *Z. Physik. Chem.*, 1930, **150A,** 203.
5. R. Marcus, *Ann. Rev. Phys. Chem.*, 1964, **15**, 155 and references therein.
6. V. G. Levich, *Advances in electrochemistry and electrochemical engineering*, ed. P. Delahay and C. W. Tobias, Wiley, New York, Vol. 4, 1966, pp. 249–371.
7. R. R. Dogonadze, *Reactions of molecules at electrodes*, ed. N. H. Hush, Wiley-Interscience, New York, 1971, Chapter 3.
8. A. M. Kuznetsov, *Faraday Disc. Chem. Soc.*, 1982, **74,** 49.
9. B. E. Conway, *Modern aspects of electrochemistry*, Plenum, New York, Vol. 16, 1985, ed. B. E. Conway, R. E. White, and J. O'M. Bockris, pp. 103–188.
10. P. J. Holmes ed., *The electrochemistry of semiconductors*, Academic Press, London, 1962.
11. S. R. Morrison, *Electrochemistry at semiconductor and oxidised metal electrodes*, Plenum, New York, 1980.
12. K. Uosaki and H. Kita, *Modern aspects of electrochemistry*, Plenum, New York, Vol. 18, 1986, ed. R. E. White, J. O'M. Bockris, and B. E. Conway, pp. 1–60.
13. A. Hamnett, *Comprehensive chemical kinetics*, Elsevier, Amsterdam, Vol. 27, 1987, ed. R. G. Compton, Chapter 2.
14. Ref. 11 p. 5.
15. H. Reiss, *J. Phys. Chem.*, 1985, **89,** 3873; S. U. M. Khan, R. C. Kainthla, and J. O'M. Bockris, *J. Phys. Chem.*, 1987, **91,** 5974; H. Reiss, *J. Electrochem. Soc.*, 1988, **135,** 247C.

5

MASS TRANSPORT

5.1 Introduction

In the last chapter it became clear that in the expression for the rate of an electrode reaction

$$\text{rate} = k_a[R]_* - k_c[O]_*$$

the values of k_a and k_c (electrode kinetics) and of $[O]_*$ and $[R]_*$ are both of extreme importance. These, in turn, are affected not only by the electrode reaction itself but also by the transport of species to and from bulk solution. This transport can occur by diffusion, convection, or migration (Section 2.8). Normally, conditions are chosen in which migration effects can be neglected, this is the effects of the electrode's electric field are limited to very small distances from the electrode, as described in Chapter 3. These conditions correspond to the presence of a large quantity (>0.1 M) of an inert electrolyte (supporting electrolyte), which does not interfere in the electrode reaction. Using a high concentration of inert electrolyte, and concentrations of 10^{-3} M or less of electroactive species, the electrolyte also transports almost all the current in the cell, removing problems of solution resistance and contributions to the total cell potential—an exception to this is ultramicroelectrodes, where the currents are so low that higher solution resistances can be tolerated. In these conditions we need to consider only diffusion and convection.

Diffusion is due to the thermal movement of charged and neutral species in solution, without electric field effects. Forced convection considerably increases the transport of species, as will be demonstrated, and in many cases can be described mathematically. Natural convection, due to thermal gradients, also exists, but conditions where this movement is negligible are generally used.

In this chapter we consider systems under conditions in which the kinetics of the electrode reaction is sufficiently fast that the control of the electrode process is totally by mass transport. This situation can, in principle, always be achieved if the applied potential is sufficiently positive (oxidation) or negative (reduction). First we consider the case of pure diffusion control, and secondly systems where there is a convection component.

5.2 Diffusion control

As mentioned previously, diffusion is the natural movement of species in solution, without the effects of the electric field. Thus the species can be charged or neutral. The rate of diffusion depends on the concentration gradients. *Fick's first law* expresses this:

$$J = -D \frac{\partial c}{\partial x} \qquad (5.1)$$

where J is the flux of species, $\partial c/\partial x$ the concentration gradient in direction x—a plane surface is assumed—and D is the proportionality constant known as the *diffusion coefficient*. Its value in aqueous solution normally varies between 10^{-5} and 10^{-6} cm^2 s^{-1}, and can, in general, be determined through application of the equations for the current–voltage profiles of the various electrochemical methods. Alternatively, the Nernst–Einstein or Stokes–Einstein relations discussed in Chapter 2 may be used to estimate values of D.

The next question is: what is the variation of concentration with time due to diffusion? The variation is described by *Fick's second law* which, for a one-dimensional system, is

$$\frac{\partial c}{\partial t} = D \frac{\partial^2 c}{\partial x^2} \qquad (5.2)$$

This is easily obtained from Fick's first law by the following reasoning. Consider an element of width dx (Fig. 5.1). The change in concentration

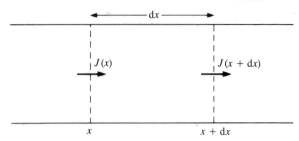

Fig. 5.1. Diffusion in one dimension. Diffusion is in the direction opposing the concentration gradient.

is given by

$$\frac{\partial c}{\partial t} = \frac{J(x) - J(x + dx)}{dx} \tag{5.3}$$

We know that

$$J(x + dx) = J(x) + \frac{\partial J(x)}{\partial x} \, dx$$

$$= J(x) - \frac{\partial}{\partial x} D \frac{\partial c}{\partial x} \, dx$$

$$= J(x) - D \frac{\partial^2 c}{\partial x^2} \, dx \tag{5.4}$$

assuming that D is constant. Substituting, we reach Fick's second law:

$$\frac{\partial c}{\partial t} = D \frac{\partial^2 c}{\partial x^2} \tag{5.2}$$

Table 5.1. The Laplace operator in various coordinate systems

Coordinates	Laplace operator
Cartesian	$\dfrac{\partial}{\partial x} + \dfrac{\partial}{\partial y} + \dfrac{\partial}{\partial z}$
Cylindrical	$\dfrac{\partial}{\partial r} + \dfrac{1}{r}\dfrac{\partial}{\partial \phi} + \dfrac{\partial}{\partial x}$
Spherical	$\dfrac{\partial}{\partial r} + \dfrac{1}{r}\dfrac{\partial}{\partial \theta} + \dfrac{1}{r \sin \theta}\dfrac{\partial}{\partial \phi}$

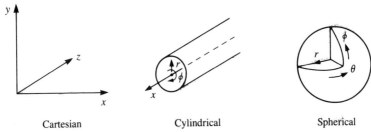

Cartesian   Cylindrical   Spherical

Fig. 5.2. Definition of the coordinates used in Table 5.1.

For any coordinate system

$$J = -D \, \nabla c \tag{5.5}$$

$$\frac{\partial c}{\partial t} = D \, \nabla^2 c \tag{5.6}$$

where ∇ is called the *Laplace operator* and has the forms shown in Table 5.1; the respective coordinates are defined in Fig. 5.2.

The solution of Fick's second law gives the variation of flux, and thence diffusion-limited current, with time, it being important to specify the conditions necessary to define the behaviour of the system (boundary conditions). Since the second law is a partial differential equation it has to be transformed into a total differential equation, solved, and the transform inverted[1]. The Laplace transform permits this (Appendix 1).

In the next two sections we use the Laplace transform to solve Fick's second law for two important cases under conditions of pure diffusion control:

● Determination of the diffusion-limited current, I_d, following the application of a potential step: change in potential from a value where there is no electrode reaction to a value where all the electroactive species react.

● Determination of the variation of potential with time resulting from the application of a constant current to the electrode.

Note that in the first case the potential is controlled and the current response and its variation with time is registered, *chronoamperometry*, and in the second case the value of the current is controlled and the variation of potential with time is registered, *chronopotentiometry*.

5.3   Diffusion-limited current: planar and spherical electrodes

The experiment leading to the diffusion-limited current involves application of a potential step at $t = 0$ to an electrode, in a solution containing

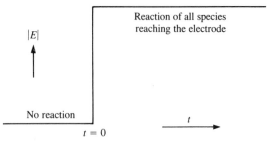

Fig. 5.3. Potential step to obtain a diffusion-limited current of the electroactive species.

either oxidized or reduced species, from a value where there is no electrode reaction to the value where all electroactive species that reach the electrode react, as shown in Fig. 5.3. This gives rise to a diffusion-limited current whose value varies with time. For a planar electrode, which is uniformly accessible, this is called semi-infinite linear diffusion, and the current is

$$I = nFAD\left(\frac{\partial c}{\partial x}\right)_0 \tag{5.7}$$

where $I = nFAJ$, x is the distance from the electrode, and we consider, for simplicity, an oxidation (anodic current) with $c = [\mathrm{R}]$. If it were a reduction a minus sign would be introduced into (5.7).

We solve Fick's second law

$$\frac{\partial c}{\partial t} = D\frac{\partial^2 c}{\partial x^2} \tag{5.8}$$

with the boundary conditions corresponding to our experiment, which are

$$t = 0, \qquad c_0 = c_\infty \qquad \text{(no electrode reaction)} \tag{5.9a}$$

$$t \geqslant 0 \qquad \lim_{x \to \infty} c = c_\infty \qquad \text{(bulk solution)} \tag{5.9b}$$

$$\left.\begin{array}{r} t > 0 \\ x = 0 \end{array}\right\} \qquad c_0 = 0 \qquad \text{(diffusion-limited current, } I_\mathrm{d}) \tag{5.9c}$$

in which c_0 represents the concentration at the electrode and c_∞ the concentration in bulk solution.

The mathematical solution to this problem is facilitated by using a dimensionless concentration

$$\gamma = \frac{c - c_\infty}{c_\infty} \tag{5.10}$$

and then (5.8) is transformed with respect to t using the Laplace transform, leading to

$$s\bar{\gamma} = D\frac{\partial^2 \bar{\gamma}}{\partial x^2} \tag{5.11}$$

The general solution to this equation is well known:

$$\bar{\gamma} = A'(s)\exp\left[-(s/D)^{1/2}x\right] + B'(s)\exp\left[(s/D)^{1/2}x\right] \tag{5.12}$$

Since the second term of the right-hand side does not satisfy the second boundary condition $(x \to \infty, \bar{\gamma} \to 0)$, then $B'(s) = 0$. The third boundary condition in Laplace space is

$$x = 0 \qquad \bar{\gamma} = -1/s \tag{5.13}$$

and we obtain from (5.12) for $x = 0$,

$$A'(s) = -1/s \tag{5.14}$$

Substituting,

$$\bar{\gamma} = \frac{1}{s}\exp\left[-(s/D)^{1/2}x\right] \tag{5.15}$$

and differentiating,

$$\left(\frac{\partial \bar{\gamma}}{\partial x}\right) = (sD)^{-1/2}\exp\left[-(s/D)^{1/2}x\right] \tag{5.16}$$

Inversion of (5.15) leads to the variation of concentration with distance from the electrode according to the value of t:

$$c = c_\infty\left\{1 - \mathrm{erfc}\left[\frac{x}{x(Dt)^{1/2}}\right]\right\} \tag{5.17}$$

which is represented in Fig. 5.4 for various values of t.

From (5.16) we obtain the current by putting $x = 0$ and inverting the transform (see Appendix 1, Table 1):

$$\left(\frac{\partial c}{\partial x}\right)_0 = \frac{1}{(\pi Dt)^{1/2}} \tag{5.18}$$

and so

$$I(t) = I_d(t) = \frac{nFAD^{1/2}c_\infty}{(\pi t)^{1/2}} \tag{5.19}$$

This is known as the *Cottrell equation*[2] (Fig. 5.5).

The current decreases with $t^{1/2}$, which means that after a certain time we cannot have confidence in the measured currents owing to the

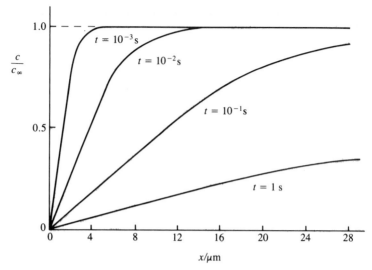

Fig. 5.4. Variation of concentration with distance at a planar electrode for various values of t after the application of a potential step, following (5.17), $D = 10^{-5}\,\mathrm{cm^2\,s^{-1}}$.

contribution of natural convection, etc., that perturbs the concentration gradients. This critical time can vary between some seconds and several minutes depending on the system's experimental arrangement.

It should also not be forgotten that, from a practical point of view, for small values of t there is a capacitive contribution to the current, due to double layer charging, that has to be subtracted. This contribution arises

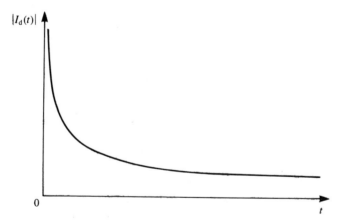

Fig. 5.5. Variation of current with time according to the Cottrell equation.

from the attraction between the electrode and the charges and dipoles in solution, and differs according to the applied potential; a rapid change in applied potential causes a very fast change in the distribution of species on the electrode surface and a large current during up to 0.1 s (see Chapter 3).

At a spherical electrode, of radius r_0, the relevant equation to solve is

$$\frac{\partial c}{\partial t} = D \left\{ \frac{\partial^2 c}{\partial r^2} + \frac{2}{r} \frac{\partial c}{\partial r} \right\} \tag{5.20}$$

with boundary conditions

$$t = 0 \quad r \geq r_0 \quad c = c_\infty \qquad \text{(no electrode reaction)} \tag{5.21a}$$

$$t \geq 0 \qquad \lim_{r \to \infty} c = c_\infty \quad \text{(bulk solution)} \tag{5.21b}$$

$$t > 0 \quad r = r_0 \quad c = 0 \qquad \text{(diffusion-limited current, } I_d) \tag{5.21c}$$

corresponding to (5.8) and (5.9). Defining a dimensionless concentration as before,

$$\gamma = \frac{c - c_\infty}{c_\infty} \tag{5.10}$$

and putting $v = r\gamma$, (5.20) becomes

$$\frac{\partial v}{\partial t} = D \frac{\partial^2 v}{\partial r^2} \tag{5.22}$$

This equation is the same as for a planar electrode and the boundary conditions are of the same form. Thus the method of solution is the same. The result is

$$I_d(t) = nFADc_\infty \left[\frac{1}{(\pi Dt)^{1/2}} + \frac{1}{r_0} \right] \tag{5.23}$$

This is the Cottrell equation, (5.19), plus a spherical correction term. Two extreme cases can be considered:

• *Small t.* The second term in (5.23) can be neglected, in other words the spherical nature of the electrode is unimportant. Diffusion at a sphere can be treated as linear diffusion. This is very important for the dropping mercury electrode (Section 8.3): for typical values of drop radius of 0.1 cm and $D = 10^{-5} \text{ cm}^2 \text{ s}^{-1}$, after $t = 3 \text{ s}$ there is only a 10 per cent error in using (5.19).

Table 5.2. Diffusion currents for planar and spherical electrodes: assuming $D_O = D_R$. r_0 = sphere radius; $\theta = [O]_*/[R]_*$

	Type of electrode	Equation	Comments
Oxidation or reduction	Planar	$\dfrac{nFAD^{1/2}c_\infty}{(\pi t)^{1/2}}$	Cottrell equation (5.19)
	Spherical	$\dfrac{nFAD^{1/2}c_\infty}{(\pi t)^{1/2}} + \dfrac{nFADc_\infty}{r_0}$	Cottrell equation plus spherical correction (5.23)
Oxidation and reduction (close to equilibrium potential)	Planar	$\dfrac{nFAD^{1/2}c_\infty}{(1+\theta)(\pi t)^{1/2}}$	
	Spherical	$\dfrac{nFAD^{1/2}c_\infty}{(1+\theta)(\pi t)^{1/2}} + \dfrac{nFADc_\infty}{(1+\theta)r_0}$	

• *Large t.* The spherical term dominates, which represents a steady-state current. However, due to the effects of natural convection this steady state is never reached at conventionally-sized electrodes. The smaller the elctrode radius, the faster the steady state is achieved. It is possible to achieve a steady state at microelectrodes. These are described further in Section 5.5.

The equations for the diffusion-limited current at planar and spherical electrodes are shown in Table 5.2 together with the expressions for the diffusion currents when the potential is not far from the equilibrium potential so that oxidation and reduction occur at the same time.

5.4 Constant current: planar electrodes

Starting at $t = 0$ a constant current is applied to the electrode in order to cause oxidation or reduction of electroactive species, and the variation of the potential of the electrode with time is measured (chronopotentiometry). Fick's second law is solved using the Laplace transform as in the previous section; the first two boundary conditions are the same, but the third is different:

$$t = 0 \quad c_0 = c_\infty \quad \text{(no electrode reaction)} \quad (5.9a)$$

$$t \geqslant 0 \quad \lim_{x \to \infty} c = c_\infty \quad \text{(bulk solution)} \quad (5.9b)$$

$$t > 0 \quad x = 0 \quad I = nFAD(\partial c/\partial x)_0 \quad (5.24)$$

The third condition expresses the fact that a concentration gradient is being imposed at the electrode surface.

As in the last section, and following the same arguments, we reach

$$\bar{\gamma} = A'(s) \exp\left[-(s/D)^{1/2}x\right] \quad (5.25)$$

given that $B'(s) = 0$. The boundary condition (5.24) after being transformed is

$$I/s = nFAD(\partial\bar{\gamma}/\partial x)_0 \qquad (5.26)$$

Differentiating (5.25) we obtain

$$\left(\frac{\partial\bar{\gamma}}{\partial x}\right)_0 = \left(\frac{s}{D}\right)^{1/2} xA'(s) \qquad (5.27)$$

and thus,

$$A'(s) = \frac{I}{s^{3/2}D^{1/2}nFAx} \qquad (5.28)$$

The equations that give the variation of concentration with time and the variation of potential with time are

$$\bar{\gamma} = \frac{I}{s^{3/2}D^{1/2}nFAx} \exp\left[-\left(\frac{s}{D}\right)^{1/2}x\right] \qquad (5.29)$$

and

$$\left(\frac{\partial\bar{\gamma}}{\partial x}\right)_0 = \frac{I}{snFAD} \qquad (5.30)$$

Inversion gives, respectively

$$c = c_\infty - \frac{I}{nFAD}\left\{2\left(\frac{Dt}{\pi}\right)^{1/2}\exp\left(\frac{-x^2}{4Dt}\right) - x\,\mathrm{erfc}\left[\frac{x}{2(Dt)^{1/2}}\right]\right\} \qquad (5.31)$$

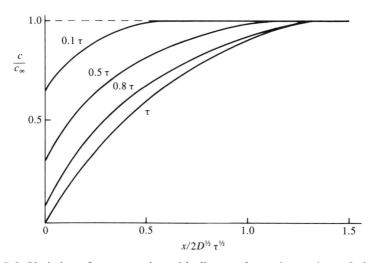

Fig. 5.6. Variation of concentration with distance for various values of t/τ in a chronopotentiometric (constant current) experiment, according to (5.31).

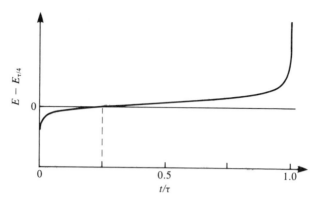

Fig. 5.7. Variation of potential with time (chronopotentiogram) in a system controlled by diffusion and with application of a constant current. τ is the transition time. $E_{\tau/4}$ is the potential when $t = \tau/4$ (Section 10.5).

and

$$c_0 = c_\infty - \frac{2It^{1/2}}{nFAD^{1/2}\pi^{1/2}} \tag{5.32}$$

When $c_0 = 0$, all species in the zone of the electrode have been consumed, as shown in Fig. 5.6. The corresponding value of t is called the *transition time*, τ.

From (5.32) and putting $c_0 = 0$ and $t = \tau$

$$\frac{I\tau^{1/2}}{c_\infty} = \frac{nFAD^{1/2}\pi^{1/2}}{2} \tag{5.33}$$

which is known as the *Sand equation*[3]. If $(I\tau^{1/2}/c_\infty)$ is not constant for several experiments with the same solution then the electrode process is not a simple electron transfer, but involves other steps. Figure 5.7 represents the theoretical variation of potential with time for this type of experiment.

It should be noted that the equation for the transition time at a spherical electrode is equal to that for a plane electrode. This result, perhaps unexpected, shows that it is only the current density that determines the transition time and not the curvature of the electrode surface.

5.5 Microelectrodes

Microelectrodes[4,5], or ultramicroelectrodes, are defined as electrodes which have at least one dimension that is a function of its size, which in practice means between 0.1 and 50 μm. Common geometries include spherical, hemispherical, disc, ring, and line. Their applications are many, and will be referred to throughout the book. They exhibit high current densities, but low total currents, so that the percentage electrolysis is small, and permit the attainment of steady states in situations, such as in the absence of added electrolyte, not possible with larger electrodes.

The diffusion-limited current at spherical and hemispherical microelectrodes follows directly from (5.23). We consider a hemispherical electrode as shown in Fig. 5.8. After a certain time, depending on the electrode size, a steady-state is reached, and the current is

$$I = \frac{nFADc_\infty}{r_0} = 2\pi nFr_0Dc_\infty \qquad (5.34)$$

This can be rewritten in terms of the surface length, d, where $d = \pi r_0$, as

$$I = 2nFdDc_\infty \qquad (5.35)$$

For a hemispherical electrode of diameter 1 μm, 95 per cent of the steady-state response is reached 0.1 s after application of a potential step. Additionally, and in general, as a result of the high rate of diffusion the current density is sufficiently high that interference from natural, and even forced, convection is negligible.

Finally, we consider the case of a plane disc microelectrode. In this

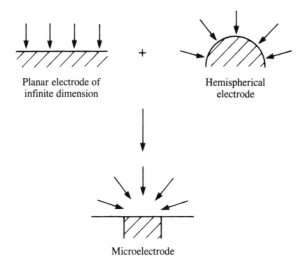

Planar electrode of
infinite dimension

Hemispherical
electrode

Microelectrode

Fig. 5.8. Schematic diagram showing the uniform current density for planar and spherical/hemispherical electrodes, and the non-uniform accessibility of the disc microelectrode.

case the solution of (5.7) is not sufficient, since we have to include the effects of radial diffusion, owing to the small dimension of the electrode. The equation to solve is

$$\frac{\partial c}{\partial t} = D\left(\frac{\partial^2 c}{\partial r^2} + \frac{1}{r}\frac{\partial c}{\partial r} + \frac{\partial^2 c}{\partial x^2}\right) \tag{5.36}$$

which does not have an analytical solution. Numerical analysis shows that for large t the current is numerically equal to that of a hemisphere of radius $a = \pi r_0/2$, showing the increase in mass transfer to and from the electrode caused by the radial diffusion component:

$$I = 4nFaDc_\infty \tag{5.37}$$

If once again we define the surface length, here as $d = 2a$, then the current is the same as at the hemispherical electrode:

$$I = 2nFdDc_\infty \tag{5.38}$$

Nevertheless there is an important difference, as shown in Fig. 5.8. In these cases the current density is not uniform. However, it is easier to make disc microelectrodes of solid materials than hemispheres. In fact, the similarity of (5.36) and (5.38) indicates that, in terms of d, the theory for the more easily tractable hemisphere can be applied without significant error, at least in the steady state, to the disc equivalent—this is found to be the case.

5.6 Diffusion layer

It can be easily verified from Figs. 5.5 and 5.6 that the concentration gradient tends asymptotically to zero at large distances from the electrode, and that the concentration gradient is not linear. However, for reasons of comparison it is useful to speak of a diffusion layer defined in the following way:

$$D\left(\frac{\partial c}{\partial x}\right)_0 = D\frac{(c_\infty - c_0)}{\delta} \tag{5.39}$$

where δ is the diffusion layer thickness (Fig. 5.9). The diffusion layer results, therefore, from the extrapolation of the concentration gradient at the electrode surface until the bulk concentration value is attained. This approximation was introduced by Nernst[6]. δ is frequently related to the mass transfer coefficient k_d, since when $c_0 = 0$

$$k_d = D/\delta \tag{5.40}$$

k_d has the dimensions of a heterogeneous rate constant.

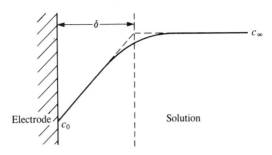

Fig. 5.9. The definition of the diffusion layer δ; $(\partial c/\partial x)_0$ is the concentration gradient at the electrode surface.

Applying (5.39) to the Cottrell equation (5.19), we can deduce the expression for the diffusion layer thickness as

$$\delta = (\pi D t)^{1/2} \tag{5.41}$$

The thickness increases with $t^{1/2}$, hence the problems of natural convection, etc. at large t. The mass transfer coefficient is (compare with (5.40)) for $c_0 = 0$

$$k_{\mathrm{d}} = (D/\pi t)^{1/2} \tag{5.42}$$

At a hemispherical microelectrode

$$k_{\mathrm{d}} = 2D/r_0 \tag{5.43}$$

The smaller δ the larger the concentration gradient at the electrode surface, leading to higher currents. It is also useful in certain investigations that δ is independent of t. Both these conditions can be satisfied by microelectrodes. They can also be obtained by imposition of forced convection at larger electrodes—hydrodynamic systems.

5.7 Convection and diffusion: hydrodynamic systems

In a fluid where there is both convection and diffusion—a hydrodynamic system—the flux is given by

$$J = c\boldsymbol{v} - D\,\nabla c \tag{5.44}$$

which is a modification of Fick's first law. The expression corresponding to Fick's second law is

$$\frac{\partial c}{\partial t} = -\boldsymbol{v}\,\nabla c + D\,\nabla^2 c \tag{5.45}$$

The forms of this equation in the three coordinate systems can be found in Table 5.3.

Table 5.3. The convective-diffusion equation in the three coordinate systems $\partial c/\partial t = D \, \nabla^2 c - \boldsymbol{v} \, \nabla c$

	Diffusion	Convection
Cartesian	$\dfrac{\partial c}{\partial t} = D\left(\dfrac{\partial^2 c}{\partial x^2} + \dfrac{\partial^2 c}{\partial y^2} + \dfrac{\partial^2 c}{\partial z^2}\right)$	$-\left(v_x \dfrac{\partial c}{\partial x} + v_y \dfrac{\partial c}{\partial y} + v_z \dfrac{\partial c}{\partial z}\right)$
Cylindrical polar	$\dfrac{\partial c}{\partial t} = D\left(\dfrac{1}{r}\dfrac{\partial}{\partial r}\left(\dfrac{\partial c}{r}\right) + \dfrac{1}{r^2}\dfrac{\partial^2 c}{\partial \phi^2} + \dfrac{\partial^2 c}{\partial x^2}\right)$	$-\left(v_r \dfrac{\partial c}{\partial r} + \dfrac{v_\phi}{r}\dfrac{\partial c}{\partial \phi} + v_x \dfrac{\partial c}{\partial x}\right)$
Spherical polar	$\dfrac{\partial c}{\partial t} = D\left(\dfrac{1}{r^2}\dfrac{\partial}{\partial r}\left(r^2 \dfrac{\partial c}{\partial r}\right) + \dfrac{1}{r^2 \sin^2 \theta}\dfrac{\partial}{\partial \theta}\right.$ $\left. \times \left(\sin \theta \dfrac{\partial c}{\partial \theta}\right) + \dfrac{1}{r^2 \sin^2 \theta}\dfrac{\partial^2 c}{\partial \phi^2}\right)$	$-\left(v_r \dfrac{\partial c}{\partial r} + \dfrac{v_\theta}{r}\dfrac{\partial c}{\partial \theta} + \dfrac{v_\phi}{r \sin \theta}\dfrac{\partial c}{\partial \phi}\right)$

Many hydrodynamic systems have been studied theoretically[7-11]. The solution to (5.45) proceeds through analysis of the velocity profile, derived from the momentum continuity equation and which is, for an incompressible fluid,

$$\nabla \boldsymbol{v} = 0 \tag{5.46}$$

and from the law of conservation of momentum

$$\frac{d\boldsymbol{v}}{dt} + \boldsymbol{v} \, \nabla \boldsymbol{v} = \frac{1}{\rho}\nabla P + \nu \, \nabla^2 \boldsymbol{v} + g \tag{5.47}$$

where P is the pressure difference through the system. The velocity profile obtained depends on whether the flow corresponds to a laminar, transition, or turbulent regime. In general, electrochemical investigations are done under laminar flow conditions. Studies of mass and heat transfer have been conducted in fluid dynamics for many types of system, including surface heterogeneous reactions. The results can often be applied directly to hydrodynamic electrochemical systems using the so-called 'similarity principle'.

An important exception to laminar flow is the calculation of velocity profiles in industrial electrochemical reactors. These often function in the turbulent regime in order to maximize the yield of the process, given that the mass transfer coefficients are higher.

In Section 5.9, we show how to solve the convective-diffusion equation for the rotating disc electrode in order to calculate the diffusion-limited current. When the forced convection is constant, then $\partial c/\partial t = 0$, which simplifies the mathematical solution.

An important assumption in these calculations is that, within the diffusion layer, there is no convection, an approximation which means that the variations in the velocity components (defining a hydrodynamic layer) must occur within distances much greater than the diffusion layer thickness—by at least a factor of 10. This assumption should be verified for any hydrodynamic system under study.

The development of convective-diffusion theories is due principally to Prandtl[9] and Schlichting[10], and their application in electrochemistry to Levich[11]. Levich was the first to solve the equations for the rotating disc electrode.

5.8 Hydrodynamic systems: some useful parameters

Besides the diffusion layer, of thickness δ, and the mass transfer coefficient, k_d, there are other parameters which are useful for describing hydrodynamic systems.

The first is the concept of a hydrodynamic layer of thickness δ_H, within which all velocity gradients occur. In practice one uses values that differ by 5 per cent from their values at infinite distance from the electrode surface, given that the components tend asymptotically to their values in bulk solution. It has been demonstrated that[11]

$$\delta \approx \left(\frac{D}{\nu}\right)^{1/3} \delta_H \qquad (5.48)$$

In aqueous solution $D \approx 10^{-5} \, \text{cm}^2 \, \text{s}^{-1}$ and $\nu \approx 10^{-2} \, \text{cm}^2 \, \text{s}^{-1}$, therefore

$$\delta \approx 0.1 \delta_H \qquad (5.49)$$

as shown in Fig. 5.10. It is therefore reasonable to suppose that there is no convection within the diffusion layer.

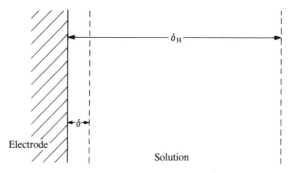

Fig. 5.10. Diagram showing the relative thicknesses of the hydrodynamic and diffusion layers in a hydrodynamic system in aqueous solution.

We now define various dimensionless groups that are useful in mass transport problems:

1. *The Reynolds number, Re*

$$Re = \frac{vl}{v} \tag{5.50}$$

where v is a characteristic velocity, l a characteristic length, and v the kinematic viscosity. Below a certain critical value of Re, Re_{crit}, the flow is laminar; above it is turbulent with a transition regime around Re_{crit}.

2. *The Schmidt number, Sc*

$$Sc = \frac{v}{D} \tag{5.51}$$

is a characteristic of the medium, v being a characteristic primarily of the solvent and D of the electrochemical species. Sc is, therefore, determined purely by the physical properties of the solution. For aqueous solution, $Sc \approx 10^3$.

3. *The Peclet number, Pe*

$$Pe = \frac{vl}{D} = Re.Sc \tag{5.52}$$

represents the relative contributions of transport by convection and by diffusion. In aqueous solution outside the diffusion layer $Pe \gg 1$, even for small Re.

4. *The Sherwood number, Sh*

$$Sh = \frac{Jl}{D(c_\infty - c_0)} = \frac{l}{\delta} = \frac{k_d l}{D} \tag{5.53}$$

where l is a characteristic length of the system. Sh is proportional to the mass transfer coefficient k_d.

5.9 An example of a convective-diffusion system: the rotating disc electrode

A rotating disc electrode[12,13] consists of a disc electrode embedded in the middle of a plane surface (theoretically of infinite extent) that rotates around its axis in a fluid, the disc being centred on the axis. In practice the electrode bodies usually have the form of a cylinder, with the sheath around the disc significantly larger than it is, so as to approximate a surface of infinite dimension (Fig. 8.2). We assume in these calculations that there are no convection effects caused by the walls of the cell where the experiment is carried out, nor effects resulting from the finite dimensions of the electrode body etc.—generically known as *edge effects*.

The first step consists in deducing the velocity profile. This fluid dynamics problem was solved by von Karman[14] and Cochran[15] and gives the velocity components:

radial	$v_r = r\omega F(\gamma)$	(5.54a)
angular	$v_\phi = r\omega G(\gamma)$	(5.54b)
perpendicular	$v_x = -(\omega v)^{1/2} H(\gamma)$	(5.54c)

where γ is the dimensionless distance from the disc

$$\gamma = \left(\frac{\omega}{v}\right)^{1/2} x \tag{5.55}$$

In these expressions ω is the disc rotation rate (rad s^{-1}) and v the kinematic viscosity. The coordinates are shown in Fig. 5.2. Figure 5.11 shows the variation of the functions F, G, and H with distance from the disc and Fig. 5.12 the corresponding streamlines. As a result of rotation, solution is sucked towards the disc and spread out sideways.

Levich[10] introduced the approximation that, close to the electrode,

$$v_{r,x\to 0} \approx Crx \tag{5.56a}$$

$$v_{\phi,x\to 0} \approx 0 \tag{5.56b}$$

$$v_{x,x\to 0} \approx -Cx^2 \tag{5.56c}$$

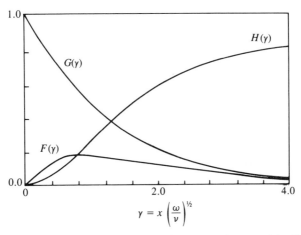

Fig. 5.11 The functions F, G, and H and their variation with dimensionless distance from a rotating disc.

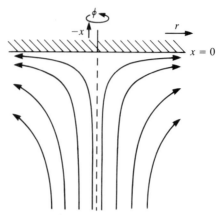

Fig. 5.12. Streamlines for a rotating disc.

where C, the convective constant, is given by

$$C = 0.510\omega^{3/2}\nu^{-1/2}D^{-1/3} \tag{5.57}$$

So, in the steady state at constant rotation speed, the convective diffusion equation is

$$Crx\frac{\partial c}{\partial r} + Cx^2\frac{\partial c}{\partial x} = D\frac{\partial^2 c}{\partial x^2} \tag{5.58}$$

neglecting radial diffusion as it is much less than radial convection. The boundary conditions for the case of the disc electrode passing the steady-state limiting current, I_L, are

$x \to \infty$	$c \to c_\infty$	(bulk solution)	(5.59a)
$r = 0$	$c_0 = 0$	(centre of disc)	(5.59b)
$x = 0$	$c_0 = 0$	(limiting current, I_L)	(5.59c)

Using the following definitions of dimensionless variables:

$$\chi = \left(\frac{3C}{D}\right)^{1/3}\left(\frac{r}{r_1}\right)x \qquad \text{distance variable} \tag{5.60}$$

$$\xi = \left(\frac{r}{r_1}\right)^3 \qquad \text{radial variable} \tag{5.61}$$

$$\gamma = \frac{c - c_\infty}{c_\infty} \qquad \text{concentration variable} \tag{5.62}$$

where r_1 is the disc radius, and the properties of partial derivatives, it is easy to reach

$$\chi \frac{\partial \gamma}{\partial \xi} = \frac{\partial^2 \gamma}{\partial \chi^2} \tag{5.63}$$

If this equation is transformed with respect to ξ by means of the Laplace transform we obtain

$$\chi s \bar{\gamma} = \frac{\partial^2 \bar{\gamma}}{\partial \chi^2} \tag{5.64}$$

which is the Airy equation (Appendix 1). The boundary conditions in Laplace space are

$$\chi \rightarrow \infty \qquad \gamma \rightarrow 0 \tag{5.65a}$$

$$\xi = 0 \qquad \bar{\gamma} = -1/s \tag{5.65b}$$

$$\chi = 0 \qquad \bar{\gamma} = -1/s \tag{5.65c}$$

The solution to the Airy equation is

$$\bar{\gamma} = A'(s) \, \text{Ai} \, (s^{1/3}\chi) \tag{5.66}$$

Boundary condition (5.65c) shows that

$$A'(s) = -\frac{1}{s \, \text{Ai} \, (0)} \tag{5.67}$$

and so

$$\bar{\gamma} = -\frac{\text{Ai} \, (s^{1/3}\chi)}{s \, \text{Ai} \, (0)} \tag{5.68}$$

Differentiating, at the electrode surface

$$\left(\frac{\partial \bar{\gamma}}{\partial \chi}\right)_0 = -\frac{\text{Ai}' \, (0)}{s^{2/3} \, \text{Ai} \, (0)} \tag{5.69}$$

On inverting

$$\left(\frac{\partial \gamma}{\partial \chi}\right)_0 = -\frac{\text{Ai}' \, (0)}{\text{Ai} \, (0)\Gamma(2/3)} \xi^{-1/3} \tag{5.70}$$

and we get

$$\left(\frac{\partial c}{\partial x}\right)_0 = 0.602 D^{-1/3} v^{-1/6} \omega^{1/2} c_\infty \tag{5.71}$$

The limiting current is given by

$$I_L = 2\pi nFD \int_0^{r_1} \left(\frac{\partial c}{\partial x}\right)_0 dr \tag{5.72}$$

which is

$$I_L = 0.620nF\pi r_1^2 D^{2/3} \nu^{-1/6} \omega^{1/2} c_\infty \tag{5.73}$$

This current is directly proportional to the electrode area, which shows that the disc is uniformly accessible. The same can be concluded from the thickness of the diffusion layer

$$\delta = 1.61 D^{1/3} \nu^{1/6} \omega^{-1/2} \tag{5.74}$$

which is independent of r. Some other hydrodynamic electrodes are not uniformly accessible such as, for example, tubular and impinging jet electrodes. However, the method for calculating the diffusion-limited current is always the same, see Section 8.2.

The various hydrodynamic electrodes and their use in investigating electrode processes are described in Chapter 8.

References

1. D. D. Macdonald, *Transient techniques in electrochemistry*, Plenum, New York, Chapter 3.
2. F. G. Cottrell, *Z. Physik. Chem.*, 1902, **42**, 385.
3. H. J. S. Sand, *Philos. Mag.*, 1901, **1**, 45.
4. M. Fleischmann, S. Pons, D. R. Rollison, and P. P. Schmidt, *Ultramicroelectrodes*, Datatech Systems Inc., Morganton, NC, 1987.
5. M. I. Montenegro, M. A. Queiros, and J. L. Daschbach (eds.), *Microelectrodes: theory and applications*, Proceedings of NATO ASI (1990), Kluwer, Dordrecht, 1991.
6. W. Nernst, *Z. Physik, Chem.*, 1904, **47**, 52.
7. J. S. Newman, *Advances in electrochemistry and electrochemical engineering*, ed. P. Delahay and C. W. Tobias, Wiley, New York, Vol. 5, 1967, pp. 87–135.
8. J. S. Newman, *Electrochemical systems*, Prentice-Hall, Englewood Cliffs, NJ, 1973.
9. L. Prandtl, *Proc. Int. Math. Congr.*, Heidelberg, 1903.
10. H. Schlichting, *Boundary layer theory*, Pergamon Press, London, 1955.
11. V. G. Levich, *Physiochemical hydrodynamics*, Prentice-Hall, Englewood Cliffs, NJ, 1962.
12. A. C. Riddiford, *Advances in electrochemistry and electrochemical engineering*, ed. P. Delahay and C. W. Tobias, Vol. 4, 1966, pp. 47–116.
13. C. M. A. Brett and A. M. C. F. Oliveira Brett, *Comprehensive chemical kinetics*, ed. C. H. Bamford and R. G. Compton, Elsevier, Amsterdam, Vol. 26, 1986, Chapter 5.
14. T. von Karman, *Z. Angew. Math. Mech.*, 1921, **1**, 233.
15. W. G. Cochran, *Proc. Camb. Phil. Soc. math. phys. sci.*, 1934, **30**, 365.

6

KINETICS AND TRANSPORT IN ELECTRODE REACTIONS

6.1 Introduction

In the last two chapters the kinetics of electrode processes and mass transport to an electrode were discussed. In this chapter these two parts of the electrode process are combined and we see how the relative rates of the kinetics and transport cause the behaviour of electrochemical systems to vary[1-4].

6.2 The global electrode process: kinetics and transport

Transport to the electrode surface as described in Chapter 5 assumes that this occurs solely and always by diffusion. In hydrodynamic systems, forced convection increases the flux of species that reach a point corresponding to the thickness of the diffusion layer from the electrode. The *mass transfer coefficient* k_d describes the rate of diffusion within the diffusion layer and k_c and k_a are the *rate constants* of the electrode reaction for reduction and oxidation respectively. Thus for the simple electrode reaction $O + ne^- \rightarrow R$, without complications from adsorption,

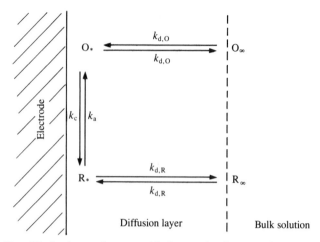

Fig. 6.1. Simplified scheme for an oxidation–reduction reaction on an electrode surface.

we can write

$$O_\infty \underset{k_{d,O}}{\overset{k_{d,O}}{\rightleftarrows}} O_* \underset{k_a}{\overset{k_c}{\rightleftarrows}} R_* \underset{k_{d,R}}{\overset{k_{d,R}}{\rightleftarrows}} R_\infty$$

as shown in Fig. 6.1, where $k_{d,O}$ and $k_{d,R}$ are the mass transfer coefficients of the species O and R. In general these coefficients differ because the diffusion coefficients differ. From Chapter 4 we have the Butler–Volmer expressions for the kinetic rate constants:

$$k_c = k_0 \exp\left[-\alpha_c nF(E - E^{\ominus\prime})/RT\right] \tag{6.1}$$

$$k_a = k_0 \exp\left[\alpha_a nF(E - E^{\ominus\prime})/RT\right] \tag{6.2}$$

Let us assume that $(\partial c/\partial t) = 0$, i.e. *steady state*—in other words the rate of transport of electroactive species is equal to the rate of their reaction on the electrode surface. The steady state also means that the applied potential has a fixed value.

The flux of electroactive species, J, is

$$J = -k_c[O]_* + k_a[R]_* \tag{6.3a}$$

$$= k_{d,O}([O]_* - [O]_\infty) \tag{6.3b}$$

$$= k_{d,R}([R]_\infty - [R]_*) \tag{6.3c}$$

When all O or R that reaches the electrode is reduced or oxidized, we obtain the *diffusion-limited cathodic or anodic current densities* $j_{L,c}$ and $j_{L,a}$:

$$j_{L,c}/nF = -k_{d,O}[O]_\infty \qquad j_{L,a}/nF = k_{d,R}[R]_\infty \tag{6.4}$$

Since $k_d = D/\delta$ (equation (5.40)), we can write

$$k_{d,O}/k_{d,R} = p = (D_O/D_R)^s \tag{6.5}$$

Making the appropriate substitutions, the concentrations can be removed from (6.3) leading to

$$j = \frac{k_c j_{L,c} + p k_a j_{L,a}}{k_{d,O} + k_c + p k_a} \tag{6.6}$$

We can point out two extreme cases for this expression:

1. Only O present in solution: $j_{L,a} = 0$ and $k_a = 0$. Thus

$$j = \frac{k_c j_{L,c}}{k_{d,O} + k_c} \tag{6.7}$$

that is

$$\frac{1}{j} = \frac{k_{d,O}}{k_c j_{L,c}} + \frac{1}{j_{L,c}} \tag{6.8}$$

$$-\frac{1}{j} = \underbrace{\frac{1}{nFk_c[O]_\infty}}_{\text{kinetics}} + \underbrace{\frac{1}{nFk_{d,O}[O]_\infty}}_{\text{transport}} \tag{6.9}$$

This result shows that the total flux is due to a transport and a kinetic term. When $k_c \gg k_{d,O}$ then

$$-\frac{1}{j} = \frac{1}{nFk_{d,O}[O]_\infty} \tag{6.10}$$

and the flux is determined by the transport. On the other hand, when $k_c \ll k_{d,O}$

$$-\frac{1}{j} = \frac{1}{nFk_c[O]_\infty} \tag{6.11}$$

and the kinetics determines the flux.

2. Only R present in solution: $j_{L,c} = 0$ and $k_c = 0$. From (6.6) we reach

$$j = \frac{k_a j_{L,a}}{k_{d,R} + k_a} \tag{6.12}$$

Rearranging,

$$\frac{1}{j} = \underbrace{\frac{1}{nFk_a[R]_\infty}}_{\text{kinetics}} + \underbrace{\frac{1}{nFk_{d,R}[R]_\infty}}_{\text{transport}} \tag{6.13}$$

The form of the expression is the same as that obtained in Case 1 with only O present in solution, and so the same comments are valid.

We now consider the factors that affect the variation of k_c (or k_a) and k_d. The kinetic rate constants depend on the applied potential and on the value of the standard rate constant, k_0. As was seen in Chapter 5, k_d is influenced by the thickness of the diffusion layer, which we can control through the type of experiment and experimental conditions, such as varying the forced convection. By altering k_c (or k_a) and k_d we can obtain kinetic information as will be described below. At the moment we note that there are two extremes of comparison between k_0 and k_d:

- $k_0 \gg k_d$ reversible system
- $k_0 \ll k_d$ irreversible system

The word reversible signifies that the system is at equilibrium at the electrode surface and it is possible to apply the Nernst equation at any potential.

The calculation of the current requires the relation between the flux and current:

$$I = \oint j \, dA = nF \oint J \, dA \qquad (6.14)$$

Normally, such as at stationary planar electrodes and at uniformly accessible hydrodynamic electrodes, for example the rotating disc, the flux over the electrode surface is constant: in this case we have the simple relation

$$I = Aj = nFAJ \qquad (6.15)$$

6.3   Reversible reactions

Reversible reactions are those where

$$k_0 \gg k_d$$

and, at any potential, there is always equilibrium at the electrode surface. The current is determined only by the electronic energy differences between the electrode and the donor or acceptor species in solution and their rate of supply. Applying the Nernst equation

$$E = E^{\ominus \prime} + \frac{RT}{nF} \ln \frac{[O]_*}{[R]_*} \qquad (6.16)$$

and given that $j/nF = k_{d,O}([O]_* - [O]_\infty)$ we have

$$\frac{j}{j_{L,c}} = \frac{[O]_* - [O]_\infty}{[O]_\infty} \tag{6.17}$$

that is

$$[O]_* = \frac{j_{L,c} - j}{j_{L,c}}[O]_\infty \tag{6.18}$$

Similarly,

$$[R]_* = \frac{j_{L,a} - j}{j_{L,a}}[R]_\infty \tag{6.19}$$

Substituting (6.18) and (6.19) in the Nernst equation, assuming the electrode is uniformly accessible $(I = Aj)$, and using (6.4) we get the steady-state expression

$$E = E^{\ominus\prime} + \frac{RT}{nF} \ln \left\{ \frac{I_{L,c} - I}{I - I_{L,a}} \cdot \frac{k_{d,R}}{k_{d,O}} \right\} \tag{6.20}$$

$$= E'_{1/2} + \frac{RT}{nF} \ln \frac{I_{L,c} - I}{I - I_{L,a}} \tag{6.21}$$

where

$$E'_{1/2} = E^{\ominus\prime} + \frac{RT}{nF} \ln \frac{k_{d,R}}{k_{d,O}} \tag{6.22}$$

$E_{1/2}$ is called the half-wave potential and corresponds to the potential when the current is equal to $(I_{L,a} + I_{L,c})/2$. Figure 6.2 shows the variation of current with applied potential, a *voltammogram*. The characteristic sigmoidal profile results from the logarithmic term in (6.21). By putting

$$\zeta = (E - E'_{1/2})\frac{nF}{RT} \tag{6.23}$$

(6.21) can be written in an alternative form

$$I = \frac{I_{L,a}}{1 + e^{-\zeta}} + \frac{I_{L,c}}{1 + e^{\zeta}} \tag{6.24}$$

which shows more clearly how the limiting current values are asymptotically approached as the potential (i.e. ζ) becomes very positive for anodic reactions or very negative for cathodic reactions.

Equation (6.21) is valid for any uniformly accessible electrode, including dropping electrodes, stationary plane electrodes, various hydro-

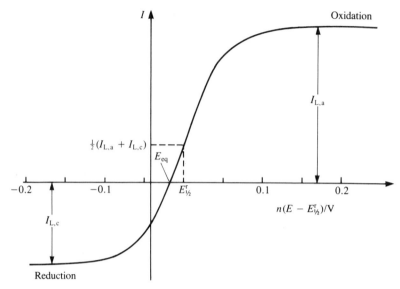

Fig. 6.2. Voltammogram for a reversible system where the solution contains O and R. Example: a mixture of Fe(II) and Fe(III) at a platinum rotating disc electrode.

dynamic systems, and hemispherical microelectrodes. In fact, the same expression is reached for non-uniformly accessible electrodes, but the reasoning is a little more complex.

In all cases

$$\frac{k_{d,R}}{k_{d,O}} = \left(\frac{D_R}{D_O}\right)^s \tag{6.25}$$

where $s = \frac{1}{2}$ (dropping or stationary electrode), $s = \frac{2}{3}$ (hydrodynamic electrodes) or $s = 1$ (microelectrodes). Even if $D_R = 1.5 D_O$, for $s = 1$, we get $E'_{1/2} - E^{\ominus'} = 10.5 \, \text{mV}$, which would be an extreme case. So, in nearly all cases, we can identify $E'_{1/2}$ with $E^{\ominus'}$ without introducing a large error.

Assuming that $D_R = D_O$ we can conclude, from (6.20), that the equilibrium potential E_{eq}, where the current is zero, is

$$E_{eq} = E^{\ominus'} + \frac{RT}{nF} \ln \frac{[O]_\infty}{[R]_\infty} \tag{6.26}$$

When $[O]_\infty = [R]_\infty$ then $E_{eq} = E'_{1/2} = E^{\ominus'}$.

From the expressions obtained above, we can write a diagnostic of reversibility:

• $E'_{1/2}$ independent of $[O]_\infty$ and $[R]_\infty$

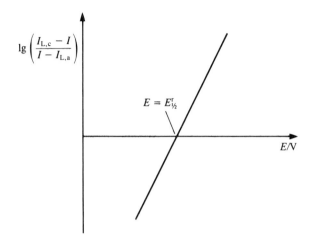

Fig. 6.3. Plot of $\lg\{(I_{L,c} - I)/(I - I_{L,a})\}$ vs. E for a reversible reaction; inverse slope is $(0.05916/n)$ V at 298 K.

• The form of the current-potential curve is independent of the diffusion layer thickness.

• A plot of $\lg[(I_{L,c} - I)/(I - I_{L,a})]$ vs. E—equation (6.21)—will give a straight line of slope $0.05916/n$ V at 298K and an intercept of $E'_{1/2}$, see Fig. 6.3.

6.4 Irreversible reactions

For irreversible reactions, $k_0 \ll k_d$. Kinetics has an important role, especially for potentials close to E_{eq}. It is necessary to apply a higher potential than for a reversible reaction in order to overcome the activation barrier and allow reaction to occur—this extra potential is called the overpotential, η. Because of the overpotential only reduction or only oxidation occurs and the voltammogram, or voltammetric curve, is divided into two parts. At the same time it should be stressed that the retarding effect of the kinetics causes a lower slope in the voltammograms than for the reversible case. Figure 6.4 shows, schematically, the curve obtained, which is explained in greater detail below.

The expressions (6.9) and (6.13) are valid for this situation. The transport term has to appear, since only at the beginning of an irreversible voltammogram can the effects of transport be neglected. This is because k_c or k_a increases on increasing the potential negatively or positively so that we finally reach the limiting current plateaux in Fig. 6.4.

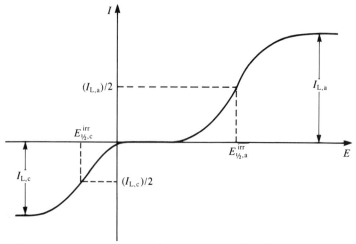

Fig. 6.4. Schematic voltammogram for an irreversible electrode reaction. Example: a mixture of anthraquinone and anthraquinol at a platinum rotating disc electrode.

The half-wave potential for reduction or oxidation varies with k_d, since there is not equilibrium on the electrode surface. For cathodic and anodic processes respectively we may write

$$E = E_{1/2}^{irr} - \frac{RT}{\alpha_c nF} \ln \frac{I_{L,c} - I}{I} \tag{6.27}$$

$$E = E_{1/2}^{irr} + \frac{RT}{\alpha_a nF} \ln \frac{I_{L,a} - I}{I} \tag{6.28}$$

where α is the charge transfer coefficient that appears in (6.1) and (6.2). $E_{1/2}^{irr}$ is not constant, but varies with the rate of transport of species to the electrode. Once more, similarly to the reversible case in Fig. 6.2, the logarithmic term in (6.27) and (6.28) explains the sigmoidal form of the curves obtained. The appearance of the factors α_c or α_a with $0 < \alpha < 1$ causes the curves to have a lower slope than for a reversible reaction. Plots of the type shown in Fig. 6.3 can be done using (6.27) and (6.28), and $(\alpha_a n)$ or $(\alpha_c n)$ calculated from the slopes of these plots.

For a uniformly accessible electrode ($I = jA$), (6.9) or (6.13) can be written in the form

$$\frac{1}{I} = \frac{1}{I_k} + \frac{1}{I_L} \tag{6.29}$$

in which I_k is the kinetic current and I_l the limiting current. A plot of I^{-1} vs. k_d^{-1} (proportional to I_L) is a straight line from which one obtains I_k

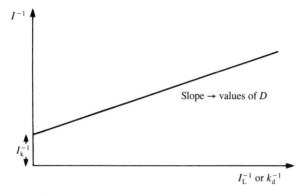

Fig. 6.5. Plot of I^{-1} vs. I_L^{-1} at a uniformly accessible electrode for an irreversible reaction.

from the intercept and the diffusion coefficient D from the slope (Fig. 6.5). For example, for the rotating disc electrode one constructs a plot of I^{-1} vs. $W^{-1/2}$, where W is the rotation speed of the electrode in Hz, since I_L is proportional to $W^{1/2}$, as demonstrated in Section 5.9.

6.5   The general case

We now consider the general case and show how reversible and irreversible systems are limiting cases of the general behaviour. For simplicity, we assume that $D_O = D_R$ ($p = 1$). Thus (6.6) becomes

$$j = \frac{k_c j_{L,c} + k_a j_{L,a}}{k_d + k_c + k_a} \tag{6.30}$$

Using again the Butler–Volmer formulation of electrode kinetics,

$$k_c = k_0 \exp\left[-\alpha_c nF(E - E^{\ominus\prime})/RT\right] \tag{6.1}$$

$$k_a = k_0 \exp\left[\alpha_a nF(E - E^{\ominus\prime})/RT\right] \tag{6.2}$$

and, in order to reduce the number of symbols, writing

$$f = \frac{F(E - E^{\ominus\prime})}{RT} \tag{6.31}$$

and substituting (6.1) and (6.2) in (6.30) we obtain

$$j = \frac{j_{L,c} \exp\left(-\alpha_c nf\right) + j_{L,a} \exp\left(\alpha_a nf\right)}{k_d/k_0 + \exp\left(-\alpha_c nf\right) + \exp\left(\alpha_a nf\right)} \tag{6.32}$$

We consider three limiting cases of (6.32)

1. $k_0 \gg k_d$: reversible system. Multiplying through by $\exp(\alpha_c nf)$ and since we are considering a simple electrode reaction where $(\alpha_a + \alpha_c) = 1$, expression (6.32) becomes

$$j = \frac{j_{L,c} + j_{L,a} \exp(nf)}{1 + \exp(nf)} \qquad (6.33)$$

or, rearranging,

$$\exp(nf) = \exp\left[\frac{nF(E - E^{\ominus\prime})}{RT}\right] = \frac{j_{L,c} - j}{j - j_{L,a}} \qquad (6.34)$$

For a uniformly accessible electrode,

$$E = E^{\ominus\prime} + \frac{RT}{nF} \ln \frac{I_{L,c} - I}{I - I_{L,a}} \qquad (6.35)$$

which is the equation for the voltammetric curve, equation (6.21), with $D_O = D_R$.

2. f large i.e. E very positive (oxidation) or very negative (reduction) in relation to $E^{\ominus\prime}$. Whatever the relative values of k_0 and k_d, we should obtain the limiting current. From (6.32):

• f large and positive, $\exp(-\alpha_c nf) \to 0$, and we obtain

$$j = j_{L,a} \qquad \text{i.e. limiting anodic current} \qquad (6.36)$$

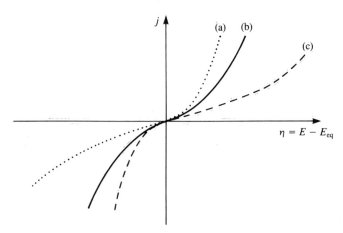

Fig. 6.6. The effect of the value of α_c on the current density, j. (a) $\alpha_c = 0.25$: oxidation favoured; (b) $\alpha_c = 0.50$: symmetric; (c) $\alpha_c = 0.75$; reduction favoured.

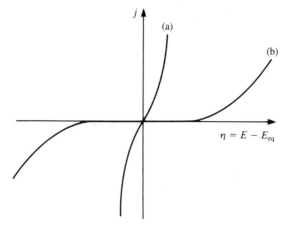

Fig. 6.7. The effect of the value of k_0 on the current density close to E_{eq}; (a) k_0 large; (b) k_0 smaller.

- f large and negative, $\exp(\alpha_a nf) \rightarrow 0$, and we obtain

$$j = j_{L,c} \quad \text{i.e. limiting cathodic current} \tag{6.37}$$

3. f close to zero, that is $E \sim E^{\ominus \prime}$. In these conditions we can approximate the exponentials by the first term in the Maclaurin expansion (Appendix 1). We obtain

$$j = \frac{j_{L,c} + j_{L,a} - j_{L,c}\alpha_c nf + j_{L,a}\alpha_a nf}{k_d/k_0 + 2} \tag{6.38}$$

In this relation j is directly proportional to E via f (see (6.31)). The proportionality constant is highly dependent on the value of α_c (and thence also on α_a), Fig. 6.6, and on the ratio k_d/k_0, Fig. 6.7.

6.6   The Tafel law

Between the limiting current plateaux of a voltammogram and the linear region close to E_{eq} described by (6.38) there is a region of potential for irreversible reactions where j depends exponentially on potential. This is the *Tafel region*. Considering a system where there is only O in bulk solution, that is there is only reduction, from the equation

$$k_c = k_0 \exp\left[-\alpha_c nF(E - E^{\ominus \prime})/RT\right] \tag{6.1}$$

it is obvious that

$$\ln k_c = \text{constant}_1 - \frac{\alpha_c nFE}{RT} \tag{6.39}$$

Since

$$j_c/nF = -k_c[O]_*$$ (6.40)

we have

$$-\ln j_c = \text{constant}_2 - \frac{\alpha_c nFE}{RT}$$ (6.41)

In a similar fashion for an oxidation,

$$\ln j_a = \text{constant}_3 + \frac{\alpha_a nFE}{RT}$$ (6.42)

These last two expressions are forms of the Tafel law. They are an example of a linear free energy relationship (linear relation between a kinetic and a thremodynamic parameter) the parameters in this case being the flux (or the current) and the potential.

Constructing plots of $\ln|j|$ vs. E we obtain Fig. 6.8. The slopes of the lines are $-\alpha_c(nF/RT)$ and $\alpha_a(nF/RT)$; the intercept at $E = E_{eq}$ gives the exchange current $I_0 = j_0 A$ and thence the standard rate constant from the expression deduced in Section 4.5:

$$I_0/nFA = j_0/nF = k_0[O]_\infty^{\alpha_a}[R]_\infty^{\alpha_c} = k_0[O]_\infty^{1-\alpha_c}[R]_\infty^{\alpha_c}$$ (6.43)

Note that close to E_{eq}, since reaction occurs in both directions there are deviations from a straight line, and far from E_{eq} transport limitations provoke alterations in concentrations, also causing deviations.

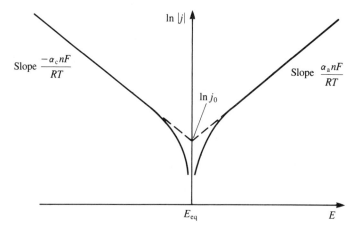

Fig. 6.8. Plot of $\ln|j|$ vs. E showing how to measure j_0 and α from the slopes of the lines.

6.7 The Tafel law corrected for transport

The Tafel law can be corrected for transport effects, enlarging the region of the voltammetric curve that can be utilized in Tafel plots.

For a reduction, from (6.9)

$$-\frac{1}{j} + \frac{1}{j_{L,c}} = \frac{1}{nFk_c[O]_\infty} = \frac{\exp(-\alpha_c nf)}{nFk_0[O]_\infty} \qquad (6.44)$$

that is

$$-\ln\left[\frac{1}{j} - \frac{1}{j_{L,c}}\right] = -\alpha_c nf + \ln k_0 + \ln(nF[O]_\infty) \qquad (6.45)$$

One does a plot of $\lg(j^{-1} - j_{L,c}^{-1})$ vs. E. For an oxidation, analogously,

$$-\ln\left[\frac{1}{j} - \frac{1}{j_{L,a}}\right] = \alpha_a nf + \ln k_0 + \ln(nF[R]_\infty) \qquad (6.46)$$

The intersection of the two lines given by (6.41) and (6.42) is when

$$\alpha_a nf + \alpha_c nf = \ln(nF[O]_\infty) - \ln(nF[R]_\infty) \qquad (6.47)$$

Since in the electrode reaction considered, $(\alpha_a + \alpha_c) = 1$, then

$$nf = \ln([O]_\infty/[R]_\infty) \qquad (6.48)$$

or

$$E_{eq} = E^{\ominus\prime} + \frac{RT}{nF} \ln \frac{[O]_\infty}{[R]_\infty} \qquad (6.49)$$

As this is the Nernst equation, the value of E can be identified with the equilibrium potential E_{eq}. Knowing $[O]_\infty$ and $[R]_\infty$ we can determine k_0 from the intersection of the two straight lines, in the same way as demonstrated in Fig. 6.8.

6.8 Kinetic treatment based on exchange current

All the equations deduced above can be formulated in terms of the exchange current I_0 (Section 4.5). In this way we have the advantage of obtaining the current as a function of the difference in applied potential and equilibrium potential, i.e. the overpotential, η; the disadvantage is that it is not directly related to the rate constants, which is important for comparison with other branches of kinetics.

As shown in Chapter 4, the exchange current density is equal to the anodic or cathodic component of the current density at equilibrium $(j_a = -j_c = j_0)$ that is

$$j_0 = -j_c = nFk_0[O]_* \exp\left[-\alpha_c nF(E_{eq} - E^{\ominus'})/RT\right] \tag{6.50}$$

On substituting the Nernst equation written as

$$\exp\left[nF(E_{eq} - E^{\ominus'})/RT\right] = [O]_\infty/[R]_\infty \tag{6.51}$$

we get

$$j_0 = nFk_0[O]_\infty^{1-\alpha_c}[R]_\infty^{\alpha_c} = nFk_0[O]_\infty^{\alpha_a}[R]_\infty^{\alpha_c} \tag{6.43}$$

From equations (6.18), (6.19), (6.31), and (6.50) it is relatively easy to reach

$$j = j_0\left\{-\frac{[O]_*}{[O]_\infty}\exp\left(-\alpha_c nf\eta\right) + \frac{[R]_*}{[R]_\infty}\exp\left(\alpha_a nf\eta\right)\right\} \tag{6.52}$$

where $\eta = E - E_{eq}$ is the overpotential. In the case of $[O]_* \sim [O]_\infty$ (<10 per cent of the limiting current)

$$j = j_0[\exp\left(\alpha_a nf\eta\right) - \exp\left(-\alpha_c nf\eta\right)] \tag{6.53}$$

In the linear region close to E_{eq} ($\exp(x) \approx x$), and writing $\alpha_a = 1 - \alpha_c$

$$j = j_0 nf\eta \tag{6.54}$$

For an irreversible reaction and for overpotentials corresponding to the exponential region, the Tafel law is

$$\ln j = \ln j_0 - \alpha_c nf\eta \qquad \text{(reduction)} \tag{6.55}$$
$$\ln j = \ln j_0 + \alpha_a nf\eta \qquad \text{(oxidation)} \tag{6.56}$$

Other information, particularly useful for multistep reactions, can be obtained from the following relations deduced from (6.43):

$$\frac{\partial \ln j_0}{\partial \ln [O]_\infty} = \alpha_a, \qquad \frac{\partial \ln j_0}{\partial \ln [R]_\infty} = \alpha_c \tag{6.57}$$

Changing $[O]_\infty$ or $[R]_\infty$, the value of j_0 varies and we can determine the value of α_a or α_c, without needing to know the value of n in the rate-determining step.

6.9   The effect of the electrolyte double layer on electrode kinetics

The electrolyte double layer affects the kinetics of electrode reactions[6]. For charge transfer to occur, electroactive species have to reach at least

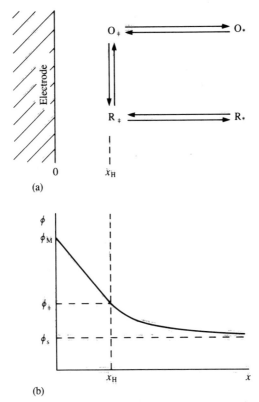

(a)

(b)

Fig. 6.9. (a) Schematic representation of the path followed by an electrode reaction. The effect of the electrode's electric field begins at the outside of the double layer, but for there to be reaction the species has to reach x_H from the electrode; (b) Variation of ϕ with distance, showing that the potential difference to cause electrode reaction is $(\phi_M - \phi_{\ddagger})$.

to the outer Helmholtz plane (distance x_H in Fig. 3.9). Hence, the potential difference available to cause reaction is $(\phi_M - \phi_{\ddagger})$ (see Fig. 6.9) and not $(\phi_M - \phi_S)$. Only when $\phi_{\ddagger} \sim \phi_S$ can we say that the double layer does not affect the electrode kinetics. Additionally, the concentration of electroactive species will be, in general, less at distance x_H from the electrode than outside the double layer in bulk solution. These assumptions can be treated quantitatively.

For a reduction a convenient representation is

$$O_* \underset{}{\overset{\text{pre-equilibrium}}{\rightleftharpoons}} O_{\ddagger}$$

outside the double layer   outer Helmholtz plane

Following the Gouy–Chapman model of the diffuse layer,

$$[O]_{\ddagger} = [O]_* \exp\left[\frac{-zF(\phi_{\ddagger} - \phi_S)}{RT}\right] \tag{6.58}$$

From the Butler-Volmer expressions, the true rate constant is then

$$k_c = k_{0,t} \exp\left[\frac{-\alpha_c nF(\phi_M - \phi_{\ddagger})}{RT}\right] \tag{6.59}$$

where $k_{0,t}$ represents the true standard rate constant of the electrode reaction. The reaction rate is

$$k_c[O]_{\ddagger} = k_{0,t} \exp\left[\frac{-\alpha_c nF(\phi_M - \phi_{\ddagger})}{RT}\right] f_{DL}[O]_* \tag{6.60}$$

where

$$f_{DL} = \exp\left[\frac{(\alpha_c n - z)F(\phi_{\ddagger} + \phi_S)}{RT}\right] \tag{6.61}$$

Clearly f_{DL} is also given by

$$f_{DL} = k_0/k_{0,t} \tag{6.62}$$

the ratio between the apparent standard rate constant, k_0, and the true standard rate constant, $k_{0,t}$, known as the *Frumkin correction*, f_{DL}. The practical consequence is variation of k_0 with potential. A more rigorous deduction from the expression for f_{DL} is obtained through the use of electrochemical potentials.

To apply the correction it is necessary to know the value of $\phi_{\ddagger}$. This value has been calculated for the mercury electrode from electrocapillary curves, calculating σ_M and thence $\phi_{\ddagger}$ using the Stern model. As this model does not include specific adsorption, the calculated values of $\phi_{\ddagger}$ are either too positive (cations) or too negative (anions) besides differences caused by the blocking effect of the ions.

As is perhaps to be expected, the double layer can also affect the values of the measured, i.e. apparent, charge transfer coefficients manifested in the slopes of the Tafel plots (Section 6.6). It is possible to show that, for the Stern model, and considering a cathodic transfer coefficient as example,

$$\alpha_c = \alpha_{c,t} + \frac{z - \alpha_{c,t}}{1 + (C_{GC}/C_H)\cosh\left(\dfrac{ze(\phi_{\ddagger} - \phi_S)}{2k_B T}\right)} \tag{6.63}$$

where α_c is the observed coefficient and $\alpha_{c,t}$ the true coefficient. There are two extreme cases:

- $[ze(\phi_{\ddagger} - \phi_S)/2k_bT]$ large, implying that $\alpha_c \sim \alpha_{c,t}$. This situation corresponds to potentials far from E_z.
- $[ze(\phi_{\ddagger} - \phi_S)/2k_BT]$ small, corresponding to potentials close to E_z.

$$\alpha_c \approx \alpha_{c,t} + \frac{z - \alpha_{c,t}}{1 + C_{GC}/C_H} \tag{6.64}$$

If, for example, $C_{GC} = C_H$, $\alpha_{c,t} = 0.5$ and $z = 1$ we obtain $\alpha_c = -0.25$. As can be seen, the apparent value of α_c can be positive or negative. This example shows the extreme importance in correcting values of α_c and α_a for double layer effects.

We can ask how effects of the double layer on electrode kinetics can be minimized and if the necessity of correcting values of α and of rate constants can be avoided? In order for this to be possible, we have to arrange for $\phi_{\ddagger} \approx \phi_S$, that is all the potential drop between electrode surface and bulk solution is confined to within the compact layer, for any value of applied potential. This can be achieved by addition of a large quantity of inert electrolyte (≈ 1.0 M), the concentration of electroactive species being much lower (<5 mM). As stated elsewhere, other advantages of inert electrolyte addition are reduction of solution resistance and minimization of migration effects given that the inert electrolyte conducts almost all the current. In the case of microelectrodes (Section 5.6) the addition of inert electrolyte is not necessary for many types of experiment as the currents are so small.

6.10 Electrode processes involving multiple electron transfer

In many reduction or oxidation half-reactions, the oxidation state changes by a value greater than 1. Examples for metallic cations are $Tl(III) \rightarrow Tl(I)$, $Cu(II) \rightarrow Cu(O)$, and examples of other species $O_2 \rightarrow H_2O_2$ ($2e^-$) or $O_2 \rightarrow H_2O$ ($4e^-$), these also involving other species in the half-reactions. In this section we consider metal ions given that, at least apparently, there are no other species involved, except for molecules of solvation etc. If the reactions are irreversible we can investigate their kinetics.

For a two-electron reduction (Fig. 6.10), we have generically

$$A + e^- \rightarrow B$$

$$B + e^- \rightarrow C$$

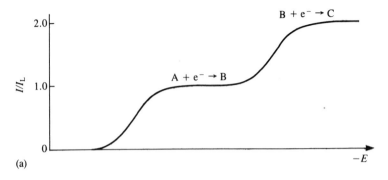

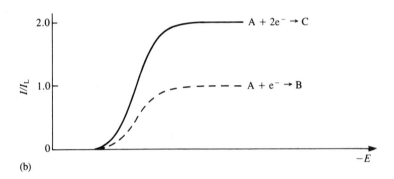

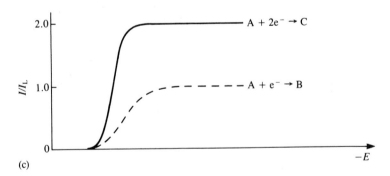

Fig. 6.10. Voltammograms for the reduction of species A following $A + e^- \rightarrow$ $B + e^- \rightarrow C$ according to the relative rates of the two steps. $I_L = I_L(A \rightarrow B)$. (a) Second step much more difficult than the first; (b) First step rate-determining; second step fast; (c) First step pre-equilibrium; second step rate-determining.

the electron transfers being consecutive. The kinetics of these two steps is conditioned by the medium where they occur and this will determine the type of voltammetric wave that is observed. The electrode reaction scheme can be written

$$
\begin{array}{ccccc}
\text{Solution} & A_\infty & & B_\infty & & C_\infty \\
& \updownarrow k_{d,A} & & \updownarrow k_{d,B} & & \updownarrow k_{d,C} \\
\text{Electrode} & A_* + e^- & \underset{k_{a,1}}{\overset{k_{c,1}}{\rightleftharpoons}} & B_* + e^- & \underset{k_{a,2}}{\overset{k_{c,2}}{\rightleftharpoons}} & C_*
\end{array}
$$

Qualitatively we can distinguish three limiting cases:

  1. Second step occurs at a more negative potential than the first:

$$k_{c,2} - k_{a,2} \ll k_{d,B}$$

We observe a one-electron reduction until the applied potential is sufficiently negative for reduction of the second electron. In other words we observe two separated voltammetric waves (Fig. 6.10a).

  2. First step rate-determining:

$$k_{c,2} - k_{a,2} \gg k_{d,B}$$

$$k_{c,2} \gg k_{a,1}$$

This situation corresponds to

$$A + e^- \rightarrow B \quad \text{rate-determining step}$$
$$B + e^- \rightarrow C \quad \text{fast}$$

The form of the voltammetric wave is the same as for $A + e^- \rightarrow B$, but the current is multiplied by 2 (Fig. 6.10b).

  3. Second step rate-determining:

$$k_{c,2} - k_{a,2} \ll k_{d,B}$$

$$k_{a,1} \ll k_{c,2}$$

This corresponds to

$$A + e^- \rightarrow B \quad \text{pre-equilibrium}$$
$$B + e^- \rightarrow C \quad \text{rate-determining step}$$

Due to pre-equilibrium, the voltammetric wave is steeper than in Case 2, Fig. 6.10c. The activated complex is more sensitive than in Case 2 to changes in applied potential and is situated roughly halfway between B and C on the reaction coordinate. This means that $\alpha_c n \approx 1.5$ (but does not mean that $\alpha_c \approx 0.75$). In actual fact, $\alpha_c \approx 0.5$ as $n = 1$ in the rate-determining step.

It is also of interest to consider the reverse reaction, oxidation of C to A. Supposing the reduction follows path 2:

$$A + e^- \rightarrow B \qquad \text{rate-determining step}$$
$$B + e^- \rightarrow C \qquad \text{fast}$$

it is quite possible that the oxidation is not the inverse, but is

$$C + e^- \rightarrow B \qquad \text{rate-determining step}$$
$$B + e^- \rightarrow A \qquad \text{fast}$$

This change in mechanism is due to the different potentials (i.e. different electronic energies) at which the reactions occur. We cannot ever assume *a priori* that the reverse of a multistep reaction with known mechanism will be the inverse.

It is also fairly evident that for certain combinations of rate constants we can change mechanisms by changing the applied potential.

The application of these concepts to the electroreduction of oxygen, important for fuel cells, with hydrodynamic electrodes is described in Chapter 8.

Finally in this section, we remember that multiple electron transfer has to follow the reaction coordinate and has consecutive steps, even if the first step is rate determining. The possibility of multiple electron transfer reactions without intermediate chemical steps has been questioned, with experimental evidence from, for example, the supposedly relatively simple reduction of Cd(II) and similar ions at mercury electrodes[6]. This is because solvation and interaction with the environment, adsorption, etc. are different for each oxidation state.

6.11 Electrode processes involving coupled homogeneous reactions

Interesting reactions occur when the charge transfer at the electrode is associated with homogeneous reactions in solution that can precede or follow the electron transfer reaction at the electrode. A selection of possible schemes is shown in Table 6.1. Note the presence of many organic compounds: the reduction or oxidation of these compounds involves, in many cases, the addition or removal of hydrogen, which

Table 6.1. Electrode reactions with coupled homogeneous reactions, adapted from Ref. 7

	Electrode process	Example
CE		
	Solution $\quad A_2 \underset{k_{-1}}{\overset{k_1}{\rightleftharpoons}} A_1$	$A_2 = H_2C(OH)_2$ $A_1 = H_2CO$
	Electrode $\quad A_1 \pm ne^- \to A_3$	$A_3 = CH_3OH$
EC′		
	Solution $\quad A_2 \underset{k_{-1}}{\overset{k_1}{\rightleftharpoons}} A_1$	$A_2 = Fe(III)$ $A_1 = Fe(II)$
	Electrode $\quad A_1 \pm ne^- \to A_3$	Homogeneous catalyst H_2O_2
EC		
	Electrode $\quad A_3 \pm ne^- \to A_1$	
	Solution $\quad A_2 \underset{k_{-1}}{\overset{k_1}{\rightleftharpoons}} A_1$	

$$A_3 = R_2N-\!\!\!\bigcirc\!\!\!-NR_2$$

$$A_1 = R_2\overset{+}{N}=\!\!\!\bigcirc\!\!\!=\overset{+}{N}R_2$$

$$A_2 = O=\!\!\!\bigcirc\!\!\!=O \text{ by}$$
reaction with OH^-

ECE		
	Electrode $\quad A_3 \pm ne^- \to A_1$	$A_3 = ClC_6H_4NO_2$
		$A_1 = ClC_6H_4NO_2^-$
	Solution $\quad A_1 \overset{k}{\to} A_2$	$A_2 = {}^{\cdot}C_6H_4NO_2$
	Electrode $\quad A_2 + n_2e^- \to A_4$	$A_4 = {}^{\prime\prime}C_6H_4NO_2$
		$(A_4 + H^+ \to C_6H_5NO_2)$

$A_3 =$ (xanthene structure) $\quad (Fe^{2-})$

DISP1		
	Electrode $\quad A_3 \pm ne^- \to A_1$	
	Solution $\quad A_1 \overset{k}{\to} A_2$	
	Solution $\quad A_1 + A_2 \to A_3 + A_4$	

$A_1 =$ (xanthene radical structure) $\quad (S^{\cdot 3-})$

$A_2 = SH^{\cdot 2-}$ $\qquad R = $ (structure)

DISP2		
	Electrode $\quad A_3 \pm ne^- \to A_1$	
	Solution $\quad A_1 \rightleftharpoons A_2$	
	Solution $\quad A_1 + A_2 \overset{k}{\to} A_3 + A_4$	

$A_4 =$ (xanthene structure) $\quad (L^{3-})$

pH 9–10 DISP1
pH 6 DISP2

proceeds via electron plus proton transfer. Such mechanisms are commonly summarized in the scheme of squares, shown below for two electron transfers and two protonation steps.

$$
\begin{array}{ccccc}
A & \underset{-e}{\overset{+e}{\rightleftarrows}} & B^- & \underset{-e}{\overset{+e}{\rightleftarrows}} & C^{2-} \\
+H^+ \updownarrow -H^+ & & +H^+ \updownarrow -H^+ & & +H^+ \updownarrow -H^+ \\
AH^+ & \underset{-e}{\overset{+e}{\rightleftarrows}} & BH & \underset{-e}{\overset{+e}{\rightleftarrows}} & CH^- \\
+H^+ \updownarrow -H^+ & & +H^+ \updownarrow -H^+ & & +H^+ \updownarrow -H^+ \\
AH_2^{2+} & \underset{-e}{\overset{+e}{\rightleftarrows}} & BH_2^+ & \underset{-e}{\overset{+e}{\rightleftarrows}} & CH_2
\end{array}
$$

A typical example is quinone/hydroquinone systems.

We consider three simple schemes, shown in Fig. 6.11, and examine the effect of homogeneous coupled reactions on the current at the electrode: they are CE, EC, and EC', where E represents an electrochemical step (at the electrode) and C a chemical step (in solution). The equations to calculate the rate constants from experimental measurements for the various types of electrode can be found in the specialized literature. In most studies the electrochemical step has been considered reversible—thus, in the following, the rate constant for the electrode reaction is not indicated.

CE process

Solution $\quad A_2 \underset{k_{-1}}{\overset{k_1}{\rightleftarrows}} A_1$

$$K = k_{-1}/k_1$$

Electrode $\quad A_1 \pm ne^- \rightarrow A_3$

The concentration of A_1 is less than in the absence of the chemical step, and is dependent on the value of K. Measurement of the diminution in current allows the determination of the values of k_1 and k_{-1}. The position of the voltammetric curve on the potential axis is not affected by the homogeneous step (Fig. 6.11a).

EC process

Electrode $\quad A_3 \pm ne^- \rightarrow A_1$

$$K = k_{-1}/k_1$$

Solution $\quad A_1 \underset{k_1}{\overset{k_{-1}}{\rightleftarrows}} A_2$

The chemical step reduces the quantity of A_1 at the electrode surface and consequently causes a shift of the voltammetric wave to more positive

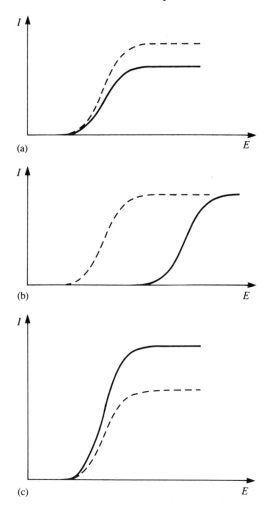

Fig. 6.11. The effect of coupled homogeneous reactions on electrode reactions illustrated for an oxidation. Mechanisms (a) CE (b) EC (c) EC′. Absence of homogeneous reaction (– – –); presence (——).

potentials (oxidation) or more negative (reduction) (Fig. 6.11*b*). This shift can be directly related to the kinetics of the homogeneous reaction.

EC′ process

$$\text{Solution } A_2 \underset{k_{-1}}{\overset{k_1}{\rightleftharpoons}} A_1$$
$$K = k_{-1}/k_1$$
$$\text{Electrode } A_1 \pm ne^- \rightarrow A_2$$

This is a catalytic process, the homogeneous reaction regenerating the reagent of the electrode reaction. The concentration of A_1 will be larger than expected and the current bigger than in the absence of the homogeneous reaction (Fig. 6.11c). Altering k_d (through the diffusion layer thickness) permits the determination of the kinetics of the homogeneous reaction.

In all these schemes for coupled homogeneous reactions, it is useful to consider in the deduction of the equations the concept of a reaction layer associated with the homogeneous reaction; all the homogeneous reaction occurs within a distance equal to the thickness of the reaction layer from the electrode. When the thickness of the layer is significantly smaller (<10 per cent) than the thickness of the diffusion layer the two layers can be considered as being independent, which simplifies the mathematical treatment. The thickness of the reaction layer depends on the values of the homogeneous rate constants k_1 and k_{-1}.

Other more complex mechanistic schemes are studied by a variety of techniques. Double hydrodynamic electrodes are particularly useful for investigating schemes involving two electron transfer steps, such as ECE and DISP schemes. Some of the applications of the different electrochemical techniques in the elucidation of these reactions are described in the following chapters.

References

1. W. J. Albery, *Electrode kinetics,* Clarendon Press, Oxford, 1975.
2. H. R. Thirsk and J. A. Harrison, *A guide to the study of electrode kinetics,* Academic Press, New York, 1972.
3. K. J. Vetter, *Electrochemical kinetics,* Academic Press, New York, 1967.
4. J. Koryta, *Principles of electrochemistry,* Wiley, London, 1987.
5. P. Delahay, *Double layer and electrode kinetics,* Wiley-Interscience, New York, 1965.
6. W. H. Reinmuth, *J. Electroanal. Chem.,* 1972, **34,** 297; C. P. M. Bongenaar, A. G. Remijnse, M. Sluyters-Rehbach, and J. H. Sluyters, *J. Electroanal. Chem.,* 1980, **111,** 155.
7. C. M. A. Brett, *Port. Electrochim. Acta,* 1985, **3,** 355.

PART II

Methods

7

ELECTROCHEMICAL
EXPERIMENTS

7.1 Introduction

The experimental aspects to be discussed in this chapter include cell design, electrode materials, construction and cleaning of electrodes, solution composition, and control instrumentation. Electrode materials specially designed for potentiometric measurements, which rely on the material selectivity, are discussed in Chapter 13.

We shall not give many practical details, but only those of greatest interest in the planning and design of electrochemical experiments. However, it is hoped that the discussion in this chapter will prove an aid to consulting more detailed expositions in this area, for example Refs. 1–6.

7.2 Electrode materials for voltammetry

The choice of an electrode material depends to a great extent on the useful potential range of the electrode in the particular solvent employed and the qualities and purity of the material. The usable potential range is limited by one or more of the following factors:

- solvent decomposition

- decomposition of the supporting electrolyte (Section 7.5)

- electrode   dissolution   or   formation   of   a   layer   of   an
insulating/semiconducting substance on its surface.

Additionally, solid electrodes can be adversely affected by poisoning
through contact with solutions containing contaminants. We now consider some frequently used materials and look at their properties as
electrodes in more detail.

Metals

Much has been written about solid metal electrodes, which have now
largely displaced liquid mercury. Those most often used as redox ('inert')
electrodes for studying electron transfer kinetics and mechanism, and
determining thermodynamic parameters are platinum, gold, and silver.
However, it should be remembered that their inertness is relative: at
certain values of applied potential bonds are formed between the metal
and oxygen or hydrogen in aqueous and some non-aqueous solutions.
Platinum also exhibits catalytic properties.

A general advantage of metal electrodes is that their high conductivity
results in low (usually negligible) background currents. It is usually fairly
easy to increase sensitivity and reproducibility at solid electrodes by
forced convection. Their surfaces can be modified by electrodeposition or
chemical modification, although the latter is more common with carbon
electrodes (see below). Another advantage of the use of metal electrodes
is the ease of construction of the electrode assembly, and ease of
polishing.

Electrodes of many metals can undergo corrosion or passivation—
formation of a salt film on the surface—and other reactions, depending
on the medium and experimental conditions. Electrochemical techniques
can be used to investigate the mechanisms of these processes.

Carbon

Carbon[7] exists in various conducting forms. Electrochemical reactions are
normally slower at carbon than at metallic electrodes, electron transfer
kinetics being dependent on structure and surface preparation[8].

Carbon has a high surface activity, which explains its susceptibility to
poisoning by organic compounds. Bonds with hydrogen, hydroxyl and
carboxyl groups, and sometimes quinones, can be formed at the carbon
surface. The presence of these groups signifies that the behaviour of these
electrodes can be very pH-sensitive. The presence of functional groups
has also been purposely used to modify the electrode surface (modified

Table 7.1. Properties of various carbon materials (from Ref. 8)

	Apparent density (g cm^{-3})	ρ $(\Omega\,\text{cm})$	L_a (nm)	L_c (nm)
Glassy carbon				
Tokai GC-10 (made at 1000°C)	1.5	4.5×10^{-3}	2^a	1.0
Tokai GC-20 (made at 2000°C)	1.5	4.2×10^{-3}	2.5^a	1.2
Tokai GC-30 (made at 3000°C)	1.5	3.7×10^{-3}	5.5	7.0
Carbon fibre^a	1.8	$(5-20) \times 10^{-4}$	>10	4.0
HOPG, a-axis	2.26	4×10^{-5}	>1000	
HOPG, c-axis		0.17		>10000
Pyrolytic graphite	2.18		100^a	100^a
Randomly oriented graphite				
(Ultracarbon UF-4s grade)	1.8	1×10^{-3}	30^a	50^a
Carbon black (Spheron-6)	1.3–2.0	0.05	2.0	1.3

^{a} Values may vary significantly with preparation procedure

electrodes, Section 14.4) with a view to obtaining new electrode properties.

Various types of carbon are used as electrodes. These include glassy carbon, carbon fibres, carbon black, various forms of graphite, and carbon paste, which consists of graphite particles in contact, incorporated in an inert matrix. They are all sp^2 carbons, and can be compared structurally by considering the length of microcrystallites, L_a, in the graphite lattice plane (a-axis), and the thickness of the microcrystallites perpendicular to the graphite planes (c-axis), L_c. These values, together with apparent density and resistivity are shown in Table 7.1.

Probably the most widely used of these is glassy carbon, which is isotropic. However, due to its hardness and fragility, electrode fabrication is difficult, which essentially limits its use to the dimensions and forms that can be acquired commercially. The manufacture of glassy carbon consists in carbonization by heating phenol/formaldehyde polymers or polyacrylonitrile between 1000°C and 3000°C under pressure. Since glassy carbon has some amorphous characteristics, as can be seen from Fig. 7.1, it is not always homogeneous.

Carbon fibres have a diameter similar to that of a hair (2–20 μm), and exhibit a stiffness greater than steel in the fibre direction. Fabrication is, in general, either from polyacrylonitrile (PAN), which gives circular concentric graphitic ribbon rings, or from pitch, which tends to give a radial structure of graphite lamellae[10]. Apart from use as microelectrodes, they are used as bundles in porous electrodes where a high electrolysis efficiency is required (Chapter 15).

Another important form of carbon is pyrolytic graphite (PG), so called

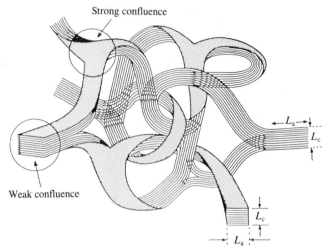

Fig. 7.1. Representation of the structure of glassy carbon, showing L_a and L_c (from Ref. 9 with permission).

because it is prepared by high-temperature decomposition of gaseous hydrocarbons on to a hot surface, which is anisotropic and has a slightly higher density than natural graphite. If this is pressure annealed at high temperature it turns into highly ordered pyrolytic graphite (HOPG), which is highly anisotropic as shown in Table 7.1, and is very reproducible. In graphite (Fig. 7.2) the choice of basal or edge plane alters the electrochemical response owing to the different structure of the exposed surface. Pores arising in graphite are sometimes impregnated with ceresin or paraffin under vacuum in order to impede the entry of solution into the electrode.

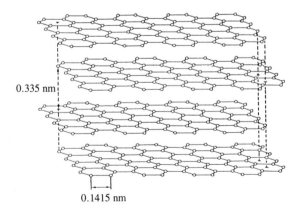

Fig. 7.2. The structure of graphite.

Electrodes made with carbon or graphite paste mixed with a hydro-phobic diluent such as Nujol, paraffin, silicone rubber, epoxy resin, Teflon, or Kel-F, have also been used. Comparative studies between the various types of carbon paste electrode have been carried out[11].

Other solid materials

Other solid electrode materials used are semiconductors, for example metal oxides[12,13], and conducting organic salts[14]. These last are of much interest at present for the immobilization of organic compounds such as enzymes, given their compatibility with these macromolecules (Chapter 17).

For spectroelectrochemical and photoelectrochemical studies, optically semi-transparent electrodes have been fabricated by vapour deposition techniques on glass or quartz substrates (Chapter 12). Tin and indium oxides, platinum, and gold have been used.

Mercury

For many years, mercury was the most used electrode material in the laboratory in the dropping mercury and hanging mercury drop elec-trodes, and more recently in the static mercury drop electrode. It possesses a very high negative overpotential for hydrogen evolution in aqueous solution, permitting a negative range of potential larger than for any other material (about $-2.0\,V$ vs. SCE instead of $-1.0\,V$ negative limit); conversely, for potentials more positive than $+0.2\,V$ vs. SCE mercury oxidation begins. Thus it is used almost always to study reduction processes, as oxidations of soluble species occurring at negative potentials are uncommon.

Mercury purity is very important, since on the one hand we have to ensure that no other element is dissolved in or amalgamated with the mercury, and on the other hand impure mercury used in a dropping mercury electrode would block the capillary.

Mercury purification processes can be summarized in four steps:

• Removal of oxides and dirt on the mercury surface by filtering through a filter paper with a pinhole (Whatman No. 40);

• Removal of dissolved basic metals (zinc and cadmium) by agitation in $2\,M$ HNO_3 for 1–3 days using a vacuum aspirator. The appearance of bubbles of mercury on the surface means that this step is finished.

• Distillation to remove noble metals (platinum, gold, silver, etc.) There are various types of distillation apparatus; care must be taken that this does not become a reservoir of impurities.

• Washing with water, drying, filtering, and distilling twice more.

There are two useful and easy methods for testing mercury purity. If basic metals are present then the mercury leaves a thin film on the wall of a glass vessel. The other method consists of placing a small quantity of mercury in a corked bottle with approximately three times its volume of distilled water and agitating: if the mercury is pure then foam will be formed, lasting for 5–15 s.

After use in an electrochemical experiment, mercury should be redistilled. It is highly toxic, especially as vapour, so care should be exercised: any drops that escape should be caught in a tray placed underneath the cells, and the laboratory must have good ventilation.

Instead of drops, liquid mercury can be used as thin films electrodeposited on some solid electrodes in order to increase the useable negative potential range of these electrodes. For example, on glassy carbon substrates the deposition can be done from a solution of 10^{-5} M Hg(II) in 0.1 M HNO_3, applying a convenient potential (-1.0 V vs. SCE) during a few minutes, and stirring the solution. Alternatively a mercury film can be formed on other substrates, such as copper, by simple immersion in liquid mercury. The special properties of these thin film electrodes are described in Section 9.10.

7.3 The working electrode: preparation and cleaning

Working electrodes are normally solid. The mercury electrode is the only liquid electrode at room temperature (with the rare exceptions of gallium and amalgam electrodes) and has been most used as a dropping electrode.

For reasons of ease of manufacture, the majority of solid electrodes have a circular or rectangular form. External links are through a conducting epoxy resin either to a wire or to a solid rod of a metal such as brass, and the whole assembly is introduced by mechanical pressure into an insulating plastic sheath (Kel-F, Teflon, Delrin, perspex, etc.) or covered with epoxy resin. It is very important to ensure that there are no crevices between electrode and sheath where solution can enter and cause corrosion. Examples of electrodes constructed by this process will be shown in Chapter 8.

Unfortunately, expansion coefficients of plastics and metals can be quite different. For this reason it is suggested that electrodes are kept in closed glass tubes in a thermostat bath when not in use, at the temperature at which they will be employed. If this is not done, there is a risk that the surfaces of the electrode and sheath will not be coplanar after a period of time. This means total repolishing, entailing considerable costs in time and electrode material.

Once the electrode is constructed, it has to be polished to obtain a smooth, brilliant surface which is free of physical defects. The polishing substance used depends on electrode material hardness, and can be used on polishing tables or with special polishing cloths. Diamond, paste or spray, and alumina powder, available containing particles of various sizes, are widely used. The process begins with large particles (perhaps 25 μm diameter), using successively smaller particles until at least 1 μm is reached, verifying the absence of scratches, etc. It is unlikely, except for very soft electrode materials, that the surface can be improved by using particles of size less than 0.3 μm diameter. During a series of experiments it is possible that the last steps of polishing have to be repeated between each experiment owing to electrode poisoning, corrosion, etc.

When carbon paste is used as electrode material an electrode body is made with a shallow hole where the paste is inserted. This electrode cannot be polished; when necessary the electrode is renewed. If it is a mixture of carbon paste with a plastic then careful polishing or surface cutting can be done.

Finally, it should be emphasized that the surface of a solid electrode is not truly clean after polishing. Particles of abrasive will be stuck in the pores of the electrode, and so on. In some experiments this can make a lot of difference. For this reason it is necessary to resort to methods such as ultrasound or electrochemical cleaning: the latter consists of applying different potentials or currents during predetermined periods of time to oxidize or reduce the impurities so that they leave the surface (see Ref. 15). At the same time there may be changes in surface properties (Section 3.5).

A *dropping mercury electrode* (DME) is constructed by linking a reservoir containing mercury of high purity through a tube of a plastic such as Tygon to a very fine glass capillary (internal diameter ≈0.05 mm). Drops of mercury are formed at the bottom of the capillary and which fall when they reach a certain size due to the action of gravity; a new drop then begins to grow. The experimental arrangement is illustrated in Fig. 8.7. By modifying the supply of mercury to the capillary it is possible to have a *hanging drop* (HMDE). A different modification that forms a hanging drop and that assures good reproducibility in drop size has been developed—the *static drop* (SMDE). This can be used in a hanging or in a quasi-dropping mode. In the latter case the drop attains its final size very rapidly, after which its surface area is constant. Electrical contact to mercury electrodes is normally made via a platinum wire inserted into the liquid mercury just above the capillary.

7.4 The cell: measurements at equilibrium

A cell to make measurements at equilibrium (potentiometric measurement) needs only two electrodes: an *indicator* and a *reference electrode* (Fig. 7.3). In general, the indicator electrode is an ion-selective electrode (Section 13.3) and the reference electrode (Table 2.1) is Ag | AgCl or calomel in aqueous solution. The difference in potential between the two electrodes is measured; since the reference electrode potential is constant, changes in cell potential are due only to the indicator electrode which responds logarithmically to the activity of the species in solution to which it is sensitive.

At equilibrium, there is no passage of current and, in this sense, the positioning of the electrodes relative to each other is not important. It should be borne in mind, however, that the greater the distance between the electrodes the larger the electrical noise; this problem arises particularly in flow systems. Noise leads to lack of stable readings on a real potentiometer or in a high-input impedance ($\approx 10^{15} \, \Omega$) voltmeter, owing to the fact that the indicator electrode has a low impedance. An important objective of ion-selective field effect transistors (ISFET) and other similar devices (Section 13.10) is the *in situ* conversion of the low-impedance signal from the electrode into a high-impedance signal, thus reducing the noise.

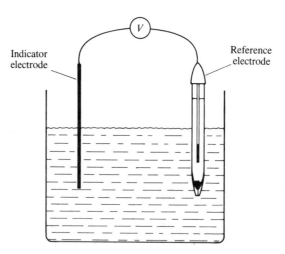

Indicator electrode

Reference electrode

Fig. 7.3. Experimental arrangement for measurements at equilibrium. The voltmeter should have high input impedance to minimize current consumption.

7.5   The cell: measurements away from equilibrium

Electrodes

Outside equilibrium there is passage of current between two electrodes, as in galvanic cells or electrolytic cells. In laboratory research we are usually interested in investigating the electrode process at one of the electrodes, the indicator or working electrode, through control of its potential (*potentiostatic control*) or the current it passes (*galvanostatic control*), the other, auxiliary, electrode being used to complete the electrical circuit. In the past, in controlled potential experiments, the auxiliary electrode was also the reference electrode, which thus had the double function of passing current and acting as a reference potential for controlling the potential of the working electrode. Sometimes this brought problems of potential stability—on passing current, the activities of the species would be slightly altered in the vicinity of the reference electrode, causing a variation of its potential. For this reason three-electrode systems were developed where the current passes from the working electrode to an auxiliary electrode (of larger area than the working electrode), the separate reference electrode serving purely as a reference potential and not passing current.

In electrochemical research, auxiliary electrodes are frequently made of platinum foil or gauze, sometimes placed in a compartment separated from the rest of the solution by a porous frit, so as to avoid contamination arising from the reaction occurring at the auxiliary electrode. This is an oxidation if reduction is occurring at the working electrode and vice versa, but the identity of the auxiliary electrode reaction is not important in this type of experiment.

Exceptions to the three-electrode system in the laboratory are now few but important. A pool of mercury at the bottom of a cell in contact with an electrolyte solution containing chloride or bromide can serve as auxiliary and reference electrodes simultaneously owing to the very large surface area exposed. Another exception is experiments with microelectrodes: the current that passes is so small that the effect on the reference/auxiliary electrode potential is negligible.

In order to control the working electrode potential accurately, it is necessary that the resistance between working and reference electrodes is as small as possible. This means close positioning, that is often not possible. In these cases a *Luggin capillary* is used, a piece of glass into which the reference electrode is inserted and whose finely drawn-out point is placed close to the working electrode. The optimal position depends on the geometry of the indicator electrode and on the electrochemical technique being employed.

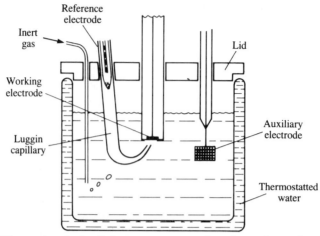

Fig. 7.4. Electrochemical cell for measurements in a three-electrode scheme at a planar electrode.

Figure 7.4 shows an electrochemical cell containing a stationary planar working electrode, a reference electrode with Luggin capillary, and an auxiliary electrode. Some other types of more specialized cells are described in subsequent chapters.

Quasi-reference electrodes can be employed in situations where the high reproducibility of potential is not necessary, such as in many voltammetric analysis experiments. Mercury pools (referred to above) or silver wires in aqueous halide media are examples. Platinum wires can also be used. The advantage of wires, apart from their small size, is in reducing the uncompensated resistance in resistive media, relative to conventional reference electrodes.

Supporting electrolyte

Normally electrode reactions take place in solutions, or sometimes in molten salts (e.g. aluminium extraction). In order to minimize the phenomenon of migration of the electroactive ions caused by the electric field (Chapter 2) and to confine the interfacial potential difference to the distance of closest approach of solvated ions to the electrode (Chapter 3), the addition of a solution containing a high concentration of inert electrolyte, called *supporting electrolyte,* is necessary. This has a concentration at least 100 times that of the electroactive species and is the principal source of electrically conducting ionic species. The concentration of supporting electrolyte varies normally between 0.01 M and 1.0 M, the concentration of electroactive species being 5 mM or less. The

Table 7.2. Some supporting electrolytes and their approximate potential ranges in water and other solvents for platinum, mercury, and carbon

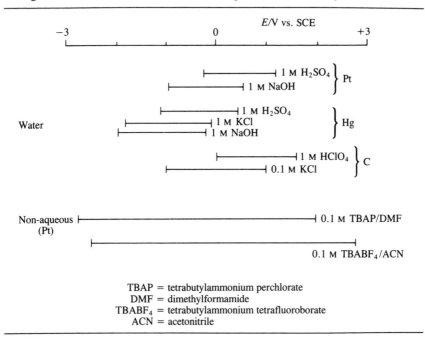

TBAP = tetrabutylammonium perchlorate
DMF = dimethylformamide
TBABF$_4$ = tetrabutylammonium tetrafluoroborate
ACN = acetonitrile

supporting electrolyte should be chosen, as well as its concentration, so that the transport numbers of the electroactive species are practically zero: it can be an inorganic or organic salt, an acid or a base, or a buffer solution such as citrate, phosphate, or acetate. The choice also has to be made in relation to the properties of the solvent employed. A description of the properties of non-aqueous solvents and usable electrolytes can be found in Ref. 16.

Table 7.2 shows some useful supporting electrolytes in aqueous and non-aqueous media. The usable potential range is limited by solvent and/or supporting electrolyte decomposition and, to a certain extent, by electrode material. Sometimes samples for laboratory analysis already contain supporting electrolyte, as is the case for sea water (0.7 M NaCl); in others acid is added to others during preparation by digestion, etc., that can serve simultaneously as supporting electrolyte. In Table 7.1 it should be noted that there are useful organic salts (tetraalkylammonium-) that are soluble in non-aqueous solvents: in these cases there is less confidence that the supporting electrolyte does not interfere in the electrode reactions[17].

All care must be taken to ensure that the supporting electrolyte is truly inert in the potential range of the experiment, not reacting with the electrode or with the products of the electrode reaction (except when desired). This also shows the importance of reagent purity for preparing an electrolyte solution of the best quality, of purification of commercial reagents if necessary, of drying and distilling pure organic solvents under the appropriate conditions, and of careful distillation of water to remove inorganic and organic impurities, these latter by photolytic degradation or reaction with permanganate.

In the case of microelectrodes where currents are sufficiently small so that the reference electrode can serve simultaneously as auxiliary electrode (see above) the solution ohmic potential drop (product of current and solution resistance) is also small. This means that measurements can be made in highly resistive media without the addition of supporting electrolyte, a fact that can be very useful.

Removal of oxygen

The atmosphere contains about 20 per cent oxygen, which is slightly heavier than air, and is dissolved appreciably ($\approx 10^{-4}$ M) in solutions open to the atmosphere. Oxygen is reduced at electrodes in two separate two-electron steps or in one four-electron step at potentials that vary between 0.05 V and -0.9 V vs. SCE, depending on pH and on electrode material. In acid solution

$$O_2 + 2H^+ + 2e^- \rightarrow H_2O_2$$
$$H_2O_2 + 2H^+ + 2e^- \rightarrow 2H_2O$$

and in alkaline solution

$$O_2 + 2H_2O + 2e^- \rightarrow 2OH^- + H_2O_2$$
$$H_2O_2 + 2e^- \rightarrow 2OH^-$$

the four-electron reduction being obtained by addition of the two two-electron steps. These half-reactions contribute to the current measured at the electrode and oxygen can oxidise the electrode surface. Besides this, oxygen, radicals such as HO_2^-, and hydrogen peroxide can react with the reagents and/or products of the electrode reaction being studied. Thus removal of oxygen from the solution is, in general, of great importance, especially for studies done at negative potentials, unless the electroreduction of oxygen itself is being investigated.

Removal of oxygen can be done chemically by the addition of the exact quantity of a compound such as hydrazine, according to the reaction

$$N_2H_4 + O_2 \rightarrow N_2 + 2H_2O$$

This procedure is used industrially to minimize corrosion.

In the laboratory solutions are saturated with an inert gas that reduces the partial pressure of oxygen to a very low value. Inert gases employed are nitrogen and argon: the latter has the advantage of being heavier than air and not escaping easily from the cell, whereas the former is lighter than air; however, nitrogen is much less expensive than argon. Gas purity is very important, and ultra-pure gases should always be preferred. If necessary the gas can be further purified to remove any remaining traces of oxygen. It will always also have to pass over a drying agent (e.g. calcium oxide) in the case of non-aqueous solvents and for all solvents bubble through a solution of supporting electrolyte to presaturate the gas before its entry into the cell. Purification methods are:

● Passage over copper-based BTS catalysts at high temperature

● Passage through Dreschler flasks containing a substance easily oxidized by oxygen, this being regenerated by a powerful reducing agent. Examples of oxidizable substances are vanadium(III), anthraquinones, pyrogallol, and the reducing agent zinc amalgam.

Note that by using the first process it is impossible to introduce impurities into the solution; however, the catalyst is expensive and regeneration with a stream of hydrogen in an oven is somewhat dangerous!

The time during which inert gas should be bubbled before an experiment is performed depends on the solvent and experimental requirements, but should never be less than 10 minutes. During the experiment gas should be passed over the solution to impede entry of oxygen—if bubbling in solution were continued there would be turbulence and possible appearance of gas bubbles on the electrode surface. This can be done simply with a two-way tap.

Another important precaution is with the tubing where the gas passes (preferably copper, glass, etc.) and, in a flow system, also with tubing through which deoxygenated solution passes. All flexible plastics are

Table 7.3. Permeabilities of some materials to oxygen[18]

	Relative permeability
Teflon TFE	0.6
Teflon FEP	0.6
Polystyrene	0.11
Polypropylene	0.02
Polyethylene	0.02

permeable to oxygen—some permeability values are given in Table 7.3. Lengths of plastic tubes should therefore be minimized.

7.6 Calibration of electrodes and cells

In investigating electrode processes it is extremely important to know whether the assumptions of the theory that is being applied are valid. Calibration of the working electrode and of the cell is thus a fundamental requirement. Since electrode response varies with time, calibration may have to be repeated fairly often.

In voltammetric experiments a normal type of calibration is the recording of voltammetric curves for a known system, constructing plots such as variation of limiting current with the transport parameter, or of current with concentration. In potentiometric experiments the equivalent would be the variation of potential with concentration. These curves are especially important in electroanalytical experiments: working curves permit the immediate conversion of a measured current or potential into a concentration.

Another useful electroanalytical procedure is the standard addition method: successive quantities of a standard solution are added to the unknown solution, the concentration of species in the unknown solution being determined from the intercept of the plot of response vs. quantity added. Note that the use of graphical methods without comparison with theoretical equations and known systems does not prove the *accuracy* of the experiments, but only their *precision*.

Finally, the detection limit of a technique, which is determined by the impossibility of separating signal from noise (blank), should be considered. In potentiometric experiments the detection limit results from a diminution down to zero variation in measured potential with concentration decrease, as discussed in Chapter 13. It is clear that reproducibility has an important effect on detection limit, and detection limits are sometimes quoted on this basis, such as three times the standard deviation. Unfortunately in the electroanalytical literature, as in many other areas, there is sometimes an incorrect use of statistical techniques that favours the authors' results or hides the degree of non-reproducibility!

7.7 Instrumentation: general

Instrumentation for electrochemical experiments has undergone rapid development, allowing much greater precision and accuracy in control and analysis parameters, lower electrical noise, and the possibility of measuring smaller currents. This is intimately linked with the impressive developments in the qualities and capabilities of electronic components. At the very least the instrument must be able to:

• Control the working electrode potential and measure the current it passes (potentiostatic control) and/or

• Control the current that the electrode passes and measure its potential (galvanostatic control). When $I = 0$ we measure the equilibrium potential.

For measurements at equilibrium, only a voltmeter with a high input impedance (about 10^{15} Ω) is necessary.

Instruments that can satisfy these criteria can be analogue or digital. We now describe these two types.

7.8 Analogue instrumentation

The basis of analogue instrumentation is the operational amplifier (OA), an integrated circuit that exists in various forms and with different characteristics according to the applications and requirements[19].

Fig. 7.5 shows a design of an OA and Table 7.4 the characteristics of an ideal OA and of three real, easily obtainable, OAs. The two input potentials should be precisely equal in the absence of an applied potential (often externally adjustable), the current consumed by the OA should be zero, and its gain infinite. In other words, the input impedance of the amplifier should be infinite and the output impedance zero. As seen from

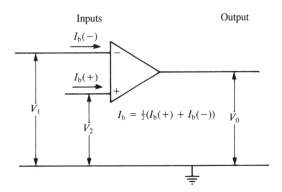

Fig. 7.5. Operational amplifier (OA), showing the two inputs and the output. Operation is by amplification of $|V_1 - V_2|$.

Table 7.4. Characteristics of operational amplifiers

	741	LM11	FET	Ideal
Maximum output current	≈ 5 mA	≈ 5 mA	≈ 5 mA	Infinite
Maximum output voltage	± 15 V	± 15 V	± 15 V	Infinite
Open circuit gain	106 dB	109 dB	106 dB	Infinite
Input impedance	2 MΩ	$10^{11}\,\Omega$	$10^{12}\,\Omega$	Infinite
Maximum offset voltage ($|V_1 - V_2|$)	1 mV	0.2 mV	3 mV	Zero
Input current	80 nA	40 pA	30 pA	Zero
Input bias current ($I_b(+) - I_b(-)$)	20 nA	1 pA	10 pA	Zero

Table 7.4 input impedances in real OAs vary from 10^6 upwards, the currents consumed from nA to less than pA and the gain between 10^4 and 10^8.

Other factors to take into account are:

• The amplifiers work from voltage sources of ± 9 V to ± 18 V; the output potential can never exceed these values:

• For high frequency there is a decline in amplifier gain and in the rate of change in output voltage to an instantaneous change in input voltage (slew rate)—generally 1V/μs at low frequency;

• The circuits need some time to stabilize at the new voltages after a voltage step, perhaps as much as 100 μs. This limitation is especially important in transient techniques (Chapters 9–11).

By choosing adequate amplifiers and using the feedback principle it is possible to construct circuits, making use of Ohm's and Kirchoff's laws to relate input voltage, V_i, with output voltage, V_o. Some of the circuit components are illustrated in Fig. 7.6 with the respective relations indicated. The gain of these components must always be less than the gain of the OA at open circuit.

For a measurement at equilibrium, it is sufficient to use a voltage follower made from a high quality OA at the voltmeter inputs for each of the two electrodes. Modern good-quality pH and ion-selective electrode meters already come with these requirements satisfied.

We now need to know how to combine the circuit components of Fig. 7.6 in order to form a potentiostat or a galvanostat. Current and/or voltage boosters can be incorporated in the schemes shown below when higher currents or voltage ranges are required.

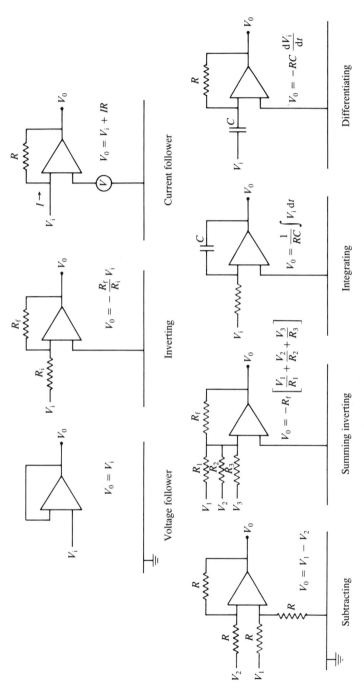

Fig. 7.6. Components of electrical circuits constructed from operational amplifiers using the feedback principle.

Potentiostat

A potentiostat controls the potential applied to the working electrode, and permits the measurement of the current it passes. Combination of some of the components in Fig. 7.6 gives the potentiostat of Fig. 7.7. In

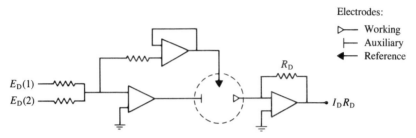

Fig. 7.7. Potentiostat circuit for control of working electrode potential. All resistances are equal, except R_D which is variable.

the circuit illustrated it is possible to apply two signals simultaneously that are added before reaching the working electrode—an example would be a voltage ramp and a sinusoidal signal. We have

$$E_w - E_{ref} = V_1 + V_2 \tag{7.1}$$

where E_w and E_{ref} are the working and reference electrode potentials respectively, and

$$V_0 = I_R R_D \tag{7.2}$$

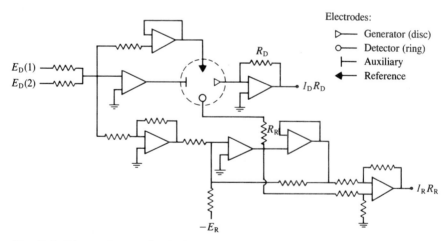

Fig. 7.8. Bipotentiostat circuit for control of the potential of two working electrodes. All resistances are equal, except R_D and R_R which are variable.

A bipotentiostat controls the potential of two working electrodes independently, and measures the current that they pass. A typical circuit is shown in Fig. 7.8. Bipotentiostats are necessary in performing studies with double hydrodynamic electrodes (Sections 8.5–8.7).

Galvanostat

A galvanostat permits control of the current that the working electrode passes. Figure 7.9 illustrates the scheme of a galvanostat circuit. The current passed at the working electrode is

$$I = \frac{V_1}{R_1} + \frac{V_2}{R_2} \tag{7.3}$$

The reference electrode is not part of the circuit. It is there in order to have a reference potential when we wish to measure the working electrode potential, for example in chronopotentiometry (Section 5.4).

In studies involving double hydrodynamic electrodes (Section 8.5), it is sometimes useful to control the current passed by the first, upstream, working electrode (*generator electrode*) and control the potential of the second, downstream, working electrode (*detector electrode*). In collection efficiency measurements, the fraction of species, *B*, produced at the generator electrode under galvanostatic control which reaches the detector electrode is measured by the current at the detector electrode caused by reaction of *B* at a potential chosen so as to give the limiting current. Plots of generator vs. detector electrode current are constructed, as in Fig. 8.11. Addition of extra components to the simple galvanostat permits this experiment to be carried out, see Fig. 7.10.

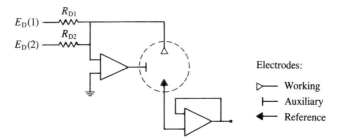

Fig. 7.9. Galvanostat circuit.

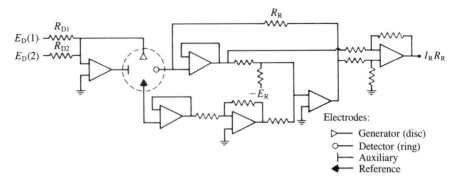

Fig. 7.10. Circuit for measuring collection efficiencies at double hydrodynamic electrodes. All resistances are equal except R_1, R_2, and R_R, which are variable.

Compensation of cell solution resistance

With a Luggin capillary (Section 7.3), one can never reduce the effects of solution resistance to zero. In certain experiments where compensation is necessary in order to increase sensitivity in the response signal, compensation can be done electronically. However, compensation can never be exact—if it is, then the electrical system becomes unstable and starts to oscillate. More details may be found in Ref. 20.

7.9 Digital instrumentation

In the past decade there has been an increasing use of digital instrumentation[21,22]; this involves controlling experiments with a microprocessor inside the instrument or by an external microcomputer. These can also be used for direct analysis of the data obtained.

Since a digital instrument functions at fixed points, that is discontinuously, any direct microprocessor control of an experiment has to be in steps. For example, a linear sweep appears as a staircase instead of a continuous ramp. To minimize these effects there are two possibilities:

● Transform the digital signal into an analogue signal using a DAC with appropriate filtering, doing the inverse with the response through an ADC;

● Use a sufficiently powerful microprocessor so that the differences between the fixed points are so small that there is no visible difference (in theoretical and practical terms) between the signal applied digitally or analogically.

The second of these options, although involving fewer steps, is only now becoming important with the advent of 16-bit and 32-bit microprocessors and microcomputers at reasonably accessible prices.

Digital instrumentation is especially useful where it is necessary to apply pulses of potential to the working electrode, i.e. a succession of steps, with current sampling (the microprocessor's internal clock is used). There has recently been a lot of progress in this area of pulse voltammetry (Section 10.9).

Finally, we remember that, due to its nature, digital instrumentation has a tendency to increase signal noise (there is damping on analogue signals). The only way to solve this problem is repeat the measurements several times and take the average values, or use other strategies such as higher-quality electronic components or improved instrument shielding[23].

References

1. D. T. Sawyer and J. L. Roberts, *Experimental electrochemistry for chemists*, Wiley, New York, 1974.
2. P. T. Kissinger and W. R. Heinemann (ed.), *Laboratory techniques in electroanalytical chemistry*, Dekker, New York, 1984.
3. D. J. G. Ives and G. J. Janz (ed.), *Reference electrodes*, Academic Press, New York, 1961.
4. L. Meites, *Polarographic techniques*, Interscience, New York, 1965.
5. J. Heyrovsky and P. Zuman, *Practical polarography*, Academic Press, New York, 1968.
6. E. Gileadi, E. Kirowa-Eisner, and J. Penciner, *Interfacial electrochemistry: an experimental approach*, Addison-Wesley, Reading, MA, 1975.
7. K. Kinoshita, *Carbon, electrochemical and physicochemical properties*, Wiley, New York, 1988.
8. R. L. McCreery, *Electroanalytical chemistry*, ed. A. J. Bard, Dekker, New York, Vol. 17, 1991, pp. 221–374.
9. G. M. Jenkins and K. Kawamura, *Nature*, 1971, **231**, 175.
10. I. L. Kalnin and H. Jaeger in *Carbon fibres and their composites*, ed. E. Fitzer, Springer-Verlag, Berlin, 1985.
11. J. Lindquist, *J. Electroanal. Chem.*, 1974, **52**, 37.
12. L. D. Burke and M. F. G. Lyons, *Modern aspects of electrochemistry*, Plenum, New York, Vol. 18, 1986, ed. R. E. White, J. O'M. Bockris, and B. E. Conway, pp. 169–248.
13. E. J. M. O'Sullivan and E. J. Calvo, *Comprehensive chemical kinetics*, Elsevier, Amsterdam, Vol. 27, 1987, ed. R. G. Compton, Chapter 3.
14. e.g. W. J. Albery and P. N. Bartlett, *J. Chem. Soc. Chem. Commun.*, 1984, 234.
15. e.g. Ref. 6, pp. 311–312.

16. C. K. Mann, *Electroanalytical chemistry*, ed. A. J. Bard, Dekker, New York, Vol. 3, 1969, pp. 57–134.
17. C. M. A. Brett and A. M. Oliveira Brett, *J. Electroanal. Chem.*, 1988, **255**, 199.
18. Ref. 1, p. 130.
19. R. Kalvoda, *Operational amplifiers in chemical instrumentation*, Ellis Horwood, Chichester, 1975.
20. Ref. 1, Chapter 5.
21. K. Schwabe, H. D. Suschke, and G. Wachler, *Electrochim. Acta*, 1980, **25**, 59.
22. P. He, J. P. Avery, and L. R. Faulkner, *Anal. Chem.*, 1982, **54**, 1313A.
23. S. G. Weber and J. T. Long, *Anal. Chem.*, 1988, **60**, 903A.

8

HYDRODYNAMIC ELECTRODES

8.1 Introduction

Hydrodynamic electrodes[1] are electrodes which function in a regime of forced convection. The advantage of these electrodes is increased transport of electroactive species to the electrode, leading to higher currents and thence a greater sensitivity and reproducibility. Most of the applications of these electrodes are in steady-state conditions, i.e. constant forced convection and constant applied potential or current. In this case $\partial c / \partial t = 0$, which simplifies the solution of the convective diffusion equation (Section 5.6)

$$\frac{\partial c}{\partial t} = D \, \nabla^2 c - \boldsymbol{v} \, \nabla c \qquad (8.1)$$

Even when $\partial c / \partial t \neq 0$, i.e. in applications of these electrodes in combination with transient techniques, forced convection is useful at the very least for improving the reproducibility of the results, owing to the weak dependence of the electrode response on the physical properties of the solution, such as viscosity (Section 8.8).

The method of resolution of (8.1) was indicated in Sections 5.7–5.9, showing as an example the calculation of the limiting current at the rotating disc electrode. In this chapter we discuss this and other hydrodynamic electrodes used in the study of electrode processes. The rotating disc electrode has probably been the hydrodynamic electrode most used for investigating the kinetics and mechanism of electrochemical reactions[2-8]. The characteristics of the dropping mercury electrode (DME) are also discussed[9]; it operation is cyclic and, as a first approximation, at the same point in successive cycles we can invoke the steady-state. Voltammetric studies at the DME are, for historical reasons, referred to as *polarography*.

In Table 8.1 all the hydrodynamic electrodes in common use are listed.

Table 8.1. Limiting currents, I_L, at hydrodynamic electrodes under laminar flow conditions[a]. Adapted from Ref. 1

A. Uniformly accessible electrodes[b]
$$(I = knFc_\infty D^{2/3} v^{-1/6} \pi r_1^2 f^{1/2})$$

	Flow parameter (f)	k	Comments
Rotating disc	ω	0.620	
Rotating hemisphere	ω	$0.433 \rightarrow 0.474$	Experimentally $k = 0.451$
Rotating core	ω	$0.62(\sin\theta)^{-1/2}$	θ is angle between rotation axis and electrode surface
Wall-tube	$V_f/(0.5a)$	0.61	
Stationary disc in uniformly laminar flow	$U/2r_1$	0.753 or 0.780	Experimental deviations of ± 4 per cent

B. Dropping mercury electrode[c]
$$(I_L = knc_\infty D^{1/2} m_1^{2/3} t^{1/6})$$

	Flow parameter (f)	k	Comments
	m_1	709	
		607	Av. current $(\bar{I}_L)$

C. Non-uniformly accessible electrodes

	Flow parameter	Expression for limiting current
Tube	V_f	$5.43 nFc_\infty D^{2/3} x^{2/3} V_f^{1/3}$
Channel	V_f	$0.925 nFc_\infty D^{2/3} (h^2 d)^{-1/3} wx^{2/3} V_f^{1/3}$
Wall-jet[b]	V_f	$1.59 nFc_\infty D^{2/3} v^{-5/12} a^{-1/2} r_1^{3/4} V_f^{3/4}$
Stationary disc in uniformly rotating fluid[b,d]	ω'	$0.761 nFc_\infty D^{2/3} v^{-1/6} \pi r_1^2 \omega'^{1/2}$
Stationary disc in fluid rotating due to a rotating disc[b,d]	ω''	$0.422 nFc_\infty D^{2/3} v^{-1/6} \pi r_1^2 \omega''^{1/2}$

[a] For coordinates see Fig. 8.1.
[b] The expression for I_L at the ring electrode is obtained by putting $(r_3^{3n/2} - r_2^{3n/2})^{2/3}$ instead of r_1^n.
[c] This is the Ilkovic equation which assumes uniform accessibility.
[d] These electrodes are not uniformly accessible, despite appearances, since the convective flow is towards and not from the centre of the disc.

Symbols (flow parameters):

ω rotation speed of electrode (rad s^{-1})	V_f volume flow rate of solution (cm^3 s^{-1})
ω' rotation speed of fluid (rad s^{-1})	U linear velocity of solution (cm s^{-1})
ω'' rotation speed of rotating disc (rad s^{-1})	a diameter of impinging jet (cm)

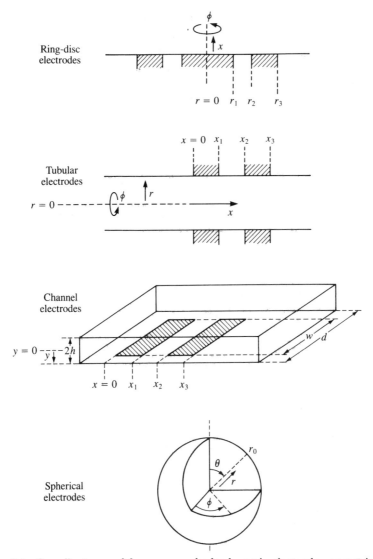

Fig. 8.1. Coordinates used for common hydrodynamic electrode geometries.

The coordinates necessary to define the parameters in Table 8.1 are shown in Fig. 8.1.

The great majority of studies have been undertaken in the laminar regime, partly because the mathematical analysis is simpler. Studies in other flow regimes are not numerous, except for the evaluation of electrochemical reactors[10], so we focus here on the laminar regime.

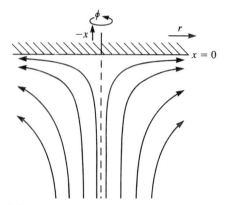

Fig. 8.2a. Schematic streamlines at a rotating disc.

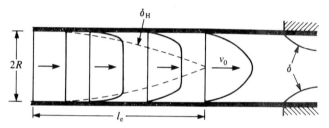

Fig. 8.2b. Establishment of Poiseuille flow and diffusion layer in a tube or channel of infinite width.

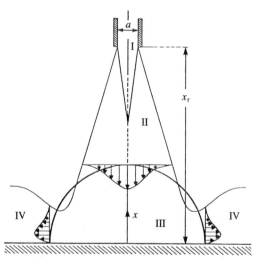

Fig. 8.2c. Mass transfer at an impinging jet electrode. I, central core potential region; II, established flow region; III, stagnation region ('wall-tube' region); wall-jet region (from Ref. 13 with permission).

8.2 Limiting currents at hydrodynamic electrodes

As described in Chapter 5, forced convection leads to a thin layer of solution next to the electrode, within which it is assumed that only diffusion occurs (i.e. it is assumed that all concentration gradients occur within this layer)—the *diffusion layer* of thickness δ. At a particular point on a hydrodynamic electrode and for constant convection, δ is constant. If the value of δ is constant over the whole electrode surface then the electrode is uniformly accessible to electroactive species that arrive from bulk solution.

If the potential applied to the electrode is sufficiently negative (reduction) or positive (oxidation), all the electroactive species that reach the electrode will react and we obtain the limiting current, I_L, whatever the value of the standard rate constant, k_0. The relation between the limiting current and the diffusion layer thickness, δ, is

$$I_L = \frac{nFAD}{\delta} \tag{8.2}$$

Thus, at a uniformly accessible electrode, I_L is directly proportional to electrode area.

The expressions for the limiting currents at commonly used hydrodynamic electrodes are presented in Table 8.1. Nearly all of these have cylindrical symmetry, and the electrodes are embedded in surfaces of infinite extension. As explained in Chapter 5, calculation of I_L begins with the velocity profile in solution[11,12]. Thence, one obtains expressions for the velocity components close to the electrode surface and calculates I_L. Streamlines and schematic profiles of solution movement for some configurations are shown in Fig. 8.2. The following points should be

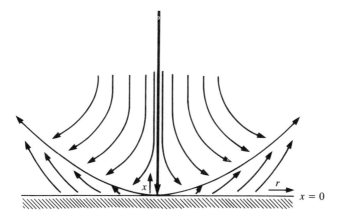

Fig. 8.2d. The wall-jet electrode: schematic streamlines.

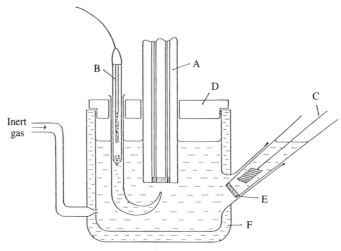

Fig. 8.3. Cell for the rotating ring-disc electrode. A, ring-disc electrode; B, reference electrode with Luggin capillary; C, auxiliary electrode; D, Teflon lid; E, porous frit; F, thermostatted water jacket (adapted from Ref. 1 with permission).

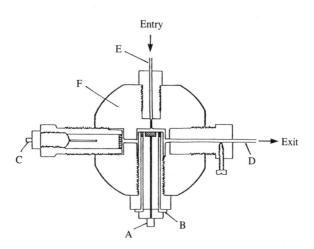

Fig. 8.4. A wall-jet cell with ring-disc electrode. A, disc electrode contact; B, ring electrode contact; C, Ag/AgCl reference electrode; D, platinum tube auxiliary electrode; E, solution entry; F, cell body in Kel-F (from Ref. 1 with permission).

noted:

• A rotating disc sucks solution from below and spreads it out sideways. The rotating disc electrode is uniformly accessible, although the rotating ring is not (Fig. 8.2*a*).

• In a tube or channel a certain entry length, l_e, is necessary before the parabolic Poiseuille flow is obtained (Fig. 8.2*b*).

• A jet impinging on a wall gives rise to four different regions. Note that region III is uniformly accessible (wall-tube region) and region IV is non-uniformly accessible (wall-jet region). To have a wall-tube or wall-jet electrode depends only on the relative radii of the solution jet and of the electrode(s) (Fig. 8.2*c*).

• The streamline profile of the wall-jet. Note the highly non-uniform accessibility and that recirculated solution can never reach the electrode a second time (Fig. 8.2*d*).

The method of calculation for I_L (except for the dropping electrode) is precisely equal to that demonstrated in Section 5.9 for the rotating disc electrode; it is always possible to arrive to (5.63) through a correct definition of the dimensionless parameters. More details of the calculations can be found in Ref. 1.

Figures 8.3, 8.4, 8.5, and 8.6 show typical designs of double hydrodynamic electrodes (two working electrodes)—rotating, wall-jet, tube, and channel. Using only one of the two working electrodes one obtains I_L as in Table 8.1. Use of the two electrodes simultaneously is described in Sections 8.5–8.8.

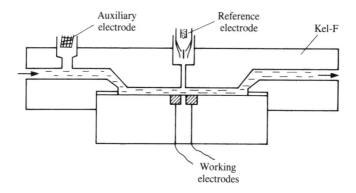

Fig. 8.5. A cell with double channel electrode.

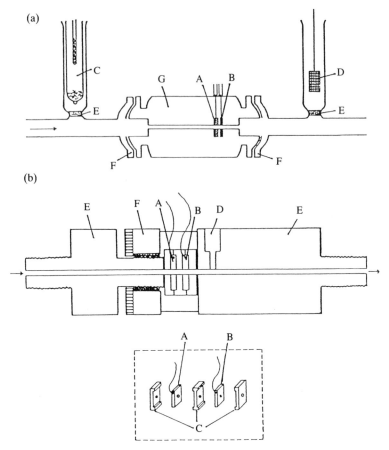

Fig. 8.6. Construction of typical tubular electrodes. (a) Integral construction: A, generator electrode; B, detector electrode; C, reference electrode; D, auxiliary electrode; E, porous frits; F, joints; G, epoxy resin. (b) Demountable construction: A, generator electrode; B, detector electrode; C, Teflon parts; D, reference electrode; E, cell body in Teflon; F, brass thread (from Ref. 1 with permission).

8.3 A special electrode: the dropping mercury electrode

The derivation of the equation for the diffusion-limited current at the dropping mercury electrode differs from that of the other electrodes described above owing to its cyclic operation. Mercury flowing down through a fine capillary forms a drop at the bottom end of the capillary. This drop, the electrode, increases in size until it falls by the force of gravity; the electrode is then renewed by formation of another drop. There are thus virtually no problems of electrode poisoning. The mode of

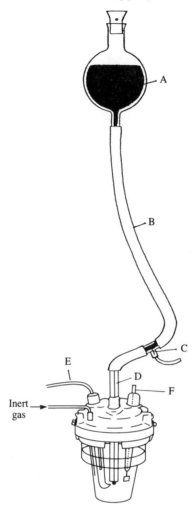

Fig. 8.7. A cell for the dropping mercury electrode. A, mercury reservoir; B, Tygon tube; C, link to mercury by platinum wire; D, capillary; E, reference electrode; F, auxiliary electrode (from Ref. 1 with permission).

operation is shown in Fig. 8.7.

One of the first to study the dropping mercury electrode mathematically was Ilkovic[14]. Following his derivation of I_L, we start from the Cottrell equation (Section 5.3) for a planar electrode

$$I_L(t) = \frac{nFAD^{1/2}c_\infty}{(\pi t)^{1/2}} \tag{8.3}$$

Remembering that the surface area of the electrode (supposed spherical)

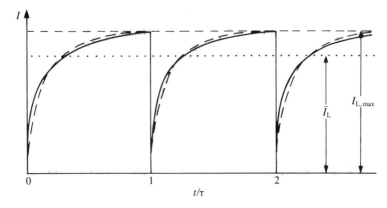

Fig. 8.8. Variation of I_L with t at the dropping electrode: (———) theoretical and (– – –) practical.

is increasing, its radius, r_0, is given by

$$r_0 = \left(\frac{3m_1 t}{4\pi\rho}\right)^{1/3} \tag{8.4}$$

where m_1 is the mass flux of liquid, given normally in mg s^{-1}, and ρ its density. Substituting expression (8.4) in (8.3), we obtain

$$I_L(t) = \frac{0.85}{\pi^{1/2}} nFc_\infty m_1^{2/3} D^{1/2} t^{1/6} \tag{8.5}$$

Drop growth causes an increase in the concentration gradient which results in a multiplicative factor of $\sqrt{(7/3)}$. The final equation for the limiting current (Ilkovic equation), with c_∞ measured in mol cm^{-3} and D in cm^2 s^{-1}, is

$$I_L(t) = 709nc_\infty D^{1/2} m_1^{2/3} t^{1/6} \tag{8.6}$$

Integrating over the lifetime of the drop, τ, the variation of I_L with time during drop life being shown in Fig. 8.8, we obtain the average value, $\bar{I}_L(t)$ which is

$$\bar{I}_L(t) = 607nc_\infty D^{1/2} m_1^{2/3} \tau^{1/6} \tag{8.7}$$

The Ilkovic equation is experimentally verified with some rigour in the form of (8.7), but not as (8.6). The current at the beginning of drop life is less than predicted and later on it is larger. Possible reasons are:

● neglect of electrode curvature

● neglect of the shielding effect of the capillary where the drop is formed, especially at the beginning of drop life

• neglect of the fact that the part of the drop linked to the capillary does not contact with the solution.

Various treatments introducting corrective terms have appeared, but few include the curvature of the electrode explicitly in the formulation of the problem. The majority of the expressions have the form

$$I = I_{\text{Ilkovic}}\{1 + A(D^{1/2}m_1^{-1/3}t^{1/6})\} \tag{8.8}$$

where the constant A varies between 17 and 39, the corrective term corresponding to spherical diffusion (see Table 5.3).

Another important factor linked with drop growth is the appearance of capacitive currents, I_C, manifested as a background current, see Fig. 8.9. We can write

$$I_C = C_i(E - E_z)\frac{dA}{dt} \tag{8.9}$$

where E_z is the point of zero charge and C_i the integral capacity of the double layer (Section 3.2). From (8.4), and since for liquid mercury $\rho = 13.4\,\text{g cm}^{-3}$, we have

$$\frac{dA}{dt} = 5.7 \times 10^{-3}m_1^{2/3}t^{-1/3} \tag{8.10}$$

Thus

$$I_C = 5.7 \times 10^{-3}C_i(E - E_z)m_1^{2/3}t^{-1/3} \tag{8.11}$$

$$\bar{I}_C = 8.6 \times 10^{-3}C_i(E - E_z)m_1^{2/3}\tau^{-1/3} \tag{8.12}$$

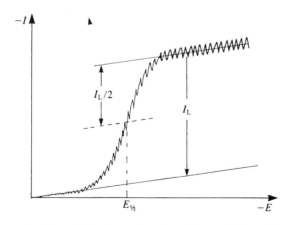

Fig. 8.9. Determination of I_L at a dropping electrode by subtraction of the background current.

So $I_C \propto t^{-1/3}$ whilst $I_L \propto t^{1/6}$, that is I_C diminishes with time whereas I_L increases. This is one of the reasons for sampling the current almost at the end of drop life in many polarographic techniques, in the hope that I_C is almost zero and can be neglected. It should also be noted that if C_i and τ do not vary with potential, then the background current varies linearly with potential, which is useful for calculating the height of polarographic peaks.

It can be shown that $\tau \propto h^3$, where h is the height of the mercury column, by considering the force of gravity and liquid flow in a capillary. So $I_C \propto h$ and $I_L \propto h^{1/2}$. This means that there is a minimum detection limit; with current sampling this is around 10^{-7} M (see Chapter 10 on pulse techniques). It is therefore important not to have too high a column of mercury.

Another important phenomenon that occurs at the dropping mercury electrode is the *polarographic maximum*[15], which occurs when, on reaching the limiting current plateau, the observed current exceeds I_L (Fig. 8.10). The causes are mass transport within the electrode and surface adsorption. Three types of maximum have been identified:

1. Variations of mercury surface tension cause movement within the mercury. Owing to the different velocity distibutions close to and far away from the capillary, there is a non-uniform current distribution. Maxima are tall and narrow.

2. Due to high flows of mercury (rapid dropping), giving small, rounded maxima.

3. Some surfactants adsorbed on the electrode surface with a compact molecular structure, i.e. liquid crystals, lead to condensed two-

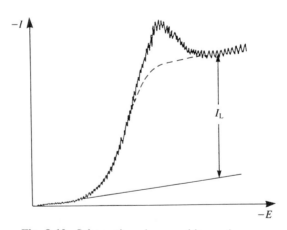

Fig. 8.10. Schematic polarographic maximum.

dimensional layers on the drop surface. This phenomenon causes turbulence. Examples are camphor and adamantol.

Maxima can be reduced to zero in many cases by addition of small quantities of certain surfactants (for example Triton-X-100 or gelatine) given the effect these have on surface tension. Another possibility consists in using very short drop lifetimes with mechanical control of drop fall. It is now realized that this is only possible for certain types of experiment; it is equivalent to the use of a microelectrode (Chapter 5).

8.4 Hydrodynamic electrodes in the study of electrode processes

In Sections 6.3–6.5 expressions for the analysis of the voltammograms corresponding to the simple electron transfer process $O + ne^- \rightarrow R$, obtained for uniformly accessible electrodes such as the rotating disc electrode, were presented. In this section these expressions will be applied to hydrodynamic electrodes in general.

Reversible reaction

In this case non-uniform accessibility has no influence. The equation for the voltammogram is then

$$E = E'_{1/2} + \frac{RT}{nF} \ln \left(\frac{I_{L,c} - I}{I - I_{L,a}} \right) \tag{8.13}$$

where

$$E'_{1/2} = E^{\ominus \prime} + \frac{RT}{nF} \ln \left(\frac{D_R}{D_O} \right)^s \tag{8.14}$$

and $s = \frac{2}{3}$ for all hydrodynamic electrodes except the dropping electrode where $s = \frac{1}{2}$. The shape of the voltammogram is shown in Fig. 6.2. An alternative way of describing the voltammetric curve, in order to separate the anodic and cathodic components, is by substituting

$$\zeta = (E - E'_{1/2})(nF/RT) \tag{8.15}$$

to obtain

$$I = \frac{I_{L,a}}{1 + \exp\{-\zeta\}} + \frac{I_{L,c}}{1 + \exp\{\zeta\}} \tag{8.16}$$

The general case

In the general case for a uniformly accessible electrode ($I_L \propto$ electrode area)

$$\frac{1}{I} = \frac{1}{I_k} + \frac{1}{I_L} \tag{8.17}$$

where $I_k = nFAk_a[R]_\infty$ (or $-nFAk_c[O]_\infty$), assuming a first-order electrode reaction. So, exemplifying for a rotating disc electrode, a plot of I^{-1} vs. $W^{-1/2}$ (see Table 8.1) gives k_0 from the intercept and D from the slope.

It has been demonstrated by Matsuda and co-workers in a series of papers[16] that is is possible to use, for first-order reactions, approximate expressions for the voltammetric curves of the form

$$I = \left(\frac{I_{L,a}}{1 + \exp\{-\zeta\}} + \frac{I_{L,c}}{1 + \exp\{\zeta\}} \right)$$

$$\times \frac{(k_0/D^{2/3})\sigma[\exp\{-\alpha_c\zeta\} + \exp\{\alpha_a\zeta\}]}{A + (k_0/D^{2/3})\sigma[\exp\{-\alpha_c\zeta\} + \exp\{\alpha_a\zeta\}]} \tag{8.18}$$

for many hydrodynamic electrodes, including some that are slightly non-uniformly accessible. In these equations

$$\zeta = (E - E^r_{1/2})\frac{nF}{RT} \tag{8.15}$$

$$D = D_O^{\alpha_a} D_R^{\alpha_c} \tag{8.19}$$

with σ an expression dependent on mass transport and A a number, constant for a uniformly accessible electrode and in other cases dependent on electrode geometry, being calculated numerically. Whilst for uniform accessibility the expression is exact, when it is non-uniform there is an error involved, the magnitude of the error increasing with the degree of non-uniform accessibility. For example, for the wall-jet disc electrode which is highly non-uniformly accessible, (8.18) cannot be applied.

When

$$A \ll (k_0/D^{2/3})\sigma[\exp\{-\alpha_c\zeta\} + \exp\{\alpha_a\zeta\}] \tag{8.20}$$

corresponding to fast electrode reactions, (8.18) reduces to the reversible case, (8.16).

Values of σ and A for various hydrodynamic electrodes are shown in Table 8.2. For these electrodes, graphical analysis of experimental results can follow (8.17), since expressions (8.17) and (8.18) have the same form.

Table 8.2. Values of σ and A for common hydrodynamic electrode geometries in the equation for the I–E curve (equation (8.18))[a]. From Ref. 1

Electrode	σ	A
Rotating disc	$v^{1/6}\omega^{-1/2}$	0.620
Rotating ring	$v^{1/6}\omega^{-1/2}$	$0.620B(r)$
Rotating ring-disc electrode	$v^{1/6}\omega^{-1/2}$	$0.799C(r, E_R)$
(reagent produced at the disc)		
Stationary disc in a uniformly	$v^{1/6}\omega^{-1/2}$	$0.761B'(r)$
rotating fluid		
Dropping mercury[b]	$\tau^{1/2}$	1.13
(expanding plane model)		
Tube	$V_f^{-1/3}Rx_1^{1/3}$	0.839
Detector of double channel electrode	$V_f^{-1/3}(h^2d)^{1/3}x_2^{1/3}$	$0.616C'(x, E_{det})$
(reagent produced at the generator)		

[a] Values of A are approximate; B and B' are functions of r; C is a function of radius and ring potential; C' is a function of electrode length and detector electrode potential.
[b] In equation (8.18) for the DME, D is raised to the power 0.5.

Equation (8.18), in the presence of only O, can be written

$$\ln\frac{I_{L,c} - I}{I} = \ln\left\{\frac{A}{(k_0/D^{2/3})\sigma}(\exp\{-\alpha_c\zeta\} + \exp\{(1-\alpha_c)\zeta\})\right\} \quad (8.21)$$

showing that logarithmic analysis to determine $E_{1/2}^r$ and $(\alpha_c n)$ can be carried out.

For a totally irreversible process, with significant overpotential,

$$E = E_{1/2}^{irr} + \frac{RT}{\alpha_c nF}\ln\frac{I_{L,c} - I}{I} \quad (8.22)$$

for a cathodic process, and

$$E = E_{1/2}^{irr} - \frac{RT}{\alpha_a nF}\ln\frac{I_{L,a} - I}{I} \quad (8.23)$$

for an anodic process, the values of $E_{1/2}^{irr}$ being, naturally, dependent on the magnitude of mass transport.

8.5 Double hydrodynamic electrodes

Double hydrodynamic electrodes have two working electrodes, the second (detector) placed following, i.e. downstream of, the first (generator) with respect to convection (see Figs. 8.2–8.5). Examples are the *rotating ring-disc electrode* (RRDE) the *wall-jet ring-disc electrode* (WJRDE) and the *tube/channel double electrode* (TDE/CDE). The second electrode can be used to detect what happened at the first or to measure what remains of the product of the reaction at the first electrode after a homogeneous reaction in solution (including decomposition). Naturally, the materials of the two electrodes can be different.

The most simple reaction scheme, in which all products are stable, is

$$\text{generator} \qquad A + n_1 e^- \rightarrow B$$

$$\text{detector} \qquad B + n_2 e^- \rightarrow C$$

C can be equal to A. The fraction of B which reaches the detector electrode is always less than unity because a part, after diffusing from the electrode to distance δ, is transported by convection to bulk solution and does not reach the detector electrode. The fraction of B reaching the detector electrode under these conditions is called the steady-state collection efficiency, N_0.

Experimentally, the generator electrode current, I_{gen}, is controlled (galvanostatic control), usually being slowly increased from zero in an anodic or cathodic direction depending on the electrode reaction, and the detector electrode is held at a potential such that all the B reaching it is converted into C, i.e. a potential in the limiting current region for B and C, and passes current I_{det}.

The steady-state collection efficiency, N_0, of the double electrode is given by

$$N_0 = \left| \frac{n_1 I_{\text{det}}}{n_2 I_{\text{gen}}} \right| = 1 - F(\alpha/\beta) + \beta^{2/3}\{1 - F(\alpha)\}$$

$$- (1 + \alpha + \beta)^{2/3}[1 - F\{(\alpha/\beta)(1 + \alpha + \beta)\}] \quad (8.24)$$

The function F is given by the expression

$$F(\theta) = \frac{3^{1/2}}{4\pi} \ln \left\{ \frac{(1 + \theta^{1/3})^3}{1 + \theta} \right\} + \frac{3}{2\pi} \arctan \left(\frac{2\theta^{1/3} - 1}{3^{1/2}} \right) + \frac{1}{4} \quad (8.25)$$

and α and β are functions only of electrode geometry and are independent of mass transport or kinetics. The non-variation of N_0 with convection rate is a very useful simplification. Values of α and β for some double hydrodynamic electrodes are given in Table 8.3.

The deduction of N_0 assumes certain conditions besides the steady state, including the non-existence of edge effects, which could occur when the cell is small and transport by convection is small (larger δ).

For fairly usual radius ratios in ring-disc electrodes of $(r_2/r_1) = 1.08$ and $(r_3/r_1) = 1.15$ where r_1 would typically be 0.35 cm at the RRDE and

Table 8.3. Values of α and β for some double hydrodynamic electrodes

Electrode	α	β
Rotating ring-disc electrode (RRDE) Wall-tube ring-disc electrode (WTRDE)	$\left(\dfrac{r_2}{r_1}\right)^3 - 1$	$\left(\dfrac{r_3}{r_1}\right)^3 - \left(\dfrac{r_2}{r_1}\right)^3$
Wall-jet ring-disc electrode (WJRDE)	$\left(\dfrac{r_2}{r_1}\right)^{9/8} - 1$	$\left(\dfrac{r_3}{r_1}\right)^{9/8} - \left(\dfrac{r_2}{r_1}\right)^{9/8}$
Tube double electrode (TDE) Channel double electrode (CDE)	$\left(\dfrac{l_2}{l_1}\right) - 1$	$\left(\dfrac{l_3}{l_1}\right) - \left(\dfrac{l_2}{l_1}\right)$

0.18 cm at the WJRDE, we obtain values of around 0.18 and 0.11 for N_0 respectively. For tubular and channel electrodes, the ratios $(l_2/l_1) = 1.08$ and $(l_3/l_1) = 1.15$ lead to $N_0 \approx 0.10$.

Another useful double electrode parameter, when the potential of both electrodes is controlled, is the reduction in the detector electrode current when the generator electrode is passing current, if the potential applied both electrodes is equal, these being of the same material. By inspection of Tables 8.1 and 8.3 it can be seen that the relation between generator limiting current, $I_{\mathrm{L,gen}}$, and detector limiting current with the generator disconnected, $I_{\mathrm{L,det}}^0$, is

$$I_{\mathrm{L,det}}^0 = \beta^{2/3} I_{\mathrm{L,gen}} \tag{8.26}$$

The limiting current at the detector electrode with the generator electrode connected, $I_{\mathrm{L,det}}$, is then

$$I_{\mathrm{L,det}} = I_{\mathrm{L,det}}^0 - N_0 I_{\mathrm{L,gen}} = I_{\mathrm{L,det}}^0 (1 - N_0 \beta^{-2/3}) \tag{8.27}$$

In fact $(N_0 \beta^{-2/3}) < 1$ and the term in brackets, the *shielding factor*, is always positive. The shielding factor is the maximum reduction in the detector electrode current that can be caused by the generator electrode: its use is to remove an unwanted electroactive species from solution that interferes with the reaction under study. To maximize the reduction in current (minimize the shielding factor) we want an electrode geometry such that $(N_0 \beta^{-2/3}) \rightarrow 1$, which corresponds to a very small interelectrode gap, $(r_2 - r_1)$, and a thin detector electrode.

We now exemplify some of the uses of these electrodes in the elucidation of various types of electrode processes.

8.6 Multiple electron transfer: the use of the RRDE

The rotating ring-disc electrode has been much used in the study of electron transfer in consecutive and parallel reactions or a mixture of both. Each of these situations is now examined in detail.

Hydrodynamic electrodes

Consecutive reactions

Consider the scheme

Disc $\quad$ A$_\infty$ $\xrightarrow{k_d}$ A$_*$ $\underset{k_1}{\overset{n_1 e}{\rightleftharpoons}}$ B $\underset{k_2}{\overset{n_2 e}{\rightleftharpoons}}$ C$_*$

$$\Bigg\downarrow$$

Ring $\qquad\qquad\qquad$ B$_*$ $\xrightarrow{n_R e}$ D

where the mass transfer coefficient is

$$k_d = 0.620 D^{2/3} v^{-1/6} \omega^{1/2} \tag{8.28}$$

and where we assume that the diffusion coefficients of all species in solution are equal. Applying the stationary state approximation to the concentrations of the various species, i.e. $dA/dt = dB/dt = 0$, we obtain

$$n_R N_0 \left| \frac{I_D}{I_R} \right| = n_1 + (n_1 + n_2) \frac{k_2}{k_d} \tag{8.29}$$

and

$$n_R N_0 \frac{I_{L,D} - I_D}{I_R} = \left[n_2 + (n_1 + n_2) \frac{k_2}{k_1} \right] (n_1 + n_2) \frac{k_d}{k_1} \tag{8.30}$$

From the first expression we conclude that for the two extremes when:

● $k_2 \gg k_d$, $I_R = 0$: A voltammetric wave is obtained at the disc corresponding to $(n_1 + n_2)$ electrons. Diffusion to the ring can be neglected.

● $k_2 \ll k_d$, $|I_R/I_D| = (n_R N_0/n_1)$: Reaction of B at the disc can be neglected relative to its diffusion to the ring electrode.

For the general case we construct plots of

● I_D/I_R vs. $\omega^{-1/2}$ (expression 8.29), obtaining k_2 from the slope and n_1 from the intercept.

● $(I_{L,D} - I_D)/I_R$ vs. $\omega^{-1/2}$ (expression 8.30), obtaining n_2 from the intercept and k_1 from the slope.

Parallel reactions

The scheme is

$$\begin{array}{c} \quad \xrightarrow[k_1]{n_1 e} D \\ \text{Disc} \quad \text{A}_\infty \longrightarrow \text{A}_* \; -\!\!\!\left[\begin{array}{c} \\ \\ \xrightarrow[k_2]{n_2 e} \text{B}_* \end{array} \right. \\ \qquad\qquad\qquad\qquad \Big\downarrow k_d \\ \text{Ring} \qquad\qquad \text{C} \xleftarrow{n_R e} \text{B} \end{array}$$

where only B is electroactive at the ring. The expressions for this case are

$$n_R N_0 \left| \frac{I_D}{I_R} \right| = \frac{n_1 k_1 + n_2 k_2}{k_2} \tag{8.31}$$

$$n_R N_0 \frac{I_{L,D} - I_D}{I_R} = (n_1 - n_2) + \frac{n_1 k_d}{k_2} \tag{8.32}$$

Consecutive and parallel reactions

This type of reaction is often called the *branched mechanism* and corresponds to many real systems such as, for example, the electroreduction of oxygen. At platinum electrodes, a mechanism that explains the experimental data is[17]

Disc $O_{2,\infty} \xrightarrow{k_{d,O_2}} O_{2,*} \xrightarrow{k_2} H_2O_{2,*} \xrightarrow{k_3} H_2O$ with k_1 branch and k_{d,H_2O_2}

Ring $H_2O \longleftarrow H_2O_2$

The calculated expressions are

$$N_0 \left| \frac{I_D}{I_R} \right| = \left[1 + 2 \frac{k_1}{k_2} \right] + \frac{2[(k_1/k_2) + 1]k_3}{k_{d,H_2O_2}} \tag{8.33}$$

$$\frac{I_{L,D}}{I_{L,D} - I_D} = 1 + \frac{k_1 + k_2}{k_{d,O_2}} \tag{8.34}$$

A plot of I_D/I_R vs, $\omega^{-1/2}$ will give straight lines of different slopes depending on the applied potential (which affects the ratio k_1/k_2 and k_3). At the same time there are extreme cases such as

$$k_1 = 0, \qquad \text{intercept } (8.34) = 1$$
$$k_1 = 0, \ k_3 = 0 \qquad \text{slope } (8.33) = 1.$$

A summary of the various mechanisms considered for the electroreduction of oxygen at platinum electrodes can be found in Ref. 17.

8.7 Hydrodynamic electrodes in the investigation of coupled homogeneous reactions

In Section 6.9 the schemes of some mechanisms involving coupled homogeneous reactions, namely EC, CE, and C'E, were shown. Hydrodynamic electrodes are useful in the study of these mechanisms owing to their high sensitivity and reproducibility. Besides this, forced convection confines the zone of homogeneous reaction to very close to the electrode permitting the rate constants of the coupled homogeneous reactions to be calculated more easily.

Instead of presenting the equations for these cases, which can be consulted in the literature[1], we consider a mechanism where the double electrode is very valuable, that is in the ECE mechanism, the chemical step being of first or second order[18]. Schematically for a double electrode, one has

$$\text{generator} \quad A + n_{gen}e^- \to B$$
$$\text{solution} \quad B\,(+\,X) \xrightarrow{k} \text{products}$$
$$\text{detector} \quad B \pm n_{det}e^- \to C \text{ (or A)}$$

The fraction of species B obtained at the detector electrode is N_k, the *kinetic collection efficiency*. An example of a first-order reaction is the bromination of anisole[19], and of a second-order reaction the bromination of some proteins[20].

Experimentally, as for the steady-state collection efficiency, the generator electrode current is controlled and the potential of the detector electrode is held at a value corresponding to mass-transport-limited conversion of B to C.

Let us see the expressions obtained for the case of a very fast second-order homogeneous reaction, where X is not electroactive and is of low concentration. It is relatively simple to show that a plot of I_{det} vs. I_{gen} has the form of Fig. 8.11[17]. Effectively B is being titrated with X. A current begins to be registered at the detector electrode only when some B reaches it, which means that there is excess of B in the reaction with X—there will be a region not containing any X around the generator electrode and which reaches to, and eventually overtakes, the detector electrode on increasing I_{gen}. After the boundary of this region reaches the end of the detector, I_{det} increases linearly, the slope of the plot of $|I_{det}|$ vs. $|I_{gen}|$ now being equal to N_0. Analysing the curve we can deduce the concentration of X. The effect of slower homogeneous kinetics is to create a zone where both B and X exist in solution, which makes the curved region of the plot larger. Analysis of the curved part of the plot leads to the rate constant of the homogeneous reaction, and we can calculate the concentration from the linear part.

The curve is defined by the equations

$$N' = G\left(\frac{\beta_J}{\alpha}\right) - \frac{1+\alpha}{(1+\alpha+\beta_J)^{1/3}} G\left(\frac{\beta_J}{\alpha(1+\alpha+\beta_J)}\right) \tag{8.35}$$

$$\frac{M}{I_{det}} = G(1/\alpha) - \frac{\beta_J^{1/3}}{(1+\alpha+\beta_J)^{1/3}} G\left(\frac{\beta_J}{\alpha(1+\alpha+\beta_J)}\right) \tag{8.36}$$

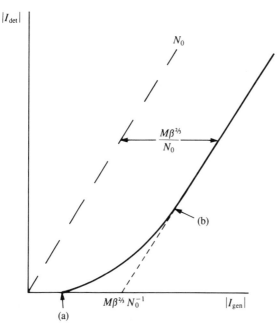

Fig. 8.11. Diffusion layer titration curve at a double hydrodynamic electrode (second order homogeneous reaction). (a) $|I_{det}|$ begins to rise when excess of B that did not react homogeneously reaches the detector electrode and $M/|I_{det}| = G(1/\alpha)$. (b) Assuming fast kinetics this is where linearity commences. From here onwards the slope is N_0.

α, β, and F have already been defined (Table 8.3 and equation (8.25), noting that

$$G(\theta) = 1 - F(1/\theta) \qquad (8.37)$$

and β_J is obtained by putting r_J instead of r_3 in the expressions for ring-disc electrodes ($r_2 < r_J < r_3$) and l_J instead of l_3 for tubular and channel electrodes ($l_2 < l_J < l_3$). M represents the limiting current of X that would be observed if it were electroactive. Usually the analysis is done from the coordinates where a line of slope N' meets the experimental curve.

Good examples are for $A = Br^-$, B being Br_2 or OBr^- depending on pH, and $X = As(III)$, amino acids or proteins, i.e. rapidly brominatable species[18,19].

8.8 Hydrodynamic electrodes and non-stationary techniques

In this chapter we have considered that $\partial c/\partial t = 0$. However, there are important applications of hydrodynamic electrodes with transient techniques, their preferential use being based on the high reproducibility of results that this type of electrode confers and the weak dependence on the physical properties of the solutions relative to stationary electrodes. These applications will be described in the chapters that follow.

Nevertheless, it is important to refer here to the fact that forced convection alters the electrode response only in the case of this being at a timescale that is long in comparison with the electrode process. For short timescales the (fast) perturbation will be confined to a very short distance from the electrode surface: the electrode reaction parameters are not affected by the convection, this being simply a way of achieving good reproducibility.

References

1. C. M. A. Brett and A. M. C. F. Oliveira Brett, *Comprehensive chemical kinetics,* Elsevier, Amsterdam, Vol. 26, 1986, ed. C. H. Bamford and R. G. Compton, Chapter 5.
2. A. C. Riddiford, *Advances in electrochemistry and electrochemical engineering,* ed. P. Delahay and C. W. Tobias, Wiley, New York, Vol. 4, 1966, pp. 47–116.
3. Yu. V. Pleskov and V. Yu. Filinovskii, *The rotating disc electrode,* Consultants Bureau, New York, 1976.
4. W. J. Albery and M. L. Hitchman, *Ring disc electrodes,* Clarendon Press, Oxford, 1971.
5. F. Opekar and P. Beran, *J. Electroanal. Chem.,* 1976, **69,** 1.
6. S. Bruckenstein and B. Miller, *Acc. Chem. Res.,* 1977, **10,** 54.
7. V. Yu. Filinovskii and Yu. V. Pleskov, *Comprehensive treatise of electrochemistry,* Plenum, New York, Vol. 9, 1984, ed. E. Yeager, J. O'M. Bockris, B. E. Conway, and S. Sarangapani, pp. 293–352.
8. R. G. Compton, M. E. Laing, D. Mason, R. J. Northing, and P. R. Unwin, *Proc. R. Soc. Lond.,* 1988, **A 418,** 113.
9. J. Heyrovsky, *Chem. Listy,* 1922, **16,** 256.
10. F. Barc, C. Bernstein and W. Vielstich, *Advances in electrochemistry and electrochemical engineering,* ed. H. Gerischer and C. W. Tobias, Wiley, New York, Vol. 13, 1984, pp. 261–353.
11. V. G. Levich, *Physicochemical hydrodynamics,* Prentice-Hall, Englewood Cliffs, NJ, 1962.
12. J. S. Newman, *Electrochemical systems,* Prentice-Hall, Englewood Cliffs, NJ, 1973.
13. D. T. Chin and C. H. Tsang, *J. Electrochem. Soc.,* 1978, **125,** 1461.
14. D. Ilkovic, *Collect. Czech. Chem. Commun.,* 1934, **6,** 498.
15. H. H. Bauer, *Elecroanalytical chemistry,* ed. A. J. Bard, Dekker, New York, Vol. 8, 1975, pp. 169–279.

16. H. Matsuda et al., *Bull. Chem. Soc. Japan*, 1955, **28,** 422; *J. Electroanal. Chem.*, 1967, **15,** 325; 1972, **35,** 77; 1972, **38,** 159; 1973, **44,** 199; 1974, **52,** 421.
17. K. L. Hsueh, D. T. Chin, and S. Srinivasan, *J. Electroanal. Chem.*, 1983, **153,** 79 and references therein.
18. Ref. 4, Chapters 7, 8, 9.
19. Ref. 4, p. 127.
20. W. J. Albery, L. R. Svanberg, and P. Wood, *J. Electroanal. Chem.*, 1984, **162,** 29.

9

CYCLIC VOLTAMMETRY AND LINEAR SWEEP TECHNIQUES

9.1 Introduction

Of all the methods available for studying electrode processes, potential sweep methods are probably the most widely used, particularly by non-electrochemists. They consist in the application of a continuously time-varying potential to the working electrode. This results in the occurrence of oxidation or reduction reactions of electroactive species in solution (faradaic reactions), possibly adsorption of species according to the potential, and a capacitive current due to double layer charging. The observed current is therefore different from that in the steady state ($\partial c/\partial t = 0$). The potential sweep technique is normally used at stationary electrodes but can also be used at hydrodynamic electrodes. Its principal use has been to diagnose mechanisms of electrochemical reactions, for the identification of species present in solution and for the semi-quantitative analysis of reaction rates[1-3]. Until recently it was difficult to determine kinetic parameters accurately from these experimental results, but new methods for the analysis and simulation of these voltammetric curves now permit much greater accuracy in the determination of rate constants.

For the cyclic voltammetry specialist, many details of the application of cyclic voltammetry to a huge variety of electrochemical systems can be found in Ref. 4.

After a description of how to control the sweep experiment and its two forms, *linear sweep voltammetry* (LSV) and *cyclic voltammetry* (CV) (where the sweep direction is inverted at a certain, chosen potential), the voltammetric waveshape obtained for slow and fast electrode reactions is analysed. Recent advances in these topics are considered. Finally, the type of curve obtained from linear sweep in a thin-layer cell is presented: thin-layer cells are important because they permit almost 100 per cent conversion of the electroactive species, and show differences in relation to electrochemical behaviour in a normal-sized cell.

9.2 Experimental basis

The basic scheme involves application of a potential sweep to the working electrode. The various parameters of interest are shown in Fig. 9.1.

In linear sweep voltammetry the potential scan is done in only one direction, stopping at a chosen value, E_f, for example at $t = t_1$ in Fig. 9.1. The scan direction can be positive or negative and, in principle, the sweep rate can have any value.

In cyclic voltammetry, on reaching $t = t_1$ the sweep direction is inverted as shown in Fig. 9.1 and swept until E_{min}, then inverted and swept to

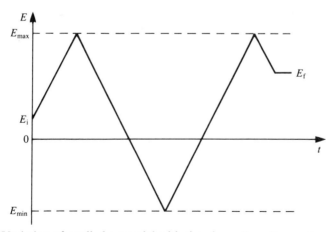

Fig. 9.1. Variation of applied potential with time in cyclic voltammetry, showing the initial potential, E_i, the final potential, E_f, maximum, E_{max}, and minimum, E_{min}, potentials. The sweep rate $|dE/dt| = v$. For linear sweep voltammetry consider only one segment. The fact that the initial sweep is positive is purely illustrative.

E_{max}, etc. The important parameters are

- the initial potential, E_i
- the initial sweep direction
- the sweep rate, v
- the maximum potential, E_{max}
- the minimum potential, E_{min}
- the final potential, E_f

It is not common, but can sometimes be convenient, to change the values of E_{max} and E_{min} between successive cycles.

A faradaic current, I_f, due to the electrode reaction, is registered in the relevant zone of applied potential where electrode reaction occurs. There is also a capacitive contribution: on sweeping the potential the double layer charge changes; this contribution increases with increasing sweep rate. The total current is

$$I = I_C + I_f = C_d \frac{dE}{dt} + I_f = v C_d + I_f \tag{9.1}$$

Thus $I_C \propto v$ and, as will be demonstrated in the following sections, $I_f \propto v^{1/2}$: this means that for very high sweep rates the capacitive current must be subtracted in order to obtain accurate values of rate constants.

In the following section we consider the equations obtained for the shape and position of the voltammetric waves according to the rate of the electrode reaction. On increasing the sweep rate there is less time to reach equilibrium at the electrode surface; reactions which appear as reversible at slow sweep rates can be quasi-reversible at high sweep rates.

9.3   Cyclic voltammetry at planar electrodes

In this section we deduce the expressions for simple electron transfer $O + ne^- \rightarrow R$, with only O initially present in solution. The initial sweep direction is therefore negative. The observed faradaic current depends on the kinetics and transport by diffusion of the electroactive species. It is thus necessary to solve the equations

$$\frac{\partial[O]}{\partial t} = D_O \frac{\partial^2[O]}{\partial x^2} \tag{9.2}$$

$$\frac{\partial[R]}{\partial t} = D_R \frac{\partial^2[R]}{\partial x^2} \tag{9.3}$$

The boundary conditions are

$$t = 0 \qquad x = 0 \qquad [O]_* = [O]_\infty \qquad [R]_* = 0 \qquad (9.4a)$$

$$t > 0 \qquad x \to \infty \qquad [O] \to [O]_\infty \qquad [R] \to 0 \qquad (9.4b)$$

$$t > 0 \qquad x = 0 \qquad D_O\left(\frac{\partial [O]}{\partial x}\right)_0 + D_R\left(\frac{\partial [R]}{\partial x}\right)_0 = 0 \qquad (9.4c)$$

$$0 < t \leqslant \lambda \qquad E = E_i - vt$$

$$t > \lambda \qquad E = E_i - v\lambda + v(t - \lambda) \qquad (9.4d)$$

where λ is the value of t when the potential is inverted. A fifth boundary condition expresses the kinetic regime of the electrode reaction. The first theoretical description of this problem was due to Randles and Sevcik in 1948[5].

If the species present in bulk solution is R and the initial sweep direction is in the positive direction, then $[O]$ and $[R]$ would be switched round in boundary conditions (9.4a) and (9.4b) and the signs in (9.4d) inverted.

The solution of (9.2) and (9.3) is carried out by using the Laplace transform (Chapter 5 and Appendix 1).

Reversible system

The final boundary condition for a reversible system is the *Nernst equation*

$$\frac{[O]_*}{[R]_*} = \exp\left[\frac{nF}{RT}(E - E^{\ominus\prime})\right] \qquad (9.5)$$

Solution of the diffusion equations leads to a result in the Laplace domain that cannot be inverted analytically, numerical inversion being necessary. The final result, after inversion, can be expressed in the form

$$I = -nFA[O]_\infty(\pi D_O\sigma)^{1/2}\chi(\sigma t) \qquad (9.6)$$

where

$$\sigma = \left(\frac{nF}{RT}\right)v \qquad (9.7)$$

and

$$\sigma t = \frac{nF}{RT}(E_i - E) \qquad (9.8)$$

Thus the current is dependent on the square root of the sweep rate. Values of $\{\pi^{1/2}\chi(\sigma(t))\}$ have been determined and are given in Table 9.1; Fig. 9.2 shows the curve obtained. Such values can be used for comparing the shapes of experimental and simulated curves.

Table 9.1. Values of the current functions $\pi^{1/2}\chi(\sigma t)$ (planar electrode) and $\phi(\sigma t)$ (spherical correction)[a] for linear sweep voltammetry and reversible charge transfer reaction $O + ne^- \rightarrow R$[6].

$n(E - E^r_{1/2})$/mV	$\pi^{1/2}\chi(\sigma t)$	$\phi(\sigma t)$	$n(E - E^r_{1/2})$/mV	$\pi^{1/2}\chi(\sigma t)$	$\phi(\sigma t)$
120	0.009	0.008	−5	0.400	0.548
100	0.020	0.019	−10	0.418	0.596
80	0.042	0.041	−15	0.432	0.641
60	0.084	0.087	−20	0.441	0.685
50	0.117	0.124	−25	0.445	0.725
45	0.138	0.146	−28.50	0.4463	0.7516
40	0.160	0.173	−30	0.446	0.763
35	0.185	0.208	−35	0.443	0.796
30	0.211	0.236	−40	0.438	0.826
25	0.240	0.273	−50	0.421	0.875
20	0.269	0.314	−60	0.399	0.912
15	0.298	0.357	−80	0.353	0.957
10	0.328	0.403	−100	0.312	0.980
5	0.355	0.451	−120	0.280	0.991
0	0.380	0.499	−150	0.245	0.997

[a] See Section 9.4: at spherical electrodes, $I = I(\text{planar}) + I(\text{spherical correction})$.

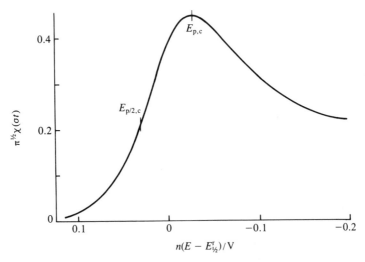

Fig. 9.2. Normalized linear sweep voltammogram for a reversible reduction at a planar electrode, using values from Table 9.1. $E = E_{p/2}$ when $I = I_{p/2}$.

The shape of the curve can be understood in the following way. On reaching a potential where the electrode reaction begins, the current rises as in a steady-state voltammogram. However, the creation of a concentration gradient and consumption of electroactive species means that, continuing to sweep the potential, from a certain value just before the maximum value of the current, the *peak current,* the supply of electroactive species begins to fall. Owing to depletion, the current then begins to decay, following a profile proportional to $t^{-1/2}$, similar to that after application of a potential step.

We now indicate some quantitative parameters in the curve, which can be deduced from data in Table 9.1. First, the current function, $\pi^{1/2}\chi(\sigma t)$, passes through a maximum value of 0.4463 at a reduction peak potential $E_{p,c}$ of

$$E_{p,c} = E^{\ominus\prime} - \frac{RT}{nF}\ln\left(\frac{D_O}{D_R}\right)^{1/2} - 0.0285/n \tag{9.9}$$

$$= E_{1/2}^r - 0.0285/n \tag{9.10}$$

Secondly, the peak current in amperes is

$$I_{p,c} = -2.69 \times 10^5 n^{3/2} A D_O^{1/2}[O]_\infty v^{1/2} \tag{9.11}$$

with A measured in cm^2, D_O in cm^2 s^{-1}, $[O]_\infty$ in mol cm^{-3} and v in V s^{-1}, substituting $T = 298$ K in (9.6) and (9.7)—an equation first obtained by Randles and Sevcik[5]. Thirdly, the difference in potential between the potential at half height of the peak, $E_{p/2,c}$ $(I = I_{p,c}/2)$, and $E_{p,c}$ is given by

$$|E_{p,c} - E_{p/2,c}| = 2.2\frac{RT}{nF} = \frac{56.6}{n}\,\text{mV} \quad \text{at 298 K} \tag{9.12}$$

If the scan direction is inverted after passing the peak for a reduction reaction, then a cyclic voltammogram, as shown schematically in Fig. 9.3, is obtained. It has been shown that, if the inversion potential, E_λ, is at least $35/n$ mV after $E_{p,c}$, then

$$E_{p,a} = E_{1/2}^r + 0.0285/n + \frac{x}{n} \tag{9.13}$$

in which $x = 0$ for $E_\lambda \ll E_{p,c}$ and is 3 mV for $|E_{p,c} - E_\lambda| = 80/n$ mV. In this case

$$|I_{p,a}/I_{p,c}| = 1 \tag{9.14}$$

The shape of the anodic curve is always the same, independent of E_λ, but the value of E_λ alters the position of the anodic curve in relation to the current axis. For this reason $I_{p,a}$ should be measured from a baseline that is a continuation of the cathodic curve, as shown in Fig. 9.3.

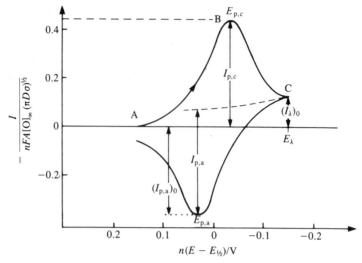

Fig. 9.3. Cyclic voltammogram for a reversible system.

We can summarize all the information in a diagnostic for linear sweep and cyclic voltammograms of reversible reactions:

- $I_p \propto v^{1/2}$
- E_p independent of v
- $|E_p - E_{p/2}| = 56.6/n$ mV

and for cyclic voltammetry alone

- $E_{p,a} - E_{p,c} = 57.0/n$ mV $(E_\lambda \ll E_{p,c} \text{ or } E_\lambda \gg E_{p,a})$
- $|I_{p,a}/I_{p,c}| = 1$

As is clear from (9.6), $I_p \propto T^{-1/2}$: so, if experiments are conducted at temperatures other than 298 K, the correction in I_p is easy to do.

Sometimes, and this is one of the disadvantages of conventional analysis of cyclic voltammograms, it is not possible to measure the baseline with sufficient precision in order to measure $I_{p,a}$. However, it is a good approximation to apply the following expression in terms of the peak current measured from the current axis $(I_{p,a})_0$ and the current at the inversion potential $(I_\lambda)_0$ (see Fig. 9.2)

$$\left| \frac{I_{p,a}}{I_{p,c}} \right| = \frac{(I_{p,a})_0}{I_{p,c}} + \frac{0.485(I_\lambda)_0}{I_{p,c}} + 0.086 \tag{9.15}$$

The capacitive contribution to the total current as given in (9.1) should also be taken into account. Writing $I_f = I_{p,c}$ we have, from (9.1) and

(9.11),

$$\left|\frac{I_{\mathrm{C}}}{I_{\mathrm{p,c}}}\right| = \frac{C_{\mathrm{d}} v^{1/2} 10^{-5}}{2.69 n^{3/2} D_{\mathrm{O}}^{1/2} [\mathrm{O}]_{\infty}} \tag{9.16}$$

Substituting typical values ($C_{\mathrm{d}} = 20\ \mu\mathrm{F\ cm}^{-2}$, $D_{\mathrm{O}} = 10^{-5}\ \mathrm{cm}^2\ \mathrm{s}^{-1}$, and $[\mathrm{O}]_{\infty} = 10^{-7}\ \mathrm{mol\ cm}^{-3}$ ($10^{-4}\ \mathrm{M}$), and $n = 1$) we obtain

$$\left|\frac{I_{\mathrm{C}}}{I_{\mathrm{p,c}}}\right| = 0.24 v^{1/2} \tag{9.17}$$

This ratio is 0.1 for $v = 0.18\ \mathrm{V\ s}^{-1}$; if $[\mathrm{O}]_{\infty}$ is an order of magnitude higher, i.e. $10^{-3}\ \mathrm{M}$, then the ratio is only 0.01. This shows the advantage of using concentrations as high as possible, millimolar concentrations representing the upper limit.

Another practical factor mentioned in Chapter 7 is the solution resistance between working and reference electrodes. This resistance leads to a shift in the potential of the working electrode of $I_{\mathrm{p}} R_{\Omega}$ where R_{Ω} is the resistance (uncompensated) of the solution. A broadening of the peaks is observed, greater separation between $E_{\mathrm{p,a}}$ and $E_{\mathrm{p,c}}$ than predicted theoretically, and the peak currents are lower. Since the peak current increases with sweep rate, this factor becomes more important for large values of v.

Irreversible system

In the case of an irreversible reaction of the type $\mathrm{O} + n e^- \to \mathrm{R}$, linear sweep and cyclic voltammetry lead to the same voltammetric profile, since no inverse peak appears on inversing the scan direction.

To solve (9.2) and (9.3), a fifth boundary condition to add to boundary conditions (9.4) is

$$D_{\mathrm{O}} \frac{\partial [\mathrm{O}]_*}{\partial x} = k_{\mathrm{c}} [\mathrm{O}]_* = k_{\mathrm{c}}' \exp\{bt\} [\mathrm{O}]_* \tag{9.18}$$

for a reduction, where

$$k_{\mathrm{c}}' = k_0 \exp\left[(-\alpha_{\mathrm{c}} n' F / RT)(E_{\mathrm{i}} - E^{\ominus\prime})\right] \tag{9.19}$$

and

$$b = \alpha_{\mathrm{c}} n' F v / RT \tag{9.20}$$

n' being the number of electrons transferred in the rate-determining step. As for the reversible case, the mathematical solution in the Laplace

Table 9.2. Values of the current functions $\pi^{1/2}\chi(bt)$ (planar electrode) and $\phi(bt)$ (spherical correction)[a] for linear sweep voltammetry and irreversible charge transfer reaction $O + ne^- \rightarrow R$[6].

Potential[b]/mV	$\pi^{1/2}\chi(bt)$	$\phi(bt)$	Potential[b]/mV	$\pi^{1/2}\chi(bt)$	$\phi(bt)$
160	0.003		15	0.457	0.323
140	0.008		10	0.462	0.396
120	0.016		5	0.480	0.482
110	0.024		0	0.492	0.600
100	0.035		−5	0.496	0.685
90	0.050		−5.34	0.4958	0.694
80	0.073	0.004	−10	0.493	0.755
70	0.104	0.010	−15	0.485	0.823
60	0.145	0.021	−20	0.472	0.895
50	0.199	0.042	−25	0.457	0.952
40	0.264	0.083	−30	0.441	0.992
35	0.300	0.115	−35	0.423	1.00
30	0.337	0.154	−40	0.406	
25	0.372	0.199	−50	0.374	
20	0.406	0.253	−70	0.323	

[a] See Section 9.4: at spherical electrodes, $I = I(\text{planar}) + I(\text{spherical correction})$.
[b] Potential scale given by $(E - E^{\ominus'})\alpha_c n' + (RT/F) \ln [(\pi D_O b)^{1/2}/k_0]$.

domain cannot be inverted analytically. Numerical inversion leads to[5]

$$I_c = -nFA[O]_\infty D_O^{1/2} v^{1/2} \left(\frac{\alpha_c n' F}{RT}\right)^{1/2} \pi^{1/2}\chi(bt) \qquad (9.21)$$

and the values of $\{\pi^{1/2}\chi(bt)\}$ are tabulated, having a maximum of 0.4958 for $E = E_p$[6], see Table 9.2. The voltammetric curve is shown in Fig. 9.4. The peak current in amperes is

$$I_{p,c} = -2.99 \times 10^5 n(\alpha_c n')^{1/2} A[O]_\infty D_O^{1/2} v^{1/2} \qquad (9.22)$$

with the units the same as in (9.11). The peak potential is given by

$$E_{p,c} = E^{\ominus'} - \frac{RT}{\alpha_c n' F}\left[0.780 + \ln \frac{D_O^{1/2}}{k_0} + \tfrac{1}{2}\ln b\right] \qquad (9.23)$$

An alternative expression for I_p is obtained from combining (9.22) and (9.23), leading to

$$I_{p,c} = -0.227 nFA[O]_\infty k_0 \exp\left[\frac{-\alpha_c n' F}{RT}(E_{p,c} - E^{\ominus'})\right] \qquad (9.24)$$

From data such as those in Table 9.2 it can be deduced that $|E_p - E_{p/2}| = 47.7/(\alpha n')$ mV and that $|dE_p/d \lg v| = 29.6/(\alpha n')$ mV.

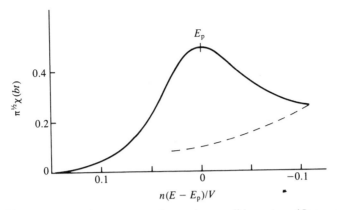

Fig. 9.4. Linear sweep voltammogram for an irreversible system $(O + ne^- \rightarrow R)$. In cyclic voltammetry, on inverting the sweep direction, one obtains only the continuation of current decay $(- - -)$.

With respect to reversible systems the waves are shifted to more negative potentials (reduction), E_p depending on sweep rate. The peaks are broader and lower.

Quasi-reversible systems

For quasi-reversible systems[7] the kinetics of the oxidation and reduction reactions have to be considered simultaneously. The mathematical solution is, therefore, more complex, but there are numerical theoretical solutions.

As a general conclusion, the extent of irreversibility increases with increase in sweep rate, while at the same time there is a decrease in the peak current relative to the reversible case and an increasing separation between anodic and cathodic peaks, shown in Fig. 9.5.

Peak shape and associated parameters are conveniently expressed by a parameter, Λ, which is a quantitative measure of reversibility, being effectively the ratio kinetics/transport,

$$\Lambda = k_0/(D_O^{\alpha_a} D_R^{\alpha_c}\sigma)^{1/2} = k_0/(D_O^{(1-\alpha_c)}D_R^{\alpha_c}\sigma)^{1/2} \tag{9.25}$$

When $D_R = D_O = D$

$$\Lambda = k_0 D^{-1/2}\sigma^{-1/2} \tag{9.26}$$

showing that small Λ corresponds to large v (i.e. large σ).

The following ranges were suggested for the different types of system

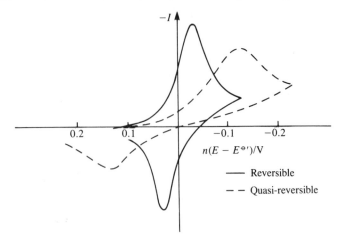

Fig. 9.5. The effect of increasing irreversibility on the shape of cyclic voltammograms.

at stationary planar electrodes[7]:

- reversible:

$$\Lambda \geqslant 15 \qquad k_0 \geqslant 0.3v^{1/2} \text{ cm s}^{-1}$$

- quasi-reversible:

$$15 > \Lambda > 10^{-2(1+\alpha)} \qquad 0.3v^{1/2} > k_0 > 2 \times 10^{-5}v^{1/2} \text{ cm s}^{-1}$$

- irreversible

$$\Lambda \leqslant 10^{-2(1+\alpha)} \qquad k_0 \leqslant 2 \times 10^{-5}v^{1/2} \text{ cm s}^{-1}$$

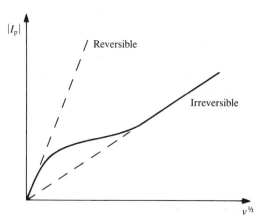

Fig. 9.6. Transition from a reversible to an irreversible system on increasing sweep rate.

Table 9.3. Variation of the difference between anodic and cathodic peak potentials with the degree of reversibility, expressed as ψ ($= \Lambda \pi^{1/2}$, see equation 9.25)[8], assuming $\alpha = 0.5$

ψ	$n(E_{p,a} - E_{p,c})/mV$	ψ	$n(E_{p,a} - E_{p,c})/mV$
20	61	0.38	117
7	63	0.35	121
6	64	0.26	140
5	65	0.25	141
3	68	0.16	176
2	72	0.14	188
1	84	0.12	200
0.91	86	0.11	204
0.80	89	0.10	212
0.75	92	0.077	240
0.61	96	0.074	244
0.54	104	0.048	290
0.50	105		

The transition between these zones is shown schematically in Fig. 9.6.

In the case of cyclic voltammograms and for $0.3 < \alpha < 0.7$ (α is α_c or α_a), E_p is almost exclusively dependent on Λ and hardly varies with α. This can be useful for the calculation of k_0 using Table 9.3, or by interpolation from a working curve drawn using these data.

Adsorbed species

If the reagent or product of an electrode reaction is adsorbed strongly or weakly on the electrode, the form of the voltammetric wave is modified[9]. There are two types of situation:

● the rate of reaction of adsorbed species is much greater than of species in solution

● it is necessary to consider the reactions of both adsorbed species and of those in solution.

From a mathematical point of view the first of these is the simpler. An adsorption isotherm has to be chosen or, alternatively, one has to assume that there is adsorption equilibrium before the experiment begins.

The details of the calculation of the voltammetric profiles can be consulted in the specialized literature[8]. Here we give the expression for a reversible reaction in which only the adsorbed species O and R contribute to the total current. The reason for this is to enable a comparison between the expressions for this situation and for thin-layer cells (Section 9.10), since they are analogous—in neither case is there any limitation from diffusion of electroactive species.

The current-potential curve for O initially adsorbed and for fast electrode kinetics is given by

$$I_c = \frac{-nF\sigma A\Gamma_{O,i}(b_O/b_R)\theta}{[1+(b_O/b_R)\theta]^2} \tag{9.27}$$

where $\Gamma_{O,i}$ is the surface concentration of adsorbed O, before the experiment begins, on an electrode of area A, $\sigma = (nF/RT)v$, b_O and b_R express the adsorption energy of O and R respectively, and

$$\theta = \exp\left[\frac{nF}{RT}(E - E^{\ominus\prime})\right] \tag{9.28}$$

The peak current for reduction, $I_{p,c}$, is obtained when $(b_O/b_R)\theta = 1$, that is

$$I_{p,c} = \frac{-n^2F^2vA\Gamma_{O,i}}{4RT} \tag{9.29}$$

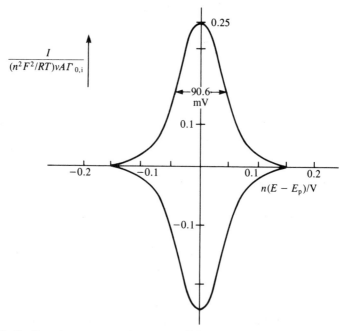

Fig. 9.7. Cyclic voltammogram for a reversible system of species adsorbed on the electrode. If O and R are adsorbed with the same strength, $E_p = E^{\ominus\prime}$.

having the same magnitude for an oxidation. The peak potential is then

$$E_p = E^{\ominus\prime} - \frac{RT}{nF} \ln\left(\frac{b_O}{b_R}\right) \tag{9.30}$$

The value of E_p is the same for oxidation and for reduction.

If the adsorption isotherm is of Langmuir type and $(b_O/b_R) = 1$, then the voltammetric profile is described by the function $\theta(1 + \theta)^{-2}$. From this it can be calculated that the peak width at half height is $90.6/n$ mV. This is all shown schematically in Fig. 9.7.

9.4 Spherical electrodes

It was demonstrated in Chapter 5 (see Table 5.2) that for a potential step the expression for the current at a spherical electrode is that of a planar electrode with a spherical correction. The same is true for potential sweep.

Considering first potential sweep for a reversible system one obtains

$$I = I_{\text{planar}} - \frac{nFAD_O[O]_\infty \phi(\sigma t)}{r_0} \tag{9.31}$$

for a reduction, in which r_0 is the electrode radius and $\phi(\sigma t)$ a current function different from $\chi(\sigma t)$. The peak current is

$$I_{p,c} = I_{p,\text{planar}} - 0.725 \times 10^5 \frac{nAD_O[O]_\infty}{r_0} \tag{9.32}$$

where r_0 is in cm and the other units are as in (9.11) for a planar electrode. Since the spherical correction does not depend on scan rate we can consider spherical electrodes as if they were planar, plus a spherical correction. The current function for the spherical correction, $\phi(\sigma t)$, is shown in Table 9.1; as can be seen it follows a sigmoidal I–E profile.

For irreversible systems the corresponding expressions are

$$I_p = I_{p,\text{planar}} - \frac{nFAD_O[O]_\infty \phi(bt)}{r_0} \tag{9.33}$$

$$I_{p,c} = I_{p,\text{planar}} - 0.670 \times 10^5 \frac{nAD_O[O]_\infty}{r_0} \tag{9.34}$$

where r_0 is in cm and the other units are as in (9.22) for a planar electrode, the spherical correction being, once more, independent of scan rate. Values of the current function for the spherical correction, $\phi(bt)$, are given in Table 9.2. As for reversible systems, the spherical correction by itself corresponds to a sigmoidal I–E profile, though of lower slope.

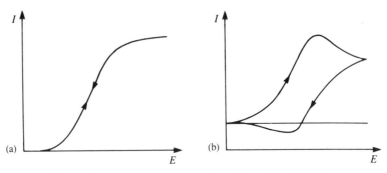

Fig. 9.8. Cyclic voltammograms at a microelectrode. (a) Low scan rate ($\sim$0.1 V s^{-1}); (b) High scan rate ($>$10 V s^{-1}). Note the similarity with hydrodynamic electrodes.

9.5 Microelectrodes

The particular advantages of microelectrodes were discussed in Section 5.5. The current density at a microelectrode is larger than that at a spherical or planar electrode of larger dimensions owing to radial and perpendicular diffusion. Mass transport is greater, and we observe differences in the experimental results obtained by the various electrochemical techniques relative to macroelectrodes.

In cyclic voltammetry we obtain the current due to perpendicular diffusion superimposed on a radial diffusion contribution, this latter being independent of scan rate. In particular, the spherical term in (9.31) and (9.33) of the previous section can easily dominate at low scan rates. Thus for small v (0.1 V s^{-1}) we observe a steady-state, scan-rate-independent, voltammogram (Fig. 9.8a); for large v ($>$10 V s^{-1}) we observe a cyclic voltammogram of the conventional type (Fig. 9.8b). Reversibility is less than that at a macroelectrode, owing to the higher mass transport: this implies that higher rate constants for electron transfer or coupled homogeneous reactions can be determined.

High-speed cyclic voltammetry[10] can easily be done at microelectrodes, increasing the range of rate constants accessible by the technique. This is because of the reduced capacitive contribution at microelectrodes— sweep rates of up to 10^6 V s^{-1} at microelectrodes have been reported. Nevertheless capacitive current subtraction is essential at these rates and instrumental artefacts can appear, which must be taken into account.

9.6 Systems containing more than one component

In solutions containing more than one electroactive species, various voltammetric waves appear. The same can happen if there is a second

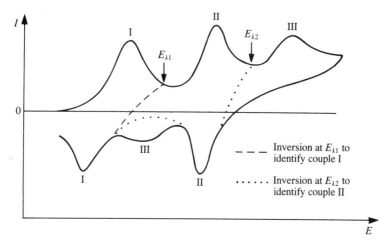

Fig. 9.9. Cyclic voltammetry in the investigation of systems of more than one component, showing the importance of the inversion potential in the identification of the peaks on the inverse scan.

step (or indeed more steps) in the electrode reaction: one or two waves will appear depending on whether the second step is easier or more difficult than the first. The formation of various species in the vicinity of the electrode permits their posterior inverse reaction on inverting the sweep direction. By choosing different inversion potentials, after the first wave and before the second, it is possible to see which waves appear on the inverse scan and which correspond to the initial sweep (Fig. 9.9). This procedure permits the identification of the species present in solution and deductions with respect to the mechanism[3].

9.7   Systems involving coupled homogeneous reactions

Cyclic voltammetry is a powerful tool for investigating electrode processes involving coupled homogeneous reactions. We exemplify with the EC mechanism:

$$\text{electrode} \quad A_3 \pm ne^- \rightarrow A_1$$

$$K = \frac{k_{-1}}{k_1}$$

$$\text{solution} \quad A_1 \underset{k_1}{\overset{k_{-1}}{\rightleftharpoons}} A_2$$

presented in Section 6.9, and consider the electrode reaction reversible. If the homogeneous step is very fast there is no current peak on inverting the sweep direction. On the other hand, increasing the sweep rate, v,

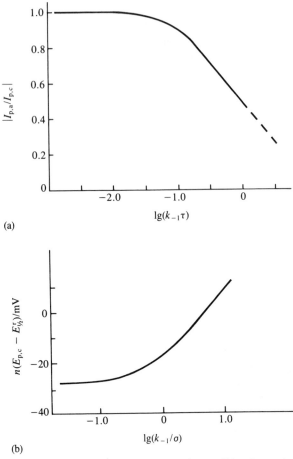

Fig. 9.10. EC mechanism in cyclic voltammetry (reversible electrode reaction and irreversible homogeneous reaction). (a) Plot of the ratio $|I_{p,a}/I_{p,c}|$ with $\lg(k_{-1}\tau)$ where $v\tau = |E_{1/2} - E_\lambda|$ (from Ref. 6 with permission). (b) Variation of cathodic peak potential, $E_{p,c}$, as a function of $\lg(k_{-1}/\sigma)$.

reduces the time between the appearance of the peak in the initial sweep and where it would appear on the inverse scan, giving less time for the homogeneous reaction to occur—the reverse peak therefore increases in size with increasing v until all the species produced initially react. The situation is represented schematically in Fig. 9.10a for a reversible electrode reaction and with $k = k_{-1}$, $k_1 = 0$ (irreversible homogeneous reaction). Analysis of the results and fitting to working curves allows k_{-1} to be determined.

An alternative strategy consists in analysing the variation of peak potential, $E_{p,c}$, as a function of scan rate (Fig. 9.10b). The horizontal region corresponds to negligible effect from the following reaction (high scan rate or low homogeneous reaction rate constant, k_{-1}).

Analysis of other mechanisms with coupled homogeneous reactions leads to plots of the same kind and which can be used for the determination of the relevant kinetic parameters. Details are given in the literature, for example Refs. 1, 2, and 4.

9.8 Convolution linear sweep voltammetry

Almost all the analysis of cyclic and linear sweep voltammograms has been done through peak currents and peak potentials. Unless digital simulation and curve-fitting by parameter adjustment is carried out, all the information contained in the rest of the wave is ignored; this brings problems of accuracy and precision. Besides this, a kinetic model has to be proposed before the results can be analysed.

An answer to this lies in the transformation of the linear sweep response into a form which is readily analysable, i.e. the form of a steady-state voltammetric wave. Two independent methods of achieving this goal have been described: the convolution technique by Saveant and co-workers[11,12], and semi-integration by Oldham[13]. In this section we describe the convolution technique, and demonstrate the equivalence of the two approaches at the end.

Convolution involves calculating integrals of the type

$$I = \left[\int_0^t I(\eta)(t-\eta)^{-1/2} \, d\eta \right] \Big/ \pi^{1/2} \tag{9.35}$$

and defining a current function

$$\psi = -I\theta^{1/2}/nFA[O]_\infty D_O^{1/2} \tag{9.36}$$

where θ is a fixed value of time. Using a dimensionless potential variable,

$$\zeta = (nF/RT)(E - E_{1/2}) \tag{9.37}$$

solution of the diffusion equation, for a reversible reduction is

$$I(\psi) = (1 + e^\zeta)^{-1} \tag{9.38}$$

that can be expressed in the form

$$E = E_{1/2} + \frac{RT}{nF} \ln \frac{I_1 - I}{I} \tag{9.39}$$

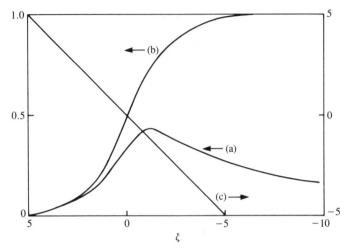

Fig. 9.11. Plots from convolution cyclic voltammetry for a reduction. (a) The current function ψ; (b) The convoluted current $I(\psi)$; (c) The logarithmic function in (9.39). All are plotted vs. ζ, the dimensionless potential, where $\zeta = (nF/RT)(E - E_{1/2})$ (from Ref. 11 with permission).

where

$$I = nFA[\mathrm{O}]_\infty D_{\mathrm{O}}^{1/2} I(\psi) \tag{9.40}$$

and I_1 is the value of I when $E \to -\infty$.

It can be seen that (9.37) indeed has the same form as the equations for voltammetric curves in the steady state (Sections 6.3–6.5). So, a plot of E vs. $\lg{(I_1 - I)/I}$ has a slope of 0.0592 V at 298 K and is a straight line if the reaction is reversible. Figure 9.11 shows the variation of I and $\lg{((I_1 - I)/I)}$ with dimensionless potential, ζ.

Electrode reactions with slower kinetics and with coupled homogeneous reactions have been considered by Saveant *et al.*[14]. A logarithmic equation of the type of (9.39) is obtained in all cases.

Semi-integration is defined by

$$\frac{\mathrm{d}^{-1/2}}{\mathrm{d}t^{-1/2}} I(t) = m(t) \tag{9.41}$$

and the theoretical results are equal to those obtained by convolution[13]. The equivalence is due to the fact that (9.35) of the convoluted integral contains a term with exponent 0.5.

If a convoluted voltammogram is to be utilized effectively and efficiently, much data transformation is required. It is necessary to use a computer, preferably linked directly to the experiment.

9.9 Linear potential sweep with hydrodynamic electrodes

Linear potential sweep at a hydrodynamic electrode can lead to two extreme situations:

• *Small* v: convection contributes much more than the sweep, no current peak appearing

• *Large* v: the sweep dominates, convection not affecting the electrode response, except to assure reproducibility. There are current peaks and the equations for stationary electrodes are applicable in the region of the peak potential.

Between the two extremes the current peak begins to appear. Figure 9.12 shows schematically the form of the voltammetric waves for different values of v.

The forms of the cyclic voltammograms have been deduced theoretically for the rotating disc[15,16], tubular[17], and wall-jet disc[18] electrodes, normally by numerical calculation. There is good agreement between theory and experiment.

Advantages of using hydrodynamic electrodes in linear sweep voltammetry are: weak dependence on the physical properties of the electrolyte, suppression of natural convection, and the possibility of obtaining values of I_p and I_L in only one experiment.

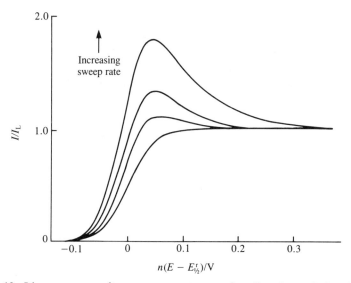

Fig. 9.12. Linear sweep voltammograms at a rotating-disc electrode for different sweep rates and for the same rotation speed—reversible reaction (from Ref. 15 with permission).

9.10   Linear potential sweep in thin-layer cells

At a hydrodynamic electrode forced convection increases the transport of species to the electrode. However, the fraction of species converted is low. For example, for a 1 mM aqueous solution of volume 100 cm³ and a rotating disc of area 0.5 cm² rotating at $W = 4\,Hz$, the quantity electrolysed in 15 minutes is approximately 1 per cent.

If the cell size is diminished until its thickness is less than that of the diffusion layer, then electrolysis is rapid and almost 100 percent. Other advantages are the necessity of only a small quantity of solution and the possibility of making simultaneous optical observations by using semi-transparent electrodes. These thin-layer cells have thicknesses ranging from 2 to 100 μm. Figure 9.13 shows two possible constructions. In the

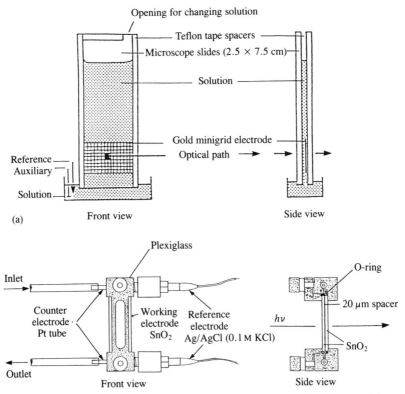

Fig. 9.13. Construction of two optically transparent thin-layer cells. (a) With minigrid electrode (from Ref. 22 with permission); (b) With semi-transparent tin dioxide electrode, and usable in a flow system.

case of minigrid electrodes, the mesh must be sufficiently fine to approximate a planar electrode in order to apply the equations which will be derived below. The auxiliary and reference electrodes have to be placed outside the thin-layer zone—this can lead to a non-uniform current distribution at the working electrode, which has somewhat limited the application of this type of cell to kinetic studies.

Thin-layer cells have been used with linear sweep, in coulometry and in chronopotentiometry. Here we limit the discussion to linear sweep—descriptions of this and other techniques can be found in Refs. 19–21.

The fact that there is total electrolysis is one of the conditions in the solution of the equation for a reduction

$$I = -nFV\frac{d[O]_*}{dt} \tag{9.42}$$

for a simple charge transfer, where V is the cell volume, and we assume that the concentrations of O and R are uniform throughout the cell. If only O is present initially in solution (concentration $[O]_i$), then at time t

$$[O]_t + [R]_t = [O]_i \tag{9.43}$$

Substituting this relation in the Nernst equation for a reversible reaction

$$E = E^{\ominus\prime} + \frac{RT}{nF}\ln\frac{[O]_*}{[R]_*} \tag{9.44}$$

we get

$$[O]_t = [O]_i\left\{1 - \left[1 + \exp\left(\frac{nF}{RT}(E - E^{\ominus\prime})\right)\right]^{-1}\right\} \tag{9.45}$$

$$= [O]_i\{1 - [1 + \theta]^{-1}\} \tag{9.46}$$

where

$$\theta = \exp\left[\frac{nF}{RT}(E - E^{\ominus\prime})\right] \tag{9.28}$$

Calculating the differential $\partial[O]_*/\partial t$ from (9.46) and substituting in (9.42), remembering that $v = -(dE/dt)$ we obtain

$$I_c = -n^2F^2vV[O]_i\frac{\theta}{(1 + \theta)^2} \tag{9.47}$$

$$= -nF\sigma V[O]_iZ(\sigma t) \tag{9.48}$$

where $\sigma = (nF/RT)v$, and $Z(\sigma t) = \theta/(1 + \theta)^2$ is the current function. It is easy to show by differentiating $Z(\sigma t)$ that the current maximum occurs for $\theta = 1$. This result means that $E_p = E^{\ominus\prime}$ and that

$$I_{p,c} = -\frac{n^2F^2vV[O]_i}{4RT} = -\frac{nF\sigma V[O]_i}{4} \tag{9.49}$$

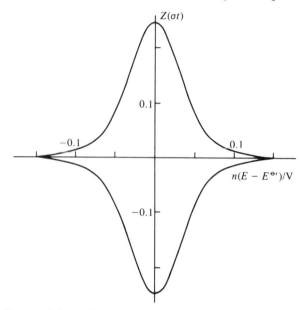

Fig. 9.14. Shape of the cyclic voltammogram obtained in a thin-layer cell for a reversible system, from (9.48).

Identical considerations are valid for an oxidation. The voltammetric curve in Fig. 9.14 shows the results obtained. Three important points should be noted in contrast with reversible reactions in normal cells:

- peak current is proportional to v, and not to $v^{1/2}$
- there is no separation between anodic and cathodic peaks
- the curve is totally symmetric round E_p

Figure 9.14 should be compared with Fig. 9.5 for adsorbed species, and (9.47) and (9.49) with (9.27) and (9.29). The form is exactly the same in the two cases and is due to the non-existence of limitations to the electrode reaction imposed by mass transfer.

For electrode reactions with slower kinetics, the first two points above are no longer valid (Fig. 9.15); however, I_p is still proportional to v. The expressions for E_p and I_p in an irreversible reduction are

$$E_{p,c} = E^{\ominus\prime} + \frac{RT}{\alpha_c n' F} \ln\left(\frac{ARTk_0}{\alpha_c n' FvV}\right) \tag{9.50}$$

$$I_{p,c} = -\frac{n\alpha_c n' F^2 vV[O]_i}{eRT} \tag{9.51}$$

Finally, we note that the mercury thin-film electrode has many of the

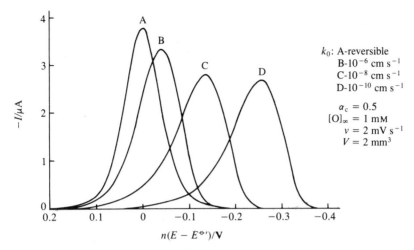

k_0: A-reversible
  B-10^{-6} cm s^{-1}
  C-10^{-8} cm s^{-1}
  D-10^{-10} cm s^{-1}

  $\alpha_c = 0.5$
  $[O]_\infty = 1$ mM
  $v = 2$ mV s^{-1}
  $V = 2$ mm^3

Fig. 9.15. Voltammograms for irreversible reactions in a thin-layer cell, for various rate constant values (from Ref. 23 with permission).

characteristics of a thin-layer cell. Anodic stripping voltammetry of metals previously deposited and present within the thin film give potential-current curves for linear sweep dissolution very similar to those of Fig. 9.15 (Section 14.4).

References

1. D. D. Macdonald, *Transient techniques in electrochemistry*, Plenum, New York, 1977, Chapter 6.
2. V. D. Parker, *Comprehensive chemical kinetics*, Elsevier, Amsterdam, Vol. 26, 1986, ed. C. H. Bamford and R. G. Compton, Chapter 2.
3. P. T. Kissinger and W. R. Heinemann, *J. Chem. Ed.*, 1983, **60,** 702.
4. M. Noel and K. I. Vasu, *Cyclic voltammetry and the frontiers of electrochemistry*, Aspect Publications, London, 1990.
5. J. E. B. Randles, *Trans. Faraday Soc.*, 1948, **44,** 327; A. Sevcik, *Collect. Czech. Chem. Commun.*, 1948, **13,** 349.
6. R. S. Nicholson and I. Shain, *Anal. Chem.*, 1964, **36,** 706.
7. H. Matsuda and Y. Ayabe, *Z. Elektrochem.*, 1955, **59,** 494.
8. R. S. Nicholson, *Anal. Chem.*, 1965, **37,** 1351; S. P. Perone, *Anal. Chem.*, 1966, **38,** 1158.
9. E. Laviron, *Electroanalytical chemistry*, ed. A. J. Bard, Dekker, New York, Vol. 12, 1982, pp. 53–157.
10. R. M. Wightman and D. O. Wipf, *Acc. Chem. Res.*, 1990, **23,** 64.
11. J. C. Imbeaux and J. M. Saveant, *J. Electroanal. Chem.*, 1973, **44,** 169.
12. L. Nadjo and J. M. Saveant, *J. Electroanal. Chem.*, 1973, **48,** 113.
13. K. B. Oldham, *Anal. Chem.*, 1972, **44,** 196; 1973, **45,** 39.

14. Ref. 1 p. 227.
15. M. Lovric and J. Osteryoung, *J. Electroanal. Chem.*, 1986, **197**, 63.
16. P. C. Andricacos and H. Y. Cheh, *J. Electrochem. Soc.*, 1980, **127**, 2153, 2385, *J. Electroanal. Chem.*, 1981, **124**, 95; G. C. Quintana, P. C. Andricacos, and H. Y. Cheh, *J. Electroanal. Chem.*, 1983, **144**, 77; 1985, **182**, 259.
17. J. Dutt and T. Singh, *J. Electroanal. Chem.*, 1985, **190**, 65; 1985, **196**, 35; 1986, **207**, 41; R. G. Compton and P. R. Unwin, *J. Electroanal. Chem.*, 1986, **206**, 57.
18. R. G. Compton, A. C. Fisher, M. H. Latham, C. M. A. Brett, and A. M. Oliveira Brett, *J. Phys. Chem.*, 1992, **96**, 8363.
19. A. T. Hubbard and F. C. Anson, *Electroanalytical Chemistry*, ed. A. J. Bard, Dekker, New York, Vol. 4, 1970, 129–214.
20. A. T. Hubbard, *CRC Crit. Rev. Anal. Chem.*, 1973, **3**, 201.
21. F. E. Woodward and C. N. Reilley, *Comprehensive treatise of electrochemistry*, Plenum, New York, Vol. 9, 1984, ed. E. Yeager, J. O'M. Bockris, B. E. Conway, and S. Sarangapani, pp. 353–392.
22. T. P. DeAngelis and W. R. Heinemann, *J. Chem. Ed.*, 1976, **53**, 594.
23. A. T. Hubbard, *J. Electroanal. Chem.*, 1969, **22**, 165.

10

STEP AND PULSE TECHNIQUES

10.1 Introduction

A step in applied potential or current[1-3] represents an instantaneous alteration to the electrochemical system. Analysis of the evolution of the system after this perturbation permits deductions about electrode reactions and their rates to be made. The equivalent in homogeneous kinetics would be a temperature or pressure jump. Potential and current step give complementary information because, whereas in the first case the potential change causes a brief capacitive current peak, in the second case a part of the applied current, the value of which probably varies with time, is always used to charge the double layer as the potential changes. Another important point is the effect of natural convection at macroscopic electrodes that begins to be felt from 20 s to 300 s after starting the experiment, depending on the care taken with the experimental arrangement.

The equations for potential and current steps in reversible systems, neglecting capacitive contributions were derived in Chapter 5. In the present chapter we show the possibilities of using these methods to elucidate electrode processes. We also consider successions of steps, that is pulses, especially with sampling of the response, which in the case of potential control has wide analytical application.

10.2 Potential step: chronoamperometry

The study of the variation of the current response with time under potentiostatic control is *chronoamperometry*. In Section 5.4 the current resulting from a potential step from a value of the potential where there is no electrode reaction to one corresponding to the mass-transport-limited current was calculated for the simple system $O + ne^- \rightarrow R$, where only O or only R is initially present. This current is the *faradaic current*, I_f, since it is due only to a faradaic electrode process (only electron transfer). For a planar electrode it is expressed by the Cottrell equation[4]

$$I_f(t) = \frac{nFAD^{1/2}c_\infty}{(\pi t)^{1/2}} \qquad (10.1)$$

However, when the potential is changed, the double layer has to be charged, giving rise to a capacitive current, I_C. The resulting I–t curve (*chronoamperogram*) is shown schematically in Fig. 10.1. Using potentiostats with good-quality components, I_C decays to zero in less than 50 μs, and so it can be neglected for longer times.

When a rapid electrode process is being studied and 50 μs is too long a timescale, the use of a microelectrode is recommended for the following reason. For a step in potential, ΔE, applied to an RC series element we obtain

$$I_C = \frac{\Delta E}{R} \exp\left(-t/RC\right) \qquad (10.2)$$

In an electrochemical cell, R is the solution resistance R_Ω, independent of electrode area, and C is the double layer capacity, C_d, directly

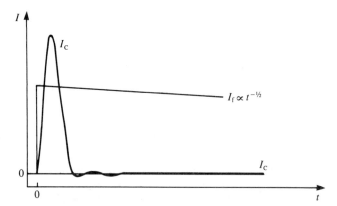

Fig. 10.1. Evolution of current with time on applying a potential step at a stationary electrode. I_f is the faradaic current and I_C the capacitive current.

dependent on electrode area. Thus I_C is proportional exponentially to the electrode area. Since, as shown in (10.1), I_f is proportional to electrode area, the ratio I_f/I_C increases with decreasing electrode area. For an electrode of area $4\,\mu\text{m}^2$ and taking extreme values of $R_\Omega = 10\,\text{k}\Omega$ and $C_d = 100\,\mu\text{F cm}^{-2}$, the double layer is 99 per cent charged in $3\,\mu\text{s}$.

For this reason, in the rest of this section we consider only the faradaic current.

We now need to calculate the current when the potential step is not sufficient to attain the limiting current. This implies considering oxidation and reduction simultaneously. For a planar electrode the diffusion equations to resolve, with only O present initially in solution, are

$$\frac{\partial [\text{O}]}{\partial t} = D_\text{O} \frac{\partial^2 [\text{O}]}{\partial x^2} \tag{10.3a}$$

$$\frac{\partial [\text{R}]}{\partial t} = D_\text{R} \frac{\partial^2 [\text{R}]}{\partial x^2} \tag{10.3b}$$

with the boundary conditions

$$t = 0 \qquad x \geqslant 0 \qquad [\text{O}]_* = [\text{O}]_\infty \qquad [\text{R}]_* = 0 \tag{10.4a}$$

$$t > 0 \qquad x \to \infty \qquad [\text{O}] \to [\text{O}]_\infty \qquad [\text{R}] \to 0 \tag{10.4b}$$

$$t > 0 \qquad x = 0 \qquad D_\text{O}\left(\frac{\partial [\text{O}]}{\partial x}\right)_0 + D_\text{R}\left(\frac{\partial [\text{R}]}{\partial x}\right)_0 = 0 \tag{10.4c}$$

$$t > 0 \qquad x = 0 \qquad I = -nFAD_\text{O}\left(\frac{\partial [\text{O}]}{\partial x}\right)_0 \tag{10.4d}$$

We use the following parameters

$$\gamma_\text{O} = [\text{O}] - [\text{O}]_\infty, \qquad \gamma_\text{R} = [\text{R}] - [\text{R}]_\infty \tag{10.5}$$

and so for $t = 0$, γ_O and $\gamma_\text{R} = 0$.

The solution of (10.2) and (10.3), after applying the Laplace transform with respect to t in a similar way as in Section 5.4, gives

$$\bar{\gamma}_\text{O} = A'(s) \exp\left[-(s/D_\text{O})^{1/2}x\right] \tag{10.6}$$

$$\bar{\gamma}_\text{R} = A''(s) \exp\left[-(s/D_\text{R})^{1/2}x\right] \tag{10.7}$$

Differentiating and considering boundary condition (10.4c) we obtain

$$\left(\frac{D_\text{O}}{s}\right)^{1/2} A'(s) + \left(\frac{D_\text{R}}{s}\right)^{1/2} A''(s) = 0 \tag{10.8}$$

that is

$$A''(s) = -pA'(s) \tag{10.9}$$

where

$$p = \left(\frac{D_O}{D_R}\right)^{1/2} \tag{10.10}$$

Equation (10.7) is transformed into

$$\bar{\gamma}_R = -pA'(s) \exp\left[-(s/D_R)^{1/2}x\right] \tag{10.11}$$

It is also necessary to apply a boundary condition corresponding to the regime of the electrode kinetics. We now consider the conditions for the various situations.

Reversible system

The boundary condition to introduce is the *Nernstian condition* at the electrode surface:

$$\theta = \frac{[O]_*}{[R]_*} = \exp\left[\frac{nF}{RT}(E - E^{\ominus\prime})\right] \tag{10.12}$$

This expression can be transformed to

$$\bar{\gamma}_{O,0} + \frac{[O]_\infty}{s} = \theta\bar{\gamma}_{R,0} \tag{10.13}$$

and combined with (10.6) and (10.11)

$$A'(s) = -\frac{[O]_\infty}{s(1+p\theta)} \tag{10.14}$$

Thus, substituting in (10.6) and differentiating

$$\left(\frac{\partial\bar{\gamma}_0}{\partial x}\right)_0 = (sD_O)^{-1/2}\frac{[O]_\infty}{(1+p\theta)} \tag{10.15}$$

On inverting the transform and removing the dimensionless variables

$$\left(\frac{\partial[O]}{\partial x}\right)_0 = \frac{[O]_\infty}{(1+p\theta)(\pi tD_O)^{1/2}} \tag{10.16}$$

and finally

$$I = -\frac{nFAD_O^{1/2}[O]_\infty}{(1+p\theta)(\pi t)^{1/2}} \tag{10.17}$$

As would be expected, when $E \ll E^{\ominus\prime}$ ($\theta \to 0$) we obtain the Cottrell equation.

Table 10.1. Currents obtained by application of a potential step to the system $O + ne^- \rightarrow R$ with only O initially present in solution

		Current (I)
Reversible	Plane	$-\dfrac{nFAD[O]_\infty}{(1+\theta)(\pi Dt)^{1/2}}$
	Sphere	$-\dfrac{nFAD[O]_\infty}{(1+\theta)}\left[\dfrac{1}{(\pi Dt)^{1/2}} + \dfrac{1}{r_0}\right]$
Irreversible	Plane	$-nFAk_c[O]_\infty \exp[k_c^2 t/D]\,\mathrm{erfc}\left(\dfrac{k_c t^{1/2}}{D^{1/2}}\right)$
	Sphere	$-nFAk_c[O]_\infty\left[\left(1 + \dfrac{D}{k_c r_0}\right)\exp\dfrac{k_c^2 t}{D}\,\mathrm{erfc}\left(\dfrac{k_c t^{1/2}}{D^{1/2}}\right) - \dfrac{D}{k_c r_0}\right]$
Irreversible (small t)	Plane and sphere	$-nFAk_c[O]_\infty\left(1 - \dfrac{2k_c t^{1/2}}{(\pi D)^{1/2}}\right)$
		$\theta = [O]_*/[R]_*$

Equation (10.17) can be written in the form

$$I(\tau) = \frac{I_{\text{Cottrell}}(\tau)}{1 + p\theta} \tag{10.18}$$

where τ represents a fixed time after application of the potential step. Taking into account the meaning of θ and p, (10.18) can be rearranged as

$$E = E_{1/2}^r + \frac{RT}{nF}\ln\frac{I_{\text{Cot}}(\tau) - I(\tau)}{I(\tau)} \tag{10.19}$$

where

$$E_{1/2}^r = E^{\ominus\prime} + \frac{RT}{nF}\ln\left(\frac{D_R}{D_O}\right)^{1/2} \tag{10.20}$$

An analogous mathematical solution can be carried out for spherical electrodes: the current is that for a planar electrode plus a spherical correction term (Table 10.1).

Quasi-reversible and irreversible systems

Instead of the Nernstian boundary condition we have to introduce

$$D_O\left(\frac{\partial[O]}{\partial x}\right)_0 = k_c[O]_* - k_a[R]_* \tag{10.21}$$

which can be transformed to

$$D_O\left(\frac{\partial\bar{\gamma}_O}{\partial x}\right)_0 = k_c\left(\bar{\gamma}_O + \frac{[O]_\infty}{s}\right) - k_a\bar{\gamma}_R \tag{10.22}$$

By differentiation of (10.6) we know that

$$\left(\frac{\partial \bar{\gamma}_O}{\partial x}\right)_0 = -A'(s)\left(\frac{s}{D_O}\right)^{1/2} \qquad (10.23)$$

Introducing (10.22) and (10.23) into (10.6) and (10.11) we obtain finally

$$A'(s) = -\frac{k_c}{D_O^{1/2}}\frac{[O]_\infty}{s(H + s^{1/2})} \qquad (10.24)$$

where

$$H = \frac{k_c}{D_O^{1/2}} + \frac{k_a}{D_R^{1/2}} \qquad (10.25)$$

Therefore

$$\bar{I}(s) = -\frac{nFAk_c[O]_\infty}{s^{1/2}(H + s^{1/2})} \qquad (10.26)$$

Inversion of the transform gives

$$I(t) = -nFAk_c[O]_\infty \exp(H^2 t)\,\mathrm{erfc}(Ht^{1/2}) \qquad (10.27)$$

If both O and R are present initially,

$$I(t) = nFA(k_a[R]_\infty - k_c[O]_\infty)\exp(H^2 t)\,\mathrm{erfc}(Ht^{1/2}) \qquad (10.28)$$

In the case of a totally irreversible reaction (consider a reduction), the expression is simplified, since

$$H = \frac{k_c}{D_O^{1/2}} \qquad (10.29)$$

and so

$$I(t) = -nFAk_c[O]_\infty \exp[k_c^2/D_O t]\,\mathrm{erfc}(k_c t^{1/2}/D_O^{1/2}) \qquad (10.30)$$

These expressions are quite complex and can be found, together with the corresponding expressions for spherical electrodes, in Table 10.1.

A very useful simplification in the calculation of kinetic parameters is the application of a small step at the foot of the voltammetric wave. Since $Ht^{1/2}$ is small we can linearize according to

$$\exp(x^2)\,\mathrm{erfc}(x) \approx 1 - \frac{2x}{\pi^{1/2}} \qquad (10.31)$$

Equation (10.27), for example, is simplified to

$$I(t) = -nFAk_c[O]_\infty\left(1 - \frac{2Ht^{1/2}}{\pi^{1/2}}\right) \qquad (10.32)$$

The value of k_c is obtained from the intercept of the plot of $I(t)$ vs. $t^{1/2}$.

More complex mechanisms

The current response to a potential step always reflects the mechanism of the electrode reaction and, in principle, can be used to distinguish mechanisms involving coupled homogeneous reactions, etc. However, sometimes the modification to the response is so small that it is impossible to differentiate it with confidence from experimental error. Recently the use of hydrodynamic electrodes, especially the rotating disc electrode, has been investigated with this aim in mind: higher transport permits easier distinction between mechanisms, besides annulling the effects of natural convection for large t. This technique has had some success[5].

10.3   Double potential step

The potential is altered between two values, perhaps repeatedly (Fig. 10.2). The second step inverts the electrode reaction. We consider an initial step from a potential where there is no electrode reaction to a value corresponding to the limiting reduction current (only O initially present in solution); at $t = \tau$ the potential reverts to its initial value and there is oxidation of R that was produced. The equations for a planar

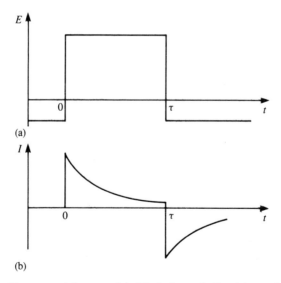

Fig. 10.2. Double potential step. (a) Variation of E with t; (b) Schematic variation of I with t.

electrode are

$$0 < t < \tau \qquad I = -nFAD_O^{1/2}[O]_\infty/(\pi t)^{1/2} \qquad (10.33)$$

$$t > \tau \qquad I = nFAD_O^{1/2}[O]_\infty\{[\pi(t-\tau)]^{-1/2} - (\pi t)^{-1/2}\} \quad (10.34)$$

Expression (10.34) shows that a conventional Cottrell response for an oxidation is obtained, superimposed on the continuation of the reduction reaction profile. Expressions for kinetic control in one of the two steps and in both steps have been derived[6].

There are various applications. If, for example, the product of the initial reaction is consumed in solution by homogeneous reaction, analysis of the reoxidation current will show its extent, and perhaps its kinetics. In the case of O being reduced to R and also to other species, the reoxidation of R (the potential would have to be very carefully chosen) gives information about the couple O | R. As a final example, if R is unstable but with a lifetime significantly greater than τ, then the generation of R *in situ* and the study of its reoxidation can lead to the calculation of the rate of decay of R.

In this last case the use of a double hydrodynamic electrode, generating R on the upstream electrode and detecting it on the downstream electrode, may be easier and more sensitive. The rotating disc electrode has also been used with success to distinguish similar mechanisms with coupled homogeneous reactions (ECE, DISP1, and DISP2)[5].

10.4   Chronocoulometry

Instead of studying the variation of current with time, we can integrate the current and study the variation of charge with time: this is *chronocoulometry*[7]. Advantages are:

● The signal usually increases with time, facilitating measurements towards the end of the transient, when the current is almost zero

● Integration is effective in reducing signal noise

● It is relatively easy to separate the capacitive charge, Q_C, from the faradaic charge, Q_f.

For a large potential step at a planar electrode and considering a reduction, we use the Cottrell equation, and by integration arrive at

$$Q_f = -\frac{2nFAD_O^{1/2}[O]_\infty t^{1/2}}{\pi^{1/2}} \qquad (10.35)$$

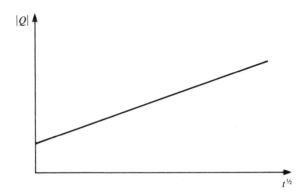

Fig. 10.3. Chronocoulometric response to a potential step.

Therefore, a plot of Q_f vs. $t^{1/2}$ is linear with zero intercept. In practice, intercepts are non-zero and correspond to a capacitive charge, Q_C, and even to reduction of adsorbed species (Fig. 10.3).

The double potential step is very powerful in identifying adsorption phenomena by chronocoulometry. From (10.34),

$$t > \tau \qquad Q = \frac{2nFAD_O^{1/2}[O]_\infty}{\pi^{1/2}}[(t-\tau)^{1/2} - t^{1/2}] \qquad (10.36)$$

Figure 10.4 illustrates how to analyse the response. It is important to remember that when $t > \tau$ there is no capacitive contribution because charge was supplied and then removed. Calling Q_R the difference between $Q(\tau)$ and $Q(t > \tau)$, the charge after $t = \tau$ is given by

$$Q_R = \frac{2nFAD_O^{1/2}[O]_\infty}{\pi^{1/2}}[t^{1/2} - \tau^{1/2} - (t-\tau)^{1/2}] \qquad (10.37)$$

$$= \frac{2nFAD_O^{1/2}[O]_\infty}{\pi^{1/2}}\theta \qquad (10.38)$$

Figure 10.4b is a plot of $Q(t < \tau)$ vs. $t^{1/2}$ and of $Q_R(t > \tau)$ vs. θ. The difference between the intercepts, $nFA\Gamma_0$, gives the amount adsorbed through the surface excess, Γ_0.

Double step chronocoulometry also gives information on the kinetics of coupled homogeneous reactions[8]. For example, any deviation, under diffusion control, from (10.35) and (10.36) implies a chemical complication, which can be compared with the responses for the various possible mechanisms.

Not much research has been done using chronocoulometry with small potential steps. Here it is necessary to integrate the expressions in Table 10.1 and consider the rate of the electrode reaction. The integrals can be

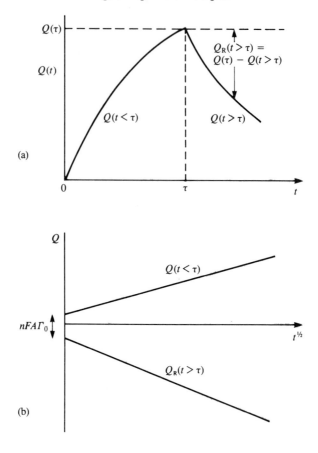

Fig. 10.4. (a) Chronocoulometric response to a double potential step; (b) Corresponding plots.

obtained directly, or in the Laplace domain before inversion through simple division by the Laplace variable. However, the possibilities for future use of the technique are many, because the signal measurement on a longer timescale is more accurate than in chronoamperometry, where measurements are more limited by current decay.

10.5 Current step: chronopotentiometry

A current step applied to an electrode provokes a change in its potential. The flux of electrons is used first to charge the double layer, and then for the faradaic reactions. The study of the variation of the potential with time is *chronopotentiometry*[2]. In Section 5.4 it was shown that, neglecting the capacitive current, the potential at a planar electrode is approximately constant until the end of the transition time, τ, which corresponds to the total consumption of the electroactive species in the neighbourhood of the electrode. The *Sand equation*[9] describes the transition time, τ, according to

$$\frac{I\tau^{1/2}}{[O]_\infty} = -\frac{nFAD_O^{1/2}\pi^{1/2}}{2} \tag{10.39}$$

for a simple system $O + ne^- \rightarrow R$ with only O initially present. The transition time is independent of the rate of charge transfer. Using hydrodynamic electrodes it is possible that $\tau \rightarrow \infty$ due to the high transport of electroactive species to the electrode.

We now consider the form of the chronopotentiogram according to the kinetics of the electrode process. We take as negligible the contribution of the capacitive current; its contribution is larger at the beginning and at the end of the chronopotentiogram owing to the greater variation of potential in those regions.

The equations for a planar electrode with only O present initially are the same as for the potential step, (10.3), as well as boundary conditions (10.4). The differences reside in the choice of the last boundary condition. To this end we need to know the concentrations of O and R on the electrode surface, $[O]_*$ and $[R]_*$ respectively.

The concentrations of O and R are given by

$$\bar{\gamma}_O = A'(s) \exp\left[-(s/D_O)^{1/2}x\right] \tag{10.6}$$

$$\bar{\gamma}_R = A''(s) \exp\left[-(s/D_R)^{1/2}x\right] \tag{10.7}$$

Differentiating, and taking account of boundary conditions (10.4c) and (10.4d), after inversion we arrive at

$$[O]_* = [O]_\infty - \frac{2It^{1/2}}{nFAD_O^{1/2}\pi^{1/2}} \tag{10.40}$$

$$[R]_* = \frac{2It^{1/2}}{nFAD_R^{1/2}\pi^{1/2}} \tag{10.41}$$

Reversible system

The final boundary condition is expressed by the Nernst equation. Substituting the Sand equation (10.39), together with (10.40) and (10.41) into the Nernst equation

$$E = E^{\ominus\prime} + \frac{RT}{nF} \ln \frac{[O]_*}{[R]_*} \tag{10.42}$$

one obtains

$$E = E_{\tau/4} + \frac{RT}{nF} \ln \frac{\tau^{1/2} - t^{1/2}}{t^{1/2}} \qquad (10.43)$$

where

$$E_{\tau/4} = E^{\ominus'} + \frac{RT}{nF} \ln \left(\frac{D_R}{D_O}\right)^{1/2} \qquad (10.44)$$

which is identifiable with $E_{1/2}^{r}$ in a conventional voltammogram. A chronopotentiometric curve is shown in Fig. 10.5a. $E_{\tau/4}$ is the quarter-wave potential, so called because $E = E^{\ominus'}$ when $t = \tau/4$, for $D_O = D_R$.

From expression (10.43) we see that a plot of E vs. $\lg [(\tau^{1/2} t^{1/2})/t^{1/2}]$ gives a straight line of slope $59.2/n$ mV at 298 K (Fig. 10.5b)—this is a

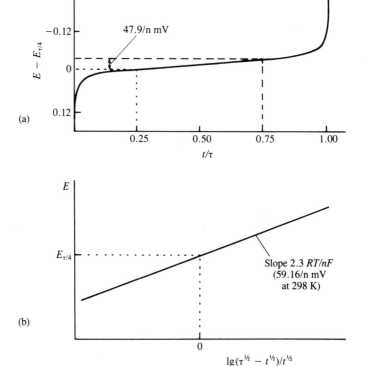

Fig. 10.5. Chronopotentiometry in a reversible system for $O + ne^- \rightarrow R$ (only O present in bulk solution). (a) Chronopotentiogram; (b) Plot of E vs. $\lg [(\tau^{1/2} - t^{1/2})/t^{1/2}]$.

diagnostic of a reversible system. On the other hand, $|E_{3\tau/4} - E_{\tau/4}| = 47.9/n$ mV at 298 K.

For an oxidation (10.43) also applies, but with a negative sign in place of the positive sign.

Quasi-reversible and irreversible systems

The final boundary condition, for only O initially present, is

$$D_O\left(\frac{\partial[O]}{\partial x}\right)_0 = k_c[O]_* - k_a[R]_* \tag{10.45}$$

substituting for $[O]_*$ and $[R]_*$ in (10.40) and (10.41) we get

$$I = nFA\{k_a P_R t^{1/2} - k_c([O]_\infty + P_O t^{1/2})\} \tag{10.46}$$

where

$$P_R = \frac{2I}{nFA(D_R\pi)^{1/2}} \qquad P_O = \frac{2I}{nFA(D_O\pi)^{1/2}} \tag{10.47}$$

Except for potentials close to E_{eq}, where it is necessary to consider the reaction in both directions, we have

$$I = nFAk_c\{P_O t^{1/2} - [O]_\infty\} \tag{10.48}$$

Substituting in the Sand equation for I, we obtain

$$k_c = \frac{\pi^{1/2} D_O^{1/2}}{2(\tau^{1/2} - t^{1/2})} \tag{10.49}$$

This expression allows the calculation of k_0, and, since only O is present in bulk solution, is equal to the expression for an irreversible system. Using

$$k_c = k_0 \exp\left[-\alpha_c nF(E - E^{\ominus'})/RT\right] \tag{10.50}$$

the result is

$$E = E^{\ominus'} + \frac{RT}{\alpha_c n'F} \ln\left[\frac{2k_0}{(\pi D_O)^{1/2}}(\tau^{1/2} - t^{1/2})\right] \tag{10.51}$$

Therefore, for a totally irreversible wave $|E_{3\tau/4} - E_{\tau/4}| = 33.8/\alpha n'$ mV at 298 K. Figure 10.6 illustrates the variation of potential with time schematically.

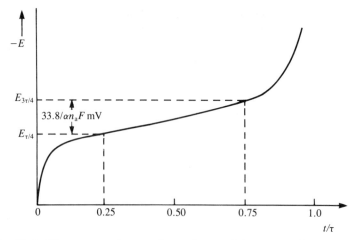

Fig. 10.6. Chronopotentiogram for an irreversible system: $O + ne^- \rightarrow R$.

10.6 Double current step

In the last section the capacitive contribution was neglected. However, for some galvanostatic experiments this procedure is not possible. Various theoretical treatments have been developed to analyse chronopotentiograms, but always with approximations that are difficult to justify, such as, for example, I_c constant!

A better alternative is to modify the experiment. The use of a double current step[10], as demonstrated in Fig. 10.7, can reduce the problem. The first step has sufficient length to charge the double layer; the current is then reduced to a lower value at which there is not capacitive component.

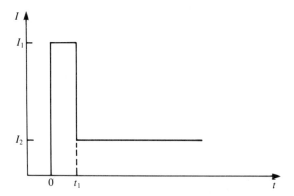

Fig. 10.7. The double current step. The first pulse is to charge the double layer.
$t_1 \approx 10^{-6}$ s.

In principle, compensation of capacitive effects can be 100 per cent. The optimal pulse duration seems to be about 1 μs.

A theoretical description of this experiment has to take into account the existence of a faradaic current during the first pulse. For small values of t, one obtains, for only O or R present, the equation

$$\eta = \frac{RT}{nF} I_2 \left\{ \frac{1}{k_0 c_\infty} + \frac{4N}{3\pi^{1/2}} t_1^{1/2} \right\} \tag{10.52}$$

where η is the overpotential. A plot of η (when $t = t_1$) vs. t_1 gives k_0 from the intercept. Knowing k_0, and that the current is purely capacitive when $t = 0$,

$$C_d = \lim_{t \to 0} \frac{nFt_1 k_0 c_\infty}{RTA} \left(\frac{I_1}{I_2} \right) \left(1 - \frac{4Nk_0 c_\infty t_1^{1/2}}{3\pi^{1/2}} \right)^{-1} \tag{10.53}$$

Double current steps, where the second step reverts to the initial value of the applied current, have been used in mechanistic studies[11].

10.7 Methods using derivatives of chronopotentiograms

If we consider reversible and irreversible systems, the derivatives dE/dt are easily obtained and are

$$\left(\frac{dE}{dt} \right)_r = \pm \frac{RT}{2nF} \frac{\tau^{1/2}}{t(\tau^{1/2} - t^{1/2})} \tag{10.54}$$

$$\left(\frac{dE}{dt} \right)_{irr} = \pm \frac{RT}{2\alpha nF} [t^{1/2}(\tau^{1/2} - t^{1/2})]^{-1} \tag{10.55}$$

respectively. The positive signs refer to oxidation and the negative signs to reduction. The minimum values of the derivatives are related to the transition times according to

$$\left(\frac{dE}{dt} \right)_{r,min} = \pm \frac{27}{8} \frac{RT}{nF\tau} \qquad t = (4\tau/9) \tag{10.56}$$

and

$$\left(\frac{dE}{dt} \right)_{irr,min} = \pm \frac{2RT}{\alpha nF\tau} \qquad t = \tau/4 \tag{10.57}$$

By measuring dE/dt we can calculate the transition time, τ. The usefulness of these expressions is that they permit the determination of τ in a region where the capacitive contribution is small (in fact at its minimum).

10.8 Coulostatic pulses

In the coulostatic pulse technique, a charge step is applied during $1\ \mu s$, the potential of the electrode being followed afterwards in open circuit[12]. Conditions are chosen such that only the double layer is charged, and even a fast reaction proceeds only to a very small extent. After the pulse the electrode relaxes to its initial state, releasing the charge. The resultant flux of electrons can be used to cause faradaic reactions. The advantage is that the solution where the measurements are made can have a high resistance so long as the measurements are done in open circuit; supporting electrolyte is not necessary.

The generic equations are

$$I_f = -I_C = -C_d \frac{d\eta}{dt} \tag{10.58}$$

and

$$\eta(t) = \eta(t = 0) + \frac{1}{C_d} \int_0^t I_f\, dt \tag{10.59}$$

We consider two special cases:

1. *Small step and without creating significant concentration gradients*: Equation (10.59) gives

$$\eta(t) = \eta(t = 0) \exp\left(\frac{-t}{\tau_C}\right) \tag{10.60}$$

where

$$\tau_C = \frac{RTC_d}{nFI_0} \tag{10.61}$$

2. *Large step, sufficient to reach the plateau of the voltammetric wave and with C_d independent of potential*:

$$\Delta E = |E(t) - E(t = 0)| = \frac{2nFAD_O^{1/2}c_\infty t^{1/2}}{\pi^{1/2}C_d} \tag{10.62}$$

from substitution of the Cottrell equation in (10.59). The plot of E vs. $t^{1/2}$ is linear, with slope proportional to concentration.

10.9 Pulse voltammetry

The potential step is the basis of pulse voltammetry[13,14]. Pulse techniques were initially developed for the dropping mercury electrode[15] (Section

8.3), the objective being to synchronize the pulses with drop growth and reduce the capacitive current contribution by current sampling at the end of drop life. After applying a pulse of potential, the capacitive current dies away faster than the faradaic current (Fig. 10.6); thus the current is measured at the end of the pulse. This type of sampling has the advantage of an increase in sensitivity and better characteristics for analytical applications. At solid electrodes there is the additional advantage of discrimination against blocking of the electrode reaction by adsorption.

We now consider some of the forms of pulse voltammetry.

Tast polarography[16]

This technique is applicable only to the dropping mercury electrode and involves current sampling immediately before drop fall, the drop lifetime being controlled mechanically to ensure good reproducibility. Although it is not strictly speaking a pulse technique, it fits well into the general scheme. Figure 10.8 shows the variation of current with time, how the

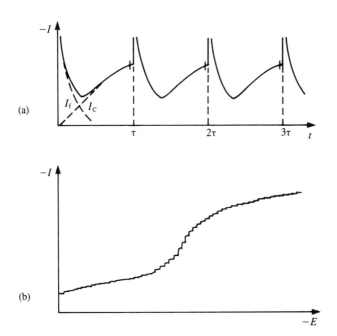

Fig. 10.8. Tast polarography showing: (a) Variation of total current ($I_c + I_f$) and sampling just before drop fall at time $t = \tau$; (b) The resulting polarogram.

sampling is done, and the polarogram, without the oscillations in current of a conventional polarogram (see Fig. 8.9). Detection limits of the order of 10^{-6} M are possible.

Normal pulse voltammetry (NPV)[15,16]

In normal pulse voltammetry a base value of potential, E_{base}, is chosen, normally where there is no faradaic reaction, and this is applied to the electrode. From this value short pulses of increasing amplitude are applied, the amplitude increment always being equal. The current is measured at the end of each pulse, the duration of which varies normally between 5 and 100 ms; the interval between pulses is 2–4 s. Figure 10.9 shows the scheme of operation and response.

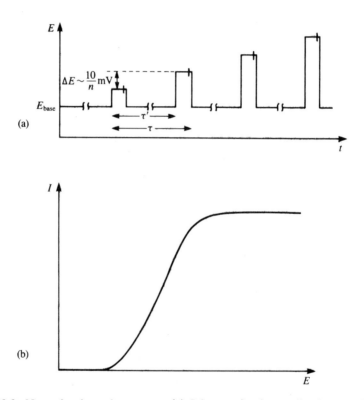

Fig. 10.9. Normal pulse voltammetry. (a) Scheme of pulse application starting at E_{base}. The current is measured at the end of the pulses and it is assumed that I at E_{base} is zero. $\tau = 2$–4 s and $(\tau - \tau')$ 5–100 ms. At the DME the end of the pulse is synchronized with drop fall; (b) Schematic I–E profile.

In relation to tast polarography, normal pulse polarography has the advantage that there is faradaic reaction only during the pulses, leading to larger currents, given that there is lower consumption of electroactive species close to the electrode. In the case of solid electrodes, problems of surface blocking by the product of the electrode reaction are reduced through the use of pulses.

The form of the response is a succession of points following the same profile as a conventional voltammogram. However, since a pulse causes greater mass transport than a steady-state technique (hydrodynamic electrode), a reaction that appears reversible in the steady state can appear quasi-reversible with this technique. On the other hand, given the short timescale, effects due to coupled homogeneous reactions may not be observed.

Differential pulse voltammetry (DPV)[15,16]

This technique is similar to NPV but with two important differences:

● The base potential is incremented between pulses, these increments being equal.

● The current is measured immediately before pulse application and at the end of the pulse: the difference between the two currents is registered.

The potential–time waveform is represented in Fig. 10.10. Pulses superimposed on a potential ramp have also been employed; for microprocessor control the staircase waveform is clearly simpler to put into operation.

Since DPV is a differential technique, the response is similar to the first derivative of a conventional voltammogram, that is a peak. The peak potential, E_p, can be approximately identified with $E_{1/2}$. With increasing irreversibility E_p moves away from $E_{1/2}$ (reversible system), at the same time as peak width increases and its height diminishes. The degree of reversibility of an electrode reaction is similar to that observed in NPV, since the timescale is the same.

Quantitative treatments for reversible systems demonstrated that, with only R (positive sign) or only O (negative sign) initially present,

$$E_{max} = E_{1/2} \pm \frac{\Delta E}{2} \tag{10.63}$$

where ΔE is the pulse amplitude. The peak current is

$$|(\delta i)_{max}| = \frac{nFAD_O^{1/2}c_\infty}{\pi^{1/2}(t - \tau')^{1/2}} \left(\frac{1 - \sigma}{1 + \sigma}\right) \tag{10.64}$$

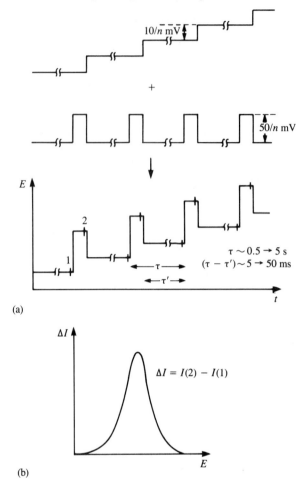

Fig. 10.10. Differential pulse voltammetry. (a) Scheme of application of potentials (sometimes superimposed on a ramp rather than a staircase); (b) Schematic
I–E profile.

τ and τ' are described in Fig. 10.10 and σ is given by

$$\sigma = \exp\left(\frac{nF}{RT}\frac{\Delta E}{2}\right) \tag{10.65}$$

The term $(1 - \sigma)/(1 + \sigma)$ describes the effect of ΔE on $(\delta i)_{\text{max}}$. This term
increases as $|\Delta E|$ increases, attaining a value of unity for very large $|\Delta E|$.
In fact the use of $|\Delta E|$ values larger than about 100 mV is not viable, as
the peak width at half height, $W_{1/2}$, also increases, leading to a loss of

resolution. At the limit, when $\Delta E \rightarrow 0$

$$W_{1/2} = 3.52 RT / nF \qquad (10.66)$$

The maximum value of the current obtained for large values of E corresponds to that obtained in NPV under identical conditions. For an electrode process without complications, DPV should be no better than NPV.

At the dropping mercury electrode the fact that DPV is better than NPV is due to the residual capacitive current contribution, which is subtracted out in the differential technique. It is relatively easy to demonstrate that the *diminution factor, f,* is given by

$$f = \left| \frac{\Delta E}{E_z - E} \right| \qquad (10.67)$$

where E_z is the potential of zero charge (Chapter 8, equation (8.9)). In general, f has a value less than 0.1. It should also be noted that with 'static electrodes' (a succession of suspended electrodes of identical, fixed area) the capacitive contribution described by equation (10.67) should also be reduced to zero.

At solid electrodes the better response from DPV than from NPV is clear in many cases, especially involving organic compounds. As these often lead to adsorption on the electrode, it is possible that a differential technique discriminates against effects that are moreless constant before and after pulse application.

Square-wave voltammetry (SWV)[17]

Square-wave voltammetry was invented in 1952 by Barker[18], but was little used at the time owing to difficulties with the controlling electronics. With advances in instrumentation it has now become an important analytical technique. The waveform is shown in Fig. 10.11, and consists of a square wave superimposed on a staircase, a full square wave cycle corresponding to the duration of one step in the staircase waveform.

Whereas NPV and DPV function with effective sweep rates of $1\text{--}10 \, \mathrm{mV \, s^{-1}}$, SWV can reach $1 \, \mathrm{V \, s^{-1}}$. There are advantages: greater speed in analysis, lower consumption of electroactive species in relation to DPV, and reduced problems with blocking of the electrode surface. Since the current is sampled in both the positive- and the negative-going pulses, peaks corresponding to the oxidation or reduction of the electroactive species at the electrode surface can be obtained in the same experiment, and by subtraction their difference. Subtraction also means that the difference current is zero for a species at a potential corresponding to the region of mass-transport limited current. In analysis, this can

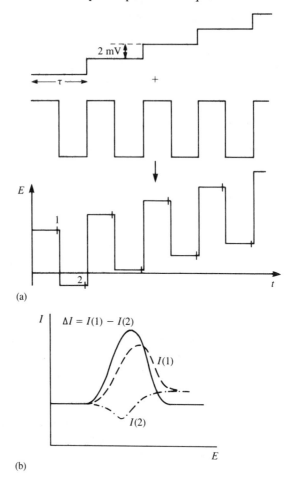

(a)

(b)

Fig. 10.11. Square wave voltammetry. (a) Scheme of application of potentials: sum of a staircase and a square wave; (b) Typical response: $E_{st} \sim 2\,\text{mV}$; $\tau_{min} = 2\,\text{ms}$. Note the similarity to DPV.

be very useful, particularly for removing the current due to reduction of dissolved oxygen.

The rapidity means that full square-wave voltammograms can be registered in quick succession, *chronovoltammograms*, an important application being in electrochemical detection of eluents from high-pressure liquid chromatography columns.

In the particular case of the static mercury drop electrode, given that the experiment can be done in 2 s, a full analysis is possible during the lifetime of one drop.

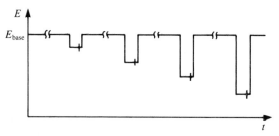

Fig. 10.12. Scheme of reverse pulse voltammetry (RPV). E_{base} corresponds to electrolysis of the electroactive species in solution.

Other pulse techniques

There are other pulse combinations, developed with particular aims besides the reduction of capacitive and adsorptive contributions. Some are described below.

Reverse Pulse Voltammetry (RPV)

This originated from a similar idea to that of the double potential step. A base potential at which all the electroactive species is electrolysed is applied, and the reverse reaction is carried out by normal pulse (Fig. 10.12). A good reason for using this technique is to diminish the problems caused by parallel electrode reactions of the initial species.

Differential Normal Pulse Voltammetry (DNPV)

This is a hybrid of DPV and NPV. Instead of applying a succession of pulses from a base potential, as in NPV, each pulse contains two pulses separated by ΔE, giving rise to a differential measurement. With the scheme in Fig. 10.13, it is possible to register the current relative to reduction or oxidation by considering pairs of alternate pulses, or, by

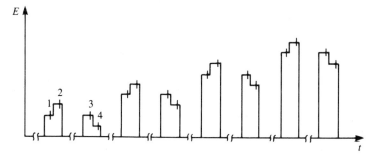

Fig. 10.13. Scheme of differential normal pulse voltammetry (DNPV). It is possible to register $(I_2 - I_1)$, $(I_4 - I_3)$, or $\{(I_2 - I_1) - (I_4 - I_3)\}$.

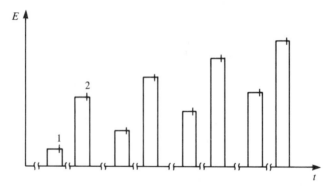

Fig. 10.14. Scheme for double differential pulse voltammetry (DDPV). $(I_2 - I_1)$ at potential $(E_1 + E_2)/2$ is registered. Pulse width ~ 50 ms.

using all pairs of pulses, the difference between them. Besides this, the reduction in the time spent at potentials where there is electrode reaction, in comparison with DPV, offers clear advantages in relation to adsorption problems.

Double Differential Pulse Voltammetry (DDPV)

The aim of this technique is precisely the same as DPV; Fig. 10.14 shows the pulse application scheme. The form of the voltammograms (peaks) was deduced theoretically[19]. As in DNPV, the great advantage is the reduction in time during which electrolysis occurs.

Applications of pulse techniques

Applications of pulse techniques in electrochemistry have been predominantly in the area of analysis, relying on the linear dependence of peak height on potential, although recently their use in mechanistic studies, particularly square-wave voltammetry, has begun to be exploited. The reason for their use in analysis is intimately linked with the low detection limits that are attainable, particularly in combination with pre-concentration techniques, as will be seen in Chapter 14. Finally, since nowadays the pulse sequences are generally controlled and responses analysed using microprocessors, the development of new waveforms for particular situations is now a much easier task than it was even a decade ago.

References

1. D. D. Macdonald, *Transient techniques in electrochemistry*, Plenum, New York, 1977, Chapter 4.

2. D. D. Macdonald, *Transient techniques in electrochemistry*, Plenum, New York, 1977, Chapter 5.
3. Z. Nagy, *Modern aspects of electrochemistry*, Plenum, New York, Vol. 21, 1990, ed. R. E. White, J. O'M. Bockris, and B. E. Conway, pp. 237–292.
4. F. G. Cottrell, *Z. Physik. Chem.*, 1902, **42,** 385.
5. R. G. Compton, M. E. Laing, R. J. Northing, and P. R. Unwin, *Proc. R. Soc. Lond.*, 1988, A**418,** 113.
6. N. M. Smit and M. D. Wijnen, *Rec. Trav. Chim.*, 1960, **79,** 5; F. Kimmerle and J. Chevalet, *J. Electroanal. Chem.*, 1969, **21,** 237.
7. F. C. Anson, *Anal. Chem.*, 1966, **38,** 54.
8. Ref. 1, pp. 93–95.
9. H. J. S. Sand, *Philos. Mag.*, 1901, **1,** 45.
10. H. Gerischer and M. Krause, *Z. Physik. Chem.*, 1957, **10,** 264; 1958, **14,** 184.
11. Ref. 2, pp. 158–176.
12. H. P. van Leeuwen, *Electroanalytical chemistry*, ed. A. J. Bard, Dekker, New York, Vol. 12, 1982, pp. 159–238.
13. P. He, J. P. Avery and L. R. Faulkner, *Anal. Chem.*, 1982, **54,** 1313A.
14. S. Borman, *Anal. Chem.*, 1982, **54,** 698A.
15. A. M. Bond, *Modern polarographic methods in analytical chemistry*, Dekker, New York, 1980.
16. J. Osteryoung and M. M. Schreiner, *CRC Crit. Rev. Anal. Chem.*, 1988, **19,** S1.
17. J. Osteryoung and J. O'Dea, *Electroanalytical chemistry*, ed. A. J. Bard, Dekker, New York, Vol. 14, 1986, pp. 209–308.
18. G. C. Barker and I. L. Jenkins, *Analyst*, 1952, **77,** 685.
19. W. J. Albery, T. W. Beck, W. N. Brooks, and M. Fillenz, *J. Electroanal. Chem.*, 1981, **125,** 205.

11

IMPEDANCE METHODS

11.1 Introduction

Electrochemical systems can be studied with methods based on impedance measurements. These methods involve the application of a small perturbation, whereas in the methods based on linear sweep or potential step the system is perturbed far from equilibrium. This small imposed perturbation can be of applied potential, of applied current or, with hydrodynamic electrodes, of convection rate. The fact that the perturbation is small brings advantages in terms of the solution of the relevant mathematical equations, since it is possible to use limiting forms of these equations, which are normally linear (e.g. the first term in the expansion of exponentials).

The response to the applied perturbation, which is generally sinusoidal, can differ in phase and amplitude from the applied signal. Measurement of the phase difference and the amplitude (i.e. the impedance) permits analysis of the electrode process in relation to contributions from diffusion, kinetics, double layer, coupled homogeneous reactions, etc. There are important applications in studies of corrosion, membranes, ionic solids, solid electrolytes, conducting polymers, and liquid/liquid interfaces.

Comparison is usually made between the electrochemical cell and an equivalent electrical circuit that contains combinations of resistances and capacitances (inductances are only important for very high perturbation frequencies, >100 kHz). The combinations normally used for faradaic reactions are considered in Section 11.3; there is a component representing transport by diffusion, a component representing kinetics (purely resistive), and another representing the double layer capacity, this for a simple electrode process. Another strategy is to choose a model for the reaction mechanism and kinetic parameters, derive the impedance expression, and compare with experiment. Given that impedance measurements at different frequencies can, in principle, furnish all the information about the electrochemical system (if we are capable of understanding all the contributions) there has been much interest in developing impedance techniques in electrochemistry[1-12].

In this chapter, after a description of specialized methods for impedance measurement, the procedure for deducing the kinetics and mechanism from the impedance spectra is demonstrated.

The principles of a.c. circuits necessary for the comprehension of some of the ideas and concepts presented here are given in Appendix 2. The impedance is the proportionality factor between potential and current; if these have different phases then we can divide the impedance into a resistive part, R, where the voltage and current are in phase, and a reactive part, $X_C = 1/\omega C$, where the phase difference between current and voltage is 90°. As shown in Appendix 2, it is often easier for posterior calculation and analysis to display the impedance vectorially in complex-plane diagrams.

11.2 Detection and measurement of impedance

There are three types of technique for the detection and measurement of impedance[6,13].

Alternating current bridge

These bridges, as shown in Fig. 11.1, use the principle of balance between the electrochemical cell under study and a variable impedance, Z_s, which in research on electrode processes normally consists of a resistance R_s in series with a capacitance C_s. Given that

$$Z_s = R_s - i/\omega C_s \tag{11.1}$$

and that

$$Z_{cell}/R_1 = Z_s/R_2 \tag{11.2}$$

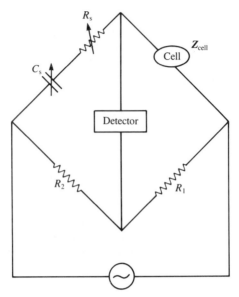

Fig. 11.1. A.c. bridge for the measurement of the impedance of electrochemical cells. The bridge is balanced when the current is zero: in this case $Z_{cell}/R_1 = Z_s/R_2$ where $Z_s = R_s - i/\omega C_s$.

we have

$$Z_{cell} = R_s R_1/R_2 - i(R_1/\omega C_s R_2) \qquad (11.3)$$
$$= R_{cell} - i\omega C_{cell} \qquad (11.4)$$

that is

$$R_{cell} = R_s R_1/R_2 \qquad (11.5)$$

and

$$C_{cell} = C_s R_2/R_1 \qquad (11.6)$$

The capacitance is due only to the working electrode, whilst the resistance includes the resistive components of the electrode process, of the solution, etc. In some cases a combination of resistance and capacitance in parallel has also been used. In these conditions the analysis is more easily carried out in terms of admittance $Y = 1/Z$: see Appendix 2.

When it is necessary to apply a d.c. potential to the cell in addition to the a.c. perturbation, it is more convenient to use a potentiostat, which simultaneously applies the d.c. potential and does the detection, rather than the conventional detector. This arrangement is called a *potentiostatic bridge,* and in this way very good stability in the applied potential and accuracy for a wide range of frequencies is possible, limited only by the characteristics of the electronic components of the potentiostat.

For very high frequencies the accuracy of bridges depends very much on cell design, and Debye–Falkenhagen effects begin to appear[14]. These occur usually above 10 MHz, and are due to the ions moving faster than the time needed for rearrangement of the ionic atmosphere: they therefore tend to leave it behind and the cell resistance drops. For normal frequencies the technique is very exact, but also very time-consuming.

A disadvantage of this type of technique is that the impedance of the whole cell is measured, whereas in the investigation of electrode processes one is interested in the properties of one of the electrodes. It is possible to reduce the contribution of the unwanted components by using an auxiliary electrode with an area large relative to that of the electrode being studied, and extrapolating the cell impedance to infinite frequency in order to remove contributions such as cell resistance.

Phase-sensitive detectors and transfer function analysers

These detectors compare the signal applied to the system with the response, giving the phase difference and the ratio of the amplitudes (i.e. impedance) (Fig. 11.2). The signal is applied by a potentiostat or galvanostat: it is necessary to subtract the resistance between the working and reference electrodes, and this can be done electronically. It is easy to register the response in terms of the alternating current as a function of applied potential, or as phase difference vs. applied potential, both these pieces of information being important in a.c. voltammetry.

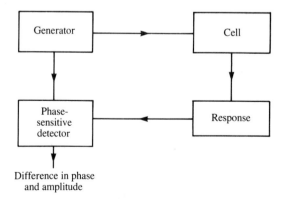

Fig. 11.2. Principle of functioning of a phase-sensitive detector.

Transfer function analysers and network analysers, both digital, have principles of electronic operation different from the conventional analogue phase-sensitive detector. However, the information obtained is similar, and for a wider range of frequencies $(10^{-4} \rightarrow 10^{6}\,\text{Hz})$. The instruments are microprocessor controlled, permitting automatic analysis, appropriate signal processing to improve signal/noise ratios, etc. It must be remembered that the processing of digital signals is complicated, and if not correctly carried out it can lead to strange and erroneous results.

In spectroscopy it is current practice to apply all frequencies at the same time through the Fourier transform. In electrochemical systems it is possible to do the same, but it is not yet clear if this is advantageous given that quite a large signal accumulation time is necessary for signal averaging; stepping through the various frequencies automatically takes more or less the same time (Section 11.12) and the results are of similar accuracy and precision.

Direct methods

If the applied signal is transmitted to a recorder (frequency <5 Hz) or an oscilloscope (frequency <5 kHz) and registered against the response signal, a *Lissajous figure* appears. The shape of this figure leads to the impedance. With a perturbing signal of

$$E(t) = \Delta E \sin \omega t \qquad (11.7)$$

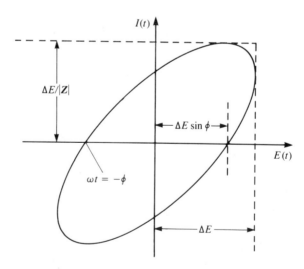

Fig. 11.3. Lissajous figure for impedance measurement (see text).

the response is

$$I(t) = \frac{\Delta E}{|\mathbf{Z}|} \sin{(\omega t + \phi)} \tag{11.8}$$

As shown in Fig. 11.3, the values of the important variables can be taken directly from the Lissajous figure.

11.3 Equivalent circuit of an electrochemical cell

Any electrochemical cell can be represented in terms of an equivalent electrical circuit that comprises a combination of resistances and capacitances (inductances only for very high high frequencies). This circuit should contain *at the very least* components to represent:

- the double layer: a pure capacitor of capacity C_d
- the impedance of the faradaic process $\mathbf{Z}_f$
- the un-compensated resistance, R_Ω, which is, usually, the solution resistance between working and reference electrodes.

The combination of these elements is shown in Fig. 11.4, with $\mathbf{Z}_f$ and C_d in parallel. The impedance $\mathbf{Z}_f$ can be subdivided in two equivalent ways:

1. Subdivision into a resistance, R_s, in series with a pseudo-capacitance, C_s, according to the scheme

which was was described in Section 11.2 on a.c. bridges.

2. Subdivision into a resistance measuring the resistance to charge transfer, R_{ct}, and an impedance that measures the difficulty of mass transport of the electroactive species, called the Warburg impedance, $\mathbf{Z}_w$:

Thus, for kinetically favoured reactions $R_{ct} \to 0$ and $\mathbf{Z}_w$ predominates, and for difficult reactions $R_{ct} \to \infty$ and R_{ct} predominates. This is called the *Randles circuit*[15].

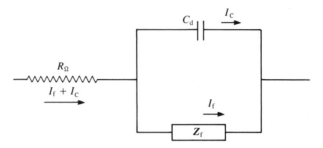

Fig. 11.4. Equivalent electrical circuit of an electrochemical cell for a simple electrode process. R_Ω is the solution resistance, of the contacts and electrode materials, $\mathbf{Z}_f$ the impedance of the electrode process, and C_d the double layer capacity.

When other steps are involved in the electrode process, homogeneous or heterogeneous, more complicated circuits have to be utilized, as indicated in Section 11.10.

11.4 The faradaic impedance for a simple electrode process

Consider the charge transfer reaction

$$O + ne^- \rightarrow R$$

at a planar electrode (semi-infinite linear diffusion), and for a sinusoidal perturbation of

$$I = I_0 \sin \omega t \tag{11.9}$$

with ω the perturbation frequency (rad s^{-1}).

For a series RC circuit we have

$$E = IR_s + \frac{Q}{C_s} \tag{11.10}$$

and by differentiation

$$\frac{dE}{dt} = R_s \frac{dI}{dt} + \frac{I}{C_s} \tag{11.11}$$

$$= R_s I \omega \cos \omega t + \frac{I}{C_s} \sin \omega t \tag{11.12}$$

The general relation reflecting the charge transfer process, that is $E = f(I, [O]_*, [R]_*)$, leads to

$$\frac{dE}{dt} = \frac{\partial E}{\partial I} \frac{dI}{dt} + \frac{\partial E}{\partial [O]_*} \frac{d[O]_*}{dt} + \frac{\partial E}{\partial [R]_*} \frac{d[R]_*}{dt} \tag{11.13}$$

The three partial derivatives describe the kinetics of the reaction and $(\partial E/\partial I)$ is the charge transfer resistance, R_{ct}. It can be shown, using Laplace transformation, that

$$\frac{d[O]_*}{dt} = \frac{I}{nFA}\left(\frac{\omega}{2D_O}\right)^{1/2}(\sin \omega t + \cos \omega t) \tag{11.14a}$$

$$\frac{d[R]_*}{dt} = \frac{-I}{nFA}\left(\frac{\omega}{2D_R}\right)^{1/2}(\sin \omega t + \cos \omega t) \tag{11.14b}$$

Since the a.c. perturbation is small, the linearized relation between current and overpotential, η (equation (6.50)), considering $\alpha_a = \alpha_c = 0.5$, may be used, that is

$$\eta = \frac{RT}{nF}\left[\frac{[O]_*}{[O]_\infty} - \frac{[R]_*}{[R]_\infty} + \frac{I}{I_0}\right] \tag{11.15}$$

Thus

$$R_{ct} = \frac{\partial E}{\partial I} = \frac{RT}{nFI_0} \tag{11.16}$$

where I_0 is the exchange current, and

$$\frac{\partial E}{\partial [O]_*} = \frac{RT}{nF[O]_\infty} \qquad \frac{\partial E}{\partial [R]_*} = \frac{-RT}{nF[R]_\infty} \tag{11.17}$$

Equation (11.13) then becomes

$$\frac{dE}{dt} = \left(R_{ct} + \frac{\sigma}{\omega^{1/2}}\right)I\omega \cos \omega t + I\sigma\omega^{1/2}\sin \omega t \tag{11.18}$$

where

$$\sigma = \frac{RT}{n^2F^2A\sqrt{2}}\left(\frac{1}{D_O^{1/2}[O]_\infty} + \frac{1}{D_R^{1/2}[R]_\infty}\right) \tag{11.19}$$

By comparison of (11.12) and (11.18) we verify that

$$R_s = R_{ct} + \sigma\omega^{-1/2} \tag{11.10}$$

and that

$$1/C_s = \sigma\omega^{1/2} \tag{11.21}$$

In this way, from (11.14), we can identify the Warburg impedance, Z_W, of the Randles circuit as consisting of a resistance and a capacitance in series and where the components in phase Z'_W and out of phase Z''_W, with $Z = Z' + iZ''$, are given by

$$Z'_W = R_W = \sigma\omega^{-1/2} \tag{11.22a}$$

$$Z''_W = -(\omega C_W)^{-1} = -\sigma\omega^{-1/2} \tag{11.22b}$$

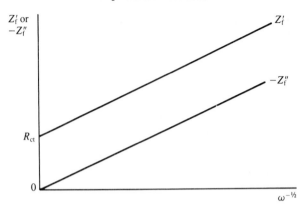

Fig. 11.5. Dependence of Z_f' and Z_f'' on the inverse square root of the frequency (Randles plot). The slope of the plots is σ.

In terms of the faradaic impedance Z_f

$$Z_f' = R_{ct} + \sigma\omega^{-1/2} \tag{11.23a}$$
$$Z_f'' = -\sigma\omega^{-1/2} \tag{11.23b}$$

A plot of Z_f' and of $-Z_f''$ vs. $\omega^{-1/2}$ should give straight lines of slope σ and of intercept R_{ct} for the in-phase component (Fig. 11.5) corresponding to infinite frequency. The physical explanation for the intercept is that at very high frequencies the time scale is so short that diffusion cannot influence the current, being dependent only on the kinetics. If the lines obtained are not parallel then either the theory cannot be applied, or the experimental accuracy is bad.

When a reaction is reversible, $R_{ct} \to 0$ and $\mathbf{Z_f} = \mathbf{Z_W} = \sigma\omega^{-1/2}(1-i)$. The observed phase angle is $\pi/4$, and the impedance is the least possible for that value of ω (σ depends not only on the transport but also on the reciprocal of the concentration of electroactive species: see (11.19)). If $R_{ct} > 0$, $\mathbf{Z_f}$ increases from its minimum value and there will be a lower phase angle. It is this phase-angle variation according to the rate of the electrode reaction that is used in the technique of a.c. voltammetry.

11.5 The faradaic impedance, Z_f, and the total impedance: how to calculate Z_f from experimental measurements

In real electrochemical systems the experimental impedance includes contributions from C_d and R_Ω as well as from $\mathbf{Z_f}$. There are several ways of attempting to eliminate these contributions:

• Do measurements in the absence of the electroactive species. In this case $Z_f = 0$ and values of C_d and R_Ω are furnished directly, so that they can be subtracted analytically or graphically from the total impedance. One assumes, which is not necessarily correct, that C_d and R_Ω are not altered by the presence of the electroactive species;

• Study the variation of Z with frequency, so that values of R_Ω, C_d, R_{ct} and Z_W can be extracted, using appropriate techniques of analysis. The disadvantage is the necessity of using an equivalent circuit chosen *a priori*;

• Other methods include varying the concentration[16] or the applied potential[17], all other parameters remaining constant.

The second method is the most used, and frequently in terms of plots of Z_f' vs. Z_f'', as will be explained in the next section. This type of analysis, developed by Sluyters *et al.*[1], is based on techniques used in electrical engineering.

Another point to consider is that we are assuming linearity between perturbation and response, owing to the fact that the perturbation is small. In a linear system Z' and Z'' are not independent, and are related by the Kramers–Kronig relations (Appendix 2). Knowing the values of Z' over all frequencies, it is possible to calculate the corresponding curve for Z''. If the theoretically obtained curve does not agree with the experimental curve, then the system is not linear and interpretation of the experimental data following the usual equations is incorrect. Unfortunately, in practice, this verification is time-consuming and quite complicated, and the whole spectrum may not be accessible.

11.6 Impedance plots in the complex plane

The experimental impedance is always obtained as if it were the result of a resistance and capacitance in series. We have already seen in (11.20) and (11.21) the relation between an RC series combination and the $R_{ct} + Z_W$ combination. It can be shown for the full Randles equivalent circuit for this simple charge transfer reaction, see Fig. 11.4, on separating the in-phase and out-of-phase components of the impedance, that

$$Z' = R_\Omega + \frac{R_{ct} + \sigma\omega^{-1/2}}{(\sigma\omega^{1/2}C_d + 1)^2 + \omega^2 C_d^2 (R_{ct} + \sigma\omega^{-1/2})^2} \quad (11.24a)$$

$$-Z'' = \frac{\omega C_d (R_{ct} + \sigma\omega^{-1/2})^2 + \sigma^2 C_d + \sigma\omega^{-1/2}}{(\sigma\omega^{1/2}C_d + 1)^2 + \omega^2 C_d^2 (R_{ct} + \sigma\omega^{-1/2})^2} \quad (11.24b)$$

These components are represented as a complex plane plot in Fig. 11.6

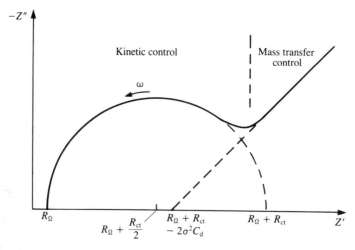

Fig. 11.6. Plot of impedance in the complex plane of a simple electrochemical system: $O + ne^- \rightarrow R$.

(Sluyters or Cole–Cole plot), in a form similar to the representation of complex numbers (Argand diagram).

It is interesting to consider two limiting forms of these equations:

1. $\omega \rightarrow 0$

$$Z' = R_\Omega + R_{ct} + \sigma\omega^{-1/2} \qquad (11.25a)$$

$$Z'' = -\sigma\omega^{-1/2} - 2\sigma^2 C_d \qquad (11.25b)$$

This low-frequency limit is a straight line of unit slope, which extrapolated to the real axis gives an intercept of $(R_\Omega + R_{ct} - 2\sigma^2 C_d)$. The line corresponds to a reaction controlled solely by diffusion, and the impedance is the Warburg impedance, the phase angle being $\pi/4$, see Fig. 11.6.

2. $\omega \rightarrow \infty$. At the high-frequency limit the control is purely kinetic, and $R_{ct} \gg Z_W$. The electrical analogy is an RC parallel combination. Thus (11.24) become

$$Z' = R_\Omega + \frac{R_{ct}}{1 + \omega^2 C_d^2 R_{ct}^2} \qquad (11.26a)$$

$$Z'' = -\frac{\omega C_d R_{ct}^2}{1 + \omega^2 C_d^2 R_{ct}^2} \qquad (11.26b)$$

Simplifying we obtain

$$\left(Z' - R_\Omega - \frac{R_{ct}}{2}\right)^2 + (Z'')^2 = \left(\frac{R_{ct}}{2}\right)^2 \qquad (11.27)$$

This last expression is the equation of a circle of radius $R_{ct}/2$ with intercepts on the Z' axis of R_{Ω} ($\omega \to \infty$) and of $R_{\Omega} + R_{ct}$ ($\omega \to 0$), see Fig. 11.6.

It is important to understand the physical reasons for the existence of this semi-circle. For very high frequency Z'' ($= -1/\omega C_d$) is very small, but rises as the frequency diminishes. For very low frequency, C_s gives a high reactance but the current passes predominantly through R_{ct}, increasing Z' and diminishing Z''.

Figure 11.6 shows the semi-circle, the straight line and a transition zone between the two. For different systems it can happen that, due to the relative values of the components R_{ct}, Z_W and C_d, only the semi-circle or only the straight line are observed.

When an electrode process involves several steps, sometimes a succession of semi-circles side by side is obtained, corresponding to RC parallel combinations in series and with different RC time constants, from which it is possible to deduce the corresponding parameters.

It is interesting to represent the impedance in three dimensions, the third axis being $\lg f$. The impedance is represented as Z, or if its value varies a lot, as $\lg Z$. This procedure is recommended by Macdonald[18] as it permits an excellent visualization of the electrode process and makes the detection of anomalies easier. Figure 11.7 shows an example.

In general, the form of complex plane plots of the type shown in Fig. 11.6 alters on changing the concentration of the electroactive species[16]. It has been shown for the various kinetic regimes that the variation of Z'' with concentration has the form in Fig. 11.8. The equation of this curve

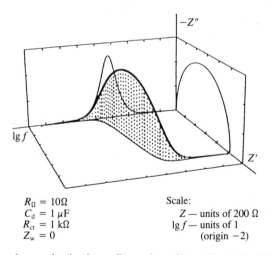

$R_{\Omega} = 10\,\Omega$
$C_d = 1\,\mu F$
$R_{ct} = 1\,k\Omega$
$Z_w = 0$

Scale:
Z — units of 200 Ω
$\lg f$ — units of 1
(origin -2)

Fig. 11.7. Impedance plot in three dimensions (from Ref. 18 with permission).

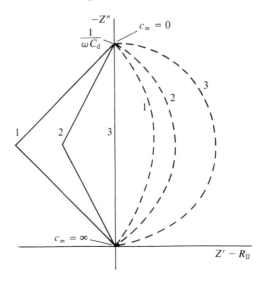

Fig. 11.8. Variation of impedance with concentration for reactions of variable kinetics: 1—reversible, $R_{ct} \to 0$; 2—quasi-reversible, $R_{ct}/\sigma\omega^{-1/2} \approx 1$; 3—ir-reversible, $R_{ct} \to \infty$.

can be obtained from (11.24), and is

$$\left(Z' - R_\Omega + \frac{1}{2\omega C_d (R_{ct}/\sigma\omega^{-1/2})}\right)^2 + \left(-Z'' - \frac{1}{2\omega C_d}\right)^2$$

$$= \frac{1}{4\omega^2 C_d^2}\left(1 + \frac{1}{((R_{ct}/\sigma\omega^{-1/2}) + 1)^2}\right) \quad (11.28)$$

which is a circle centred on $[R_\Omega - (2\omega C_d)^{-1}(R_{ct}/\sigma\omega^{-1/2} + 1)^{-1}]$ on the Z' axis and on $-(2\omega C_d)^{-1}$ on the Z'' axis. A segment of the circle is traced from $(Z' = R_\Omega, Z'' = -1/\omega C_d)$ at zero concentration to $(Z' = R_\Omega, Z'' = 0)$ at infinite concentration. The curvature diminishes from irreversible to reversible reactions. The centre of the circle traced by the experimental points linked to the points for $c_\infty = 0$ and $c_\infty = \infty$ gives the phase angle: $\pi/4$ for a reversible reaction (see Fig. 11.8) and zero for a totally irreversible reaction. These plots furnish the same information as complex plane plots and are an alternative to these in the interpretation of the results of impedance measurements.

11.7 Admittance and its use

Admittance is the inverse of impedance, and is represented by the symbol Y. In certain circumstances, admittance is very useful since elements in parallel in electrical circuits are added. The components of the experimental faradaic admittance, as is easily deduced from (11.23), are

$$Y_f' = \frac{R_{ct} + \sigma\omega^{-1/2}}{(R_{ct} + \sigma\omega^{-1/2})^2 + \sigma^2\omega^{-1}} \qquad (11.29a)$$

$$Y_f'' = \frac{-\sigma\omega^{-1/2}}{(R_{ct} + \sigma\omega^{-1/2})^2 + \sigma^2\omega^{-1}} \qquad (11.29b)$$

and the theoretical deduction from the Randles circuit for the capacitive contribution, Y_C, shows that

$$Y_C' = 0 \qquad (11.30a)$$

$$Y_C'' = \omega C_d \qquad (11.30b)$$

The limiting behaviour, depending on the value of R_{ct}, is:

• $R_{ct} \rightarrow 0$ (reversible reaction). $Y' = (2\sigma\omega^{-1/2})^{-1}$; $Y'' = Y' + \omega C_d$.

• $R_{ct} \rightarrow \infty$ (totally irreversible reaction). $Y' = 1/R_{ct}$; $Y'' = \sigma\omega^{-1/2}/R_{ct}^2 + \omega C_d$. This is a vertical line in the complex plane.

These equations show the possibility of determination of C_d in the presence of the electroactive species. A good strategy is;

• Measurement of Z' and Z'' and subtraction of the value of R_Ω from the value of Z' at infinite frequency, as in Fig. 11.6.

• Calculation of Y_f' from the formula (11.29a) for each frequency.

• Determination of R_{ct} and σ from the values of Y' (since $Y' = Y_f'$).

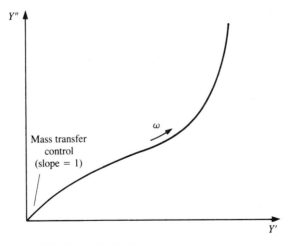

Fig. 11.9. Plot of admittance in the complex plane.

• Insertion of the values for R_{ct} and σ in the expression for Y'', obtaining the value of C_d directly.

One assumes that the double layer admittance is that of a pure capacitor. Thus its value should be independent of frequency. The graphical variation of Y' with Y'' is shown schematically in Fig. 11.9. More details may be found in Ref. 9.

As a final point we call attention to the fact that a plot of Y' vs. E has the same form as an a.c. voltammogram, i.e. it comprises a peak centred on the half-wave potential—this because the measured current is inversely proportional to R_{ct}.

11.8 A.c. voltammetry

If a fixed frequency sinusoidal perturbation is applied at a succession of fixed d.c. potentials, we can register the resulting alternating current and its phase angle vs. the d.c. potential. The form of the curves obtained gives information about the kinetics, and can also be used for analytical purposes[5,19–21].

The theoretical treatment is simplified when the perturbation frequency is sufficiently high that the contributions of E_{ac} and E_{dc} can be separated, owing to the diffusion layer derived from the d.c. potential being much larger than that derived from the a.c. perturbation. As in other techniques involving measurements for various potentials, the d.c. potential can be swept slowly. The use of hydrodynamic electrodes, including the dropping mercury electrode (*a.c. polarography*) allows a greater reproducibility and sensitivity in the results obtained.

Here we only consider a reversible system; treatments for other kinetic regimes and more complicated electrode processes can be found in Refs. 5 and 6. For a reversible system we know, from Section 11.6, that the phase angle is $\pi/4$. Theoretical considerations show that, for the reduction of O,

$$i = -\frac{n^2 F^2 A \omega^{1/2} D_O^{1/2}[O]_\infty \Delta E}{4RT \cosh^2{(p/2)}} \tag{11.31}$$

where $p = (D_O/D_R)^{1/2}$ and ΔE is the r.m.s. amplitude of the potential perturbation. The peak current is

$$I_p = -\frac{n^2 F^2 A \omega^{1/2} D_O^{1/2}[O]_\infty \Delta E}{4RT} \tag{11.32}$$

at a potential $E_{dc} = E_{1/2}$. From (11.31) and (11.32) it is easy to show that

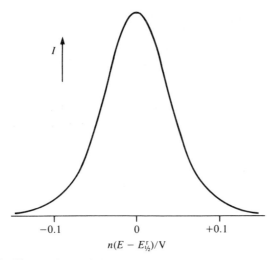

Fig. 11.10. Shape of a peak in a.c. voltammetry for a reversible system.

Table 11.1. Parameters for reversible and quasi-reversible a.c. voltammograms with O initially present (reversible d.c. behaviour)

Reversible

$\cot \phi$	1
$\lvert I \rvert$	$\dfrac{\xi \omega^{1/2} D_{O}^{1/2}}{4 \cosh^2 (q/2)}$
$\lvert I_{p} \rvert$	$\dfrac{\xi \omega^{1/2} D_{O}^{1/2}}{4}$
E_{dc}	$E^{r}_{1/2} + \dfrac{2RT}{nF} \ln \left[\left(\dfrac{I_{p}}{I} \right)^{1/2} - \left(\dfrac{I_{p} - I}{I} \right)^{1/2} \right]$

Quasi-reversible

$\cot \phi$	$1 + \dfrac{(2 D_{O}^{\alpha_a} D_{R}^{\alpha_c} \omega)^{1/2}}{k_0} \left[\dfrac{1}{e^{\alpha_a} + e^{q}} \right]$
$\lvert I \rvert$	$k_0 \xi p^{\alpha_c} \left(\dfrac{e^{\alpha_a}}{1 + e^{-q}} \right)$
$\lvert I_{p} \rvert$	$k_0 \xi p^{\alpha_c} \alpha_a^{\alpha_a} \alpha_c^{\alpha_c}$
$E_{dc,max}$	$E^{r}_{1/2} + \dfrac{RT}{nF} \ln \dfrac{\alpha_a}{\alpha_c}$

$$\xi = \frac{n^2 F^2 A \, \Delta E [O]_{\infty}}{RT} \qquad q = \frac{nF}{RT}(E_{dc} - E^{r}_{1/2})$$

$$p = (D_O/D_R)^{1/2} \qquad E(t) = E_{dc} - \Delta E \sin (\omega t)$$

the voltammogram shape is described by

$$E_{dc} = E_{1/2}^r + \frac{2RT}{nF} \ln \left[\left(\frac{I_p}{I} \right)^{1/2} - \left(\frac{I_p - I}{I} \right)^{1/2} \right] \qquad (11.33)$$

as shown in Fig. 11.10. At a dropping mercury electrode without current sampling the characteristic variation of current with drop size is superimposed on this curve.

The equations for reversible and quasi-reversible systems are given in Table 11.1. The table shows, as mentioned in Section 11.4, that the phase angle is lower the slower the electrode reaction.

11.9 Second-order effects

Higher harmonics

A musician learns that on playing a note at a frequency of x Hz, sounds of reduced intensity at frequencies $2x$, $3x$, etc. appear. In fact, any periodic excitation leads to the same type of behaviour.

In an electrochemical system, Taylor's expansion shows that

$$\Delta I = \left(\frac{\partial I}{\partial E} \right)_{E_0, I_0} \Delta E + \frac{1}{2} \left(\frac{\partial^2 I}{\partial E^2} \right)_{E_0, I_0} (\Delta E)^2 + \cdots \qquad (11.34)$$

Until now we have used conditions for which the second and following terms are (or are supposed to be) negligible, in other words I is linear in E; this corresponds to small sinusoidal perturbations. The terms in the Taylor expansion are called first harmonic or fundamental, second harmonic, third harmonic, etc. If we look at the form of a normal voltammogram (Fig. 6.2) the approximation of a linear system is valid close to $E_{1/2}$ much more than in other parts of the voltammogram where the curvature of the I–E profile is more pronounced.

If we register the second harmonic current vs. d.c. potential, this will have the same form as the second derivative of the voltammetric curve, as Fig. 11.11 shows. One of the advantages of the use of the second harmonic is that, since the double layer capacity is essentially linear, it contributes much less to the second harmonic than to the fundamental frequency and the calculation of accurate kinetic parameters is much facilitated.

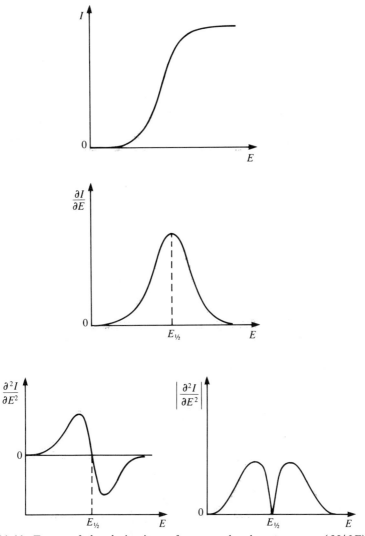

Fig. 11.11. Forms of the derivatives of a normal voltammogram. $(\partial I/\partial E)$ and $(\partial^2 I/\partial E^2)$ have the forms of the a.c. voltammograms of the fundamental and second harmonic respectively.

Other second-order methods

Apart from the second harmonic there are other second-order effects, which are developed in the techniques of faradaic rectification and demodulation. Both these techniques are utilized to study systems with very fast kinetics.

Faradaic rectification[22]

A sinusoidal perturbation gives rise to second harmonics that, for a signal of the type $I_0 \sin \omega t$, will have terms in $\sin^2 \omega t$. If we remember that

$$\sin^2 \omega t = \tfrac{1}{2}(1 - \cos 2\omega t) \tag{11.35}$$

we see that the response has a d.c. component. Figure 11.12 shows visually the origin of this component in the curvature of the I–E profile.

The use of perturbing frequencies of greater than 1 MHz is possible in this technique. The d.c. response is obtained through the application of the a.c. signal in short duration pulses (Fig. 11.13*a*).

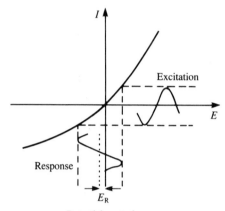

Potential control

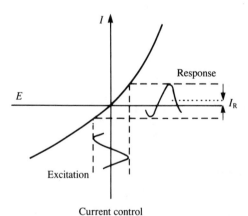

Current control

Fig. 11.12. Diagrams showing the appearance of a d.c. component in the response of an electrochemical cell to an a.c. perturbation.

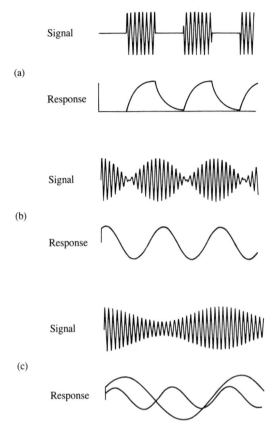

Fig. 11.13. Perturbing signals and responses for (a) faradaic rectification; (b) demodulation type 1; (c) demodulation type 2.

Demodulation[23]

Instead of applying pulses of sinusoidal perturbations, the amplitude of the perturbation can be modulated in a sinusoidal fashion (Fig. 11.13b). If ω_H is the a.c. perturbation frequency and ω_L the modulation frequency, then the applied signal can be either

$$\sin(\omega_H t) \cos(\omega_L t) \tag{11.36}$$

(Fig. 11.13b) or

$$\sin(\omega_H t)[1 + M \cos(\omega_L t)] \tag{11.37}$$

as in Fig. 11.13c. The second-order response to modulation depends on the square of these terms, which can be shown through trigonometric conversion, such as that in (11.29), to be dependent in the first case on frequencies ω_L and $2\omega_L$ and in the second case only on $2\omega_L$. The reason for the use of this technique is, once more, to reduce the relative contribution to the measurements of linear components such as C_d, thus improving the accuracy.

11.10   More complex systems, porous electrodes, and fractals

Many electrode processes are more complex than those discussed above. Besides this, the impedance of an interface is dependent on its microscopic structure which, in the case of a solid electrode, can have an important influence. Impedance measurements can be used to study complicated corrosion phenomena (Chapter 16), blocked interfaces (i.e. where there is no redox process nor adsorption/desorption), the liquid/liquid interface[24,25], transport through membranes[26], the electrode/solid electrolyte interface etc. Experimental measurements always furnish values of Z' and Z'' or their equivalents Y' and Y'', or of the complex permittivities ϵ' and ϵ'' ($\epsilon = Y/i\omega C_c$, C_c being the capacitance of the empty cell). In this section we attempt to show how to

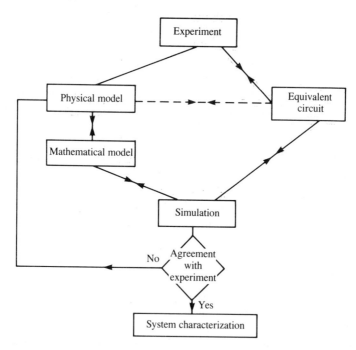

Fig. 11.14. Flow diagram for evaluating experimental impedance results.

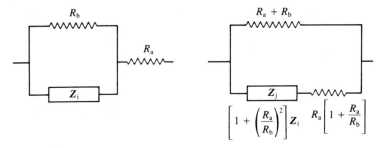

Fig. 11.15. Two circuits with the same impedance at all frequencies.

interpret the results in terms of a physical model, which should always follow reasoning such as that examplified in Fig. 11.14. The importance of the physical model is shown in Fig. 11.15, whose two circuits give precisely equal impedance values at all frequencies. Some help can be had from altering experimental conditions (concentration for example) and obtaining various series of measurements.

Two types of electrical analogy model for the interpretation of impedance data can be used: based on combinations of resistances and capacitances, or based on transmission lines. These possibilities are now described.

For consecutive or parallel electrode reactions it is logical to construct circuits based on the the Randles circuit, but with more components. Figure 11.16 shows a simulation of a two-step electrode reaction, with strongly adsorbed intermediate, in the absence of mass transport control. When the combinations are more complex it is indispensable to resort to digital simulation so that the values of the components in the simulation can be optimized, generally using a non-linear least squares method (complex non-linear least squares fitting).

Transmission lines have been much used recently to simulate interfaces, especially where the solid is rough. We consider, to illustrate, a simple transmission line (Fig. 11.17).

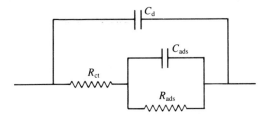

Fig. 11.16. An equivalent circuit for a two-step electrode reaction with strongly adsorbed intermediate.

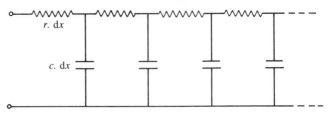

Fig. 11.17. A resistive-capacitive transmission line that describes a semi-infinite linear diffusion process.

Elementary considerations show that

$$I = -\frac{\partial V}{\partial x}\frac{1}{r} \tag{11.38}$$

$$\frac{\partial V}{\partial t} = -\frac{\partial I}{\partial x}\frac{1}{c} \tag{11.39}$$

$$= \frac{\partial^2 V}{\partial x^2}\frac{1}{rc} \tag{11.40}$$

where $r = R/x$ is the resistance per unit length of the transmission line and $c = C/x$ is the capacitance per unit length. Expression (11.40) is of the same type as Fick's second law, thence its application to electrochemistry. The impedance is

$$Z = \left[\frac{r}{(\sigma + i\omega)c}\right]^{1/2} \tag{11.41}$$

which is a line of slope $\pi/4$ in the complex plane, that is the Warburg impedance for semi-infinite diffusion (Fig. 11.18a). If the transmission line is finite there can be two situations:

1. Termination in an open circuit (corresponding to reflection) as in conducting polymers and porous electrodes, such as porous carbon in acid media (Fig. 11.18b).

2. Termination in a large resistance, i.e. a blocked interface such as a metal totally covered with oxide or a highly resistive membrane used in ion exchange selective electrodes (Section 13.6) (Fig. 11.18c). This is sometimes referred to as the *finite Warburg impedance*.

Rugosity and porosity give rise to the so-called *constant phase element* (CPE), which can be described by groups of parallel or branched transmission lines. The CPE is manifested in real systems by an impedance spectrum altered from the expected shape, especially in the

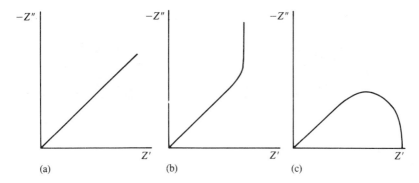

Fig. 11.18. Variation of impedance for diffusive systems: (a) Semi-infinite diffusion; (b) Reflective finite diffusion; (c) Transmissive finite diffusion.

diffusive part, where an angle of less than $\pi/4$ relative to the real axis is obtained. In the case of a blocked interface the element can be described by

$$Y_{\text{CPE}} = b(i\omega C)^{\alpha} \tag{11.42}$$

where b is a proportionality constant. Using a transmission line model, de Levie[27,28] showed, for porous electrodes, that $\alpha = 0.5$ whereas for a smooth electrode $\alpha = 1$. In this case the impedance diagrams can be corrected by calculating values of $|Z|^2$ and multiplying the phase angle by 2, as demonstrated in Fig. 11.19.

It is recognized, however, that many interfaces are rough but do not correspond to porous electrodes. This suggests that values of α between 0.5 and 1 should be considered. It has been observed that the large majority of interfaces contain the same or similar features at different

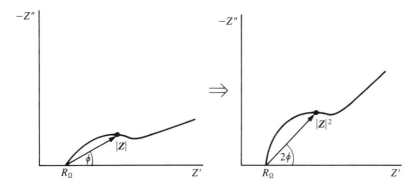

Fig. 11.19. Conversion of the impedance spectrum of a porous electrode to the equivalent smooth electrode.

$D_f = 2.500$ $D_f = 2.730$

Fig. 11.20. An example of fractal geometry.

scales (for example scratches or pits) which is called *fractal geometry* (Fig. 11.20) and which seems to be universal[29]. The fractal dimension, D_f, has been related to the roughness of the electrode/solution interface using the transmission line method[30]. For a diffusion impedance it was shown that

$$\alpha = (D_f - 1)/2 \tag{11.43}$$

D_f varying between 2, for a porous electrode, and 3 for a smooth electrode. In the case of a blocked interface, the conclusions up to now are that there is no simple relation between α and the fractal dimension. However, the analogy seems useful from an interpretative point of view. Reviews of the response at fractal and rough electrodes have recently appeared[31,32].

11.11 Hydrodynamic electrodes and impedance[7]

Hydrodynamic electrodes permit the control of the diffusion layer thickness by imposing convection. This thickess can also be modulated. Implicit functions link the current, potential and convection modulation. For the rotating disc electrode

$$F(I, E, W_E) = 0 \tag{11.44}$$

where W_E is the electrode rotation speed. So

$$\left(\frac{\partial W_E}{\partial I}\right)_E = -\left(\frac{\partial W_E}{\partial E}\right)_I \left(\frac{\partial E}{\partial I}\right)_{W_E} \tag{11.45}$$

and here

$\left(\dfrac{\partial W_E}{\partial I}\right)_E$ is the electrohydrodynamic impedance under potential control

$\left(\dfrac{\partial W_E}{\partial E}\right)_I$ is the electrohydrodynamic impedance under galvanostatic control

$\left(\dfrac{\partial E}{\partial I}\right)_{w_E}$ is   the   electrochemical   impedance   at   fixed   rotation   speed

The last of these is the impedance which has been considered throughout this chapter. We now consider *forced convection*. For low frequencies the diffusion layer thickness due to the a.c. perturbation is similar to that of the d.c. diffusion layer: in these cases convection effects will be apparent in the impedance expressions. For the rotating disc electrode these frequencies are lower than 40 Hz[33]. For higher frequencies where the two diffusion layers are of quite different thicknesses, the advantage of hydrodynamic electrodes is that transport is well defined with time, as occurs with linear sweep voltammetry.

Applications of double hydrodynamic electrodes are particularly interesting because the change of phase in the current measured at the downstream electrode permits discrimination between the electron flow and the flux of electroactive species produced at the upstream electrode[34].

Modulation of the convective flux, originally proposed by Bruckenstein[35], leads to the electrohydrodynamic impedance. It has been used for determining kinetic parameters and diffusion parameters in Newtonian and Ostwaldian fluids, and in corrosion.

11.11   Transforms and impedance

The determination of the characteristics of an electrochemical system by a.c. techniques requires measuring the impedance at various frequencies, to give a frequency spectrum. Instead of applying each frequency successively the application of all the frequencies simultaneously has been suggested, using a transform to analyse the response[36]. This procedure can be particularly advantageous at a dropping electrode, where it would permit the recording of a full spectrum during the lifetime of one drop. Possible excitation signals vary from a group of sinusoidal signals to white noise. Comparisons have been made, but owing to the time necessary to obtain a reasonable signal/noise ratio in the response in the frequency domain, it is not clear if this method is faster than the automatic sweeping of the frequency spectrum.

Whatever the excitation, the transformation of the response from the frequency to the time domain (Fig. 11.21) is done with the inverse Fourier transform, normally as the FFT (fast Fourier transform) algorithm, just as for spectra of electromagnetic radiation. Remembering that the Fourier transform is a special case of the Laplace transform with

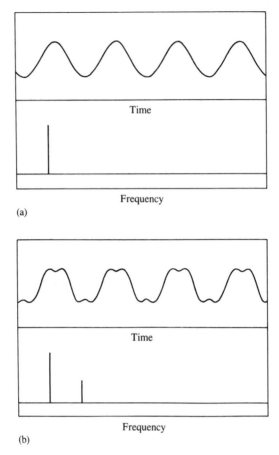

(a)

(b)

Fig. 11.21. The Fourier transform: the link between the time and frequency domain. (a) A simple sinusoidal signal; (b) The first two components of a square wave (Fig. A1.1).

$s = i\omega$ (Appendix 1) this type of method can be applied to any technique involving change in potential or current, given that any waveform can be synthesized by superposition of sinusoidal waves of variable frequency and amplitude (Fourier synthesis). The response can then be analysed in terms of the resultant impedances. Unfortunately, from an experimental point of view it is not realistic to proceed in this way, but it is important to attempt to unify the various methods at a theoretical level which are, effectively, probing different aspects of the same problem.

References

1. M. Sluyters–Rehbach and J. H. Sluyters, *Electroanalytical chemistry*, ed. A. J. Bard, Dekker, New York, Vol. 4, 1970, pp. 1–128.
2. R. D. Armstrong, M. F. Bell, and A. A. Metcalfe, *Chem. Soc. Spec. Per. Reports, Electrochemistry*, 1978, **6,** 98–127.
3. W. I. Archer and R. D. Armstrong, *Chem. Soc. Spec. Per. Reports, Electrochemistry*, 1980, **7,** 157–202.
4. I. Epelboin, *Comprehensive treatise of electrochemistry*, Plenum, New York, Vol. 4, 1981, ed. J. O'M. Bockris, B. E. Conway, E. Yeager, and R. E. White, Chapter 3.
5. D. D. Macdonald, *Transient techniques in electrochemistry*, Plenum, New York, 1977, Chapter 7.
6. D. D. Macdonald and M. C. H. McKubre, *Modern aspects of electrochemistry*, Plenum, New York, Vol. 14, 1982, ed. J. O'M. Bockris, B. E. Conway, and R. E. White, pp. 61–150.
7. C. Gabrielli, *The identification of electrochemical processes by frequency response analysis*, Solartron Instruments, UK, 1980.
8. M. Sluyters–Rehbach and J. H. Sluyters, *Comprehensive treatise of electrochemistry*, Plenum, New York, Vol. 9, 1984, ed. E. Yeager, J. O'M. Bockris, B. E. Conway, and S. Sarangapani, pp. 177–292.
9. M. Sluyters–Rehbach and J. H. Sluyters, *Comprehensive chemical kinetics*, Elsevier, Amsterdam, Vol. 26, 1986, ed. C. H. Bamford and R. G. Compton, Chapter 4.
10. J. R. Macdonald (ed.), *Impedance spectroscopy*, Wiley, New York, 1987.
11. C. Gabrielli, *Use and applications of electrochemical impedance spectroscopy*, Schlumberger Instruments, UK, 1990.
12. C. Gabrielli (ed.), *Proceedings of the first international symposium on electrochemical impedance spectroscopy*, Electrochim. Acta, 1990, **35,** No. 10.
13. Ref. 10, Chapter 3.
14. P. Debye and H. Falkenhagen, *Phys. Z.*, 1928, **29,** 121 and 401.
15. J. E. B. Randles, *Disc. Faraday soc.*, 1947, **1,** 11.
16. Ref. 1, pp. 37–38.
17. Ref. 1, pp. 38–41.
18. J. R. Macdonald, J. Schoonman, and A. P. Lehnen, *Solid state ionics*, 1981, **5,** 137.
19. D. E. Smith, *Electroanalytical chemistry*, ed. A. J. Bard, Dekker, New York, Vol. 1, 1966, pp. 1–155.
20. B. Breyer and H. H. Bauer, *A.c. polarography and tensammetry*, Wiley, New York 1963.
21. D. E. Smith, *Crit. Rev. Anal. Chem.*, 1971, **2,** 247.
22. H. P. Argawal, *Modern aspects of electrochemistry*, Plenum, New York, Vol. 20, 1989, ed. J. O'M. Bockris, R. E. White, and B. E. Conway, pp. 177–263.
23. Ref. 9, pp. 252–3.
24. J. Koryta and P. Vanýsek, *Advances in electrochemistry and electrochemical engineering*, ed. H. Gerischer and C. W. Tobias, Wiley, New York, Vol. 12, 1981, pp. 113–176.
25. P. Vanýsek, *Electrochemistry on liquid/liquid interfaces*, Lecture Notes in Chemistry Vol. 39, Springer-Verlag, Berlin, 1985.

26. e.g. R. P. Buck, *J. Electroanal. Chem.*, 1986, **210,** 1.
27. R. de Levie, *Electrochim. Acta,* 1964, **9,** 1231.
28. R. de Levie, *Advances in electrochemistry and electrochemical engineering,* ed. P. Delahay and C. W. Tobias, Wiley, New York, Vol. 6, 1967, pp. 329–397.
29. B. B. Mandelbrot, *The fractal geometry of nature,* Freeman, San Francisco, 1982.
30. A. Le Mehauté and G. Crepy, *Solid state ionics,* 1983, **9/10,** 17; S. Liu, *Phys. Rev. Lett.,* 1985, **55,** 529; L. Nyikos and T. Pajkossy, *Electrochim. Acta,* 1985, **30,** 1533, 1986, **31,** 1347; M. Keddam and H. Takenouti, *Electrochim. Acta,* 1988, **33,** 445; T. Pajkossy and L. Nyikos, *Electrochim. Acta,* 1989, **34,** 171, 181.
31. R. de Levie, *J. Electroanal. Chem.,* 1990, **281,** 1.
32. T. Pajkossy, *J. Electroanal. Chem.,* 1991, **300,** 1.
33. K. Tokuda and H. Matsuda, *J. Electroanal. Chem.,* 1977, **82,** 157; 1978, **90,** 149; 1979, **95,** 147.
34. W. J. Albery et al., *J. Chem. Soc. Faraday Trans. 1,* 1971, **67,** 2414; 1978, **74,** 1007; 1979, **75,** 1623.
35. S. Bruckenstein, M. I. Bellavance, and B. Miller, *J. Electrochem. Soc.,* 1973, **120,** 1351.
36. D. E. Smith, *Anal. Chem.,* 1976, **48,** 221A and 517A.

12

NON-ELECTROCHEMICAL
PROBES OF ELECTRODES AND
ELECTRODE PROCESSES

12.1 Introduction

There is a great variety of electrochemical experiments that can be done to elucidate electrode processes, as has become clear. However, the information obtained is often complex and interpretation may be ambiguous. Aditionally, there is no direct insight into what is happening at a microscopic or molecular level on the electrode surface. For these reasons, one has recourse to spectroscopic and microscopic techniques, many of which are used in surface science. Some of these can be employed in conjunction with the electrochemical experiment in real time as *in situ* probes. They are not electrochemical, so they give a different type of information that complements the electrochemical information. This field is expanding so rapidly that the survey in this chapter cannot be exhaustive in its description of techniques: it would become disproportionately long if this were attempted. References 1–5 give a good and more complete survey of the immense progress achieved in recent years.

Spectroscopic techniques can be carried out *in situ* (low-energy photon, etc.) and *ex situ* or *in vacuo* (high-energy photon and electron techniques). *Ex situ* microscopic techniques have been employed for many years to examine surfaces, and are now widely used tools. However, *in situ* microscopic techniques with resolution approaching the atomic scale are very recent: scanning tunnelling microscopy and atomic force microscopy are invaluable in imaging the surface structure. A particularly interesting variant of using *in situ* probes with direct electrochemical control is scanning electrochemical microscopy.

Electromagnetic radiation, besides being a probe of surface structure, can excite electrons in the species in solution (especially in organic compounds) or in the electrode itself (especially in semiconductor electrodes). This photon excitation can lead to electron transfer between electrode and solution. The study of these phenomena is *photoelectrochemistry* and can be very important in conversion of solar energy into electricity in order to convert substances (*photoelectrolysis*).

A slightly different application is where species produced electrochemically lead to photon emission in the visible spectrum, via the formation of organic radicals by homogeneous reaction from electrochemically generated precursors. The electrode controls the quantity of precursor, enabling quantitative parameters of the homogeneous reaction to be elucidated. This is known as electrogenerated chemiluminescence or *electrochemiluminescence* (ECL).

12.2 *In situ* spectroscopic techniques

In these techniques a beam of photons is directed to the electrode such that it is transmitted or reflected. The majority of the techniques are reflective, since transmission is limited to transparent or semi-transparent electrode materials.

Transmission[6,7]

Since the electrode has to be transparent, the electrode material is limited to thin films of metals or semiconductors deposited on a transparent substrate (for example a thin film of tin(IV) oxide or platinum on quartz) or to very fine grids of the electrode material, as shown in Fig. 9.13. The first of these two options is preferable, since the transmission coefficient is uniform and the electrode can be truly planar, and as such can be used as a hydrodynamic electrode, for example. The change in absorbance with time due to one of the reagents or products of the electrode reaction characterizes the mechanism.

This type of cell and electrode can be used to photochemically activate an electrochemical system, the electrode reaction being used to detect or electrolyse the species produced (e.g. photoelectrolysis of water, Section 12.5).

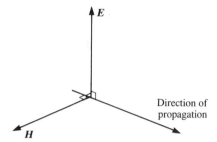

Fig. 12.1. Propagation of electromagnetic radiation: the magnetic, **H**, and electric, **E**, vectors. A beam of ordinary light contains the vectors in randomized directions. $|H| = |E|$.

Reflectance, electroreflectance, and ellipsometry[8]

These techniques depend on the fact that electromagnetic radiation comprises two perpendicular vectors mutually perpendicular to the direction of photon transmission—the magnetic vector **H** and the electric vector **E**, Fig. 12.1. In normal non-polarized radiation, the directions of these vectors are not aligned, their sum being zero. In linearly polarized radiation the vectors are all aligned in the same direction; in partially polarized radiation the alignment is partial.

Reflection at a surface of a beam of linearly polarized photons alters the direction and amplitude of the electric and magnetic vectors. It is these differences between incident and reflected beams that give information concerning surface structure, as they depend on the interaction of the beam with the electronic distribution and with the associated local electric and magnetic fields on the surface. The phase and amplitude change for the vectors is different for the component parallel to the plane of incidence the component perpendicular to it. The result is a vector that follows a spiral during its propagation, and is referred to as elliptically polarized, Fig. 12.2. A deeper treatment of these optical properties can be found in Ref. 9. Such measurements are referred to as *specular reflectance*.

In these experiments cells are designed that have a thin electrolyte layer over the working electrode covered by a window that is transparent to the incident radiation. Refraction of the beam by the window has to be taken into account in the calculations.

The reflectance, R, is the ratio between the intensities of the reflected and incident beams. Experimentally only the intensity of the reflected beam, I_R, is measured and changes in I_R caused by:

- *Modulation of photon intensity* (*reflectance*): The equation for this is

$$\frac{\Delta R}{R} = \frac{\Delta I_R}{R} \tag{12.1}$$

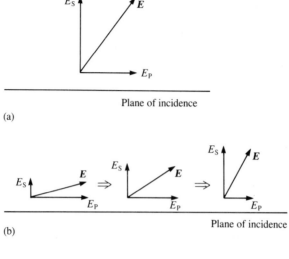

(a)

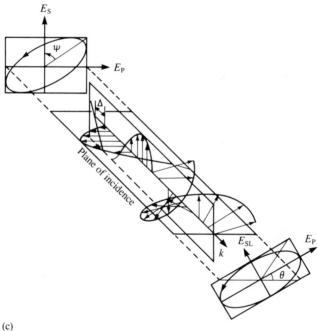

(b)

(c)

Fig. 12.2 Conversion of linearly polarized electromagnetic radiation into ellipti-cally polarized radiation by reflection. Consideration of the electric vector. (a) Decomposition of **E** in components perpendicular and parallel to the plane of incidence; (b) Example after reflection: the vector moves anticlockwise; (c) Representation in three dimensions.

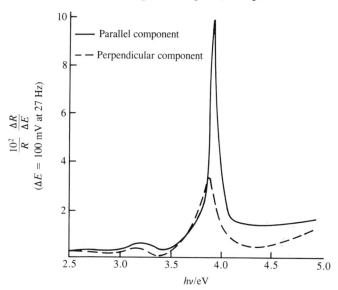

Fig. 12.3. Typical electroreflectance results: Ag in $1.0\,M\,NaClO_4$; $E_{dc} = -0.5\,V$ vs. SCE (from Ref. 10 with permission).

• *Modulation of potential (electroreflectance)*[10,11]: The equation is

$$\frac{1}{R}\frac{\Delta R}{\Delta E} = \frac{\Delta I_R}{I_R \Delta E} \qquad (12.2)$$

Potential modulation causes alterations in the electrode surface that, measured as a function of incident light frequency, permits conclusions to be taken in relation to electronic structure, adsorption, film formation, etc. The type of curve obtained is shown in Fig. 12.3.

Whereas electroreflectance is conventionally used in the UV/visible region, in the 1980s Bewick *et al.*[12–14] developed the technique in the IR region. Since IR radiation interacts with the vibrations of chemical bonds, important information regarding bonds with adsorbed species has been obtained[15], especially useful for research into electrocatalysts. New developments in signal processing, such as the Fourier transform (FT-IR), have veen very valuable.

There are three types of infrared experiment that can be conducted:

1. *Electrochemically modulated infrared spectroscopy* (EMIRS), involving potential modulation. Modulation frequencies from 1–100 Hz are employed and phase-sensitive detection used to calculate ΔR.

2. *Subtractively normalized interfacial Fourier transform infrared spectroscopy* (SNIFTRS). Using an FTIR spectrometer, multifrequency

determinations are possible, but instrumental limitations with respect to speed of data collection put too low an upper limit on the potential modulation frequency. Spectra are therefore recorded at two different potentials and subtracted.

3. *Infrared reflection-absorption spectroscopy* (IRRAS) is done at fixed potential. Electric vectors in the incident beam parallel to the metal surface do not interact with adsorbed molecules, whereas those perpendicular to the surface do. The light beam is switched successively between the two directions and the results subtracted.

We can measure the phase change of the intensities of the perpendicular and parallel components of the electric vector after reflection: this is *ellipsometry*[9,16-18]. In the most accurate instruments the reflected beam is adjusted successively by a group of polarizers until the beam is totally extinguished. The results are conveniently represented in a plot of parallel vs. perpendicular intensity in relation to the incident beam. Applications are especially to the study of film growth on electrode surfaces (Fig. 12.4).

Internal reflection[30]

An electrode (or any other transparent material such as an optical fibre), reflects radiation internally if the angle of incidence is larger than the critical angle. Prisms are used to let the radiation enter and leave (Fig.

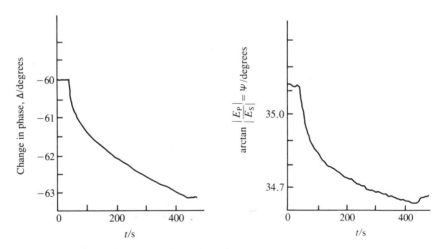

Fig. 12.4. Ellipsometry trace for polymerization of thionine on platinum, showing effect of coating between $t = 50$ s and $t = 450$ s, using light of $\lambda = 450$ nm (solution 40 μM thionine in 0.05 M H_2SO_4) (from Ref. 19 with permission).

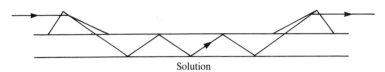

Fig. 12.5. A scheme for internal reflection.

12.5). Radiation is reflected at the solid/solution interface where it interacts optically. In fact the electric field of the wave extends into solution following the relation

$$\langle E^2 \rangle = \langle E_0^2 \rangle \exp\left(-x/\delta\right) \tag{12.3}$$

$\langle E^2 \rangle$ is the average of the squared amplitude of the electric vector at distance x inside the solution, $\langle E_0 \rangle$ is the vector's value when $x = 0$, i.e. at the interface, and δ the penetration distance. This is the *evanescent wave*, which exists in solution and interacts with species that adsorb in that region of small thickness. Absorbance measurements can give the concentration of an electroactive species, and if the process is transient, determine the development of the concentration profile at the electrode surface. The radiation can be visible or, for adsorbed species, in the IR region.

Raman spectroscopy

The Raman effect is due to the interaction of photons with the vibrations of chemical bonds. These wavelengths are in the IR region but the excitation source for observation of the Raman effect is often in the visible region. The major part of incident light passes through the system without change of photon energy (*Rayleigh effect*); if there is energy exchange with the system the *Raman effect* is observed.

The photons excite the molecules to a virtual electronic state, (Fig. 12.6*a*) from which emission occurs; emission to the ground vibrational state is *Rayleigh scattering*. If the photons have a very high energy then the virtual state is within the vibration levels of the excited electronic state. In this case there is a much greater interaction between the radiation and the molecules and an increase in intensity of the Raman effect by a factor of 10^4–10^6—the *resonance Raman effect* (Fig. 12.6*b*).

The resonance Raman effect has been applied to electrochemical cells, generally with laser excitation[21,22]. As it is possible to construct cells that are transparent to IR, it is not necessary to use transparent electrodes. The Raman results are useful for mechanistic diagnosis and for investigating the vibrational and electronic properties of the species under study.

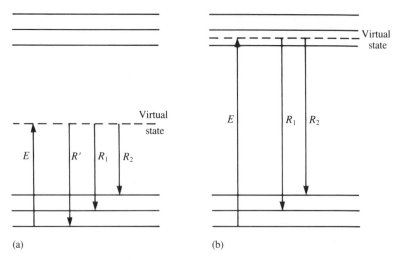

Fig. 12.6. Raman spectroscopy—radiation emission. (a) The normal Raman effect. R' represents Rayleigh scattering; (b) The resonance Raman effect.

A related technique is based on the fact that signals from adsorbed species are much larger than from the same species in solution (*surface enhanced Raman spectroscopy*, SERS)[23,24]. The phenomenon was first noted in a study of the adsorption of pyridine on silver electrodes[25], and has been extended to the investigation of the adsorption of many species such as, for example, porphyrins.

Electron spin resonance (ESR) spectroscopy[26]

Electron spin resonance spectroscopy is used for detecting and identifying paramagnetic species. ESR spectroscopy is particularly powerful when used together with other techniques. Thus in electrochemistry it is useful for studying radicals, radical ions, and certain species containing transition metals, which are produced by electrode reactions, and is valuable in elucidating mechanisms. It has been much used with stationary electrodes, but also with tubular and channel hydrodynamic electrodes[27]. These last offer the possibility of easy quantitative calculation of the radical flux, allowing determination of kinetic parameters. The electrode that produces the radicals can be positioned upstream of the spectrometer cavity, or within it, in order to increase the signal intensity from radicals with very short lifetimes.

X-ray absorption spectroscopy[28-30]

The absorption coefficient, μ, of atoms for X-rays with increasing X-ray energy (frequency) decreases until threshold values corresponding to photoionization of core electrons are reached. There is then a rapid increase in μ, called an *absorption edge*. Atoms give rise to several absorption edges due to the different core electron shells: K (one edge), L (three edges), and M (five edges). For an isolated atom, after an absorption edge is reached there is a steady decrease in μ until the next edge. However, for atoms surrounded by other atoms there is interaction of the emitted photoelectron with these leading to backscattering (Fig. 12.7a). This in turn leads to oscillations in the absorption coefficient (Fig. 12.7b). These oscillations are a probe of short-range atomic order.

Beyond the absorption edge the variation in absorption coefficient can be divided into two regions. If the photoelectron energy is sufficiently high (40–1000 eV above the absorption edge) then we can approximate that the frequency of oscillation of the absorption coefficient depends on the distance between the central absorber and its neighbour, whereas the signal amplitude depends on this and the identity and concentration of neighbours. This is the region of *extended X-ray absorption fine structure* (EXAFS). Information on bond distances, coordination numbers, and atom identification can be obtained.

For lower photoelectron energies, interaction with the electronic environment of the absorber atom is much greater and each electron cannot be considered singly. The fine structure is richer and gives information such as spin state and local symmetry, but theoretical

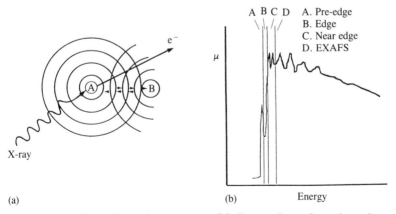

A B C D	A. Pre-edge
	B. Edge
	C. Near edge
	D. EXAFS

(a) (b) Energy

Fig. 12.7. X-ray absorption spectroscopy. (a) Interaction of a photoelectron produced by X-ray absorption with neighbouring atoms, showing backscattering; (b) Typical absorption spectrum, showing the various regions.

modelling is more difficult. This is referred to as the *X-ray absorption near edge structure* (XANES) region.

For EXAFS and particularly for XANES, data analysis is complex. The oscillation frequency/bond distance dependence means that extensive use is made of Fourier transform analysis. Most applications to date have been in the EXAFS region. In order to acquire sufficiently strong signals in a reasonable time, use has to be made of high-intensity photon fluxes, which are available at synchrotron facilities. These provide a broad-band tuneable source of high-intensity radiation, but the reduced number of facilities limits widespread dissemination of the technique. Reflection (fluorescent detection) mode is usually preferred to transmission. Experiments can be conducted in any phase, and the probing of electrode surfaces *in situ* is an important application.

Types of electrode/solution interface studied include oxide films on metals, monolayer deposits obtained by underpotential deposition, adsorption, and spectroelectrochemistry in thin-layer cells.

A useful illustrative example is shown in Fig. 12.8 of iron passivated by chromate and nitrite. Fourier transform of the EXAFS data to give distance-dependent signals shows the similarities and differences between the two passivation methods.

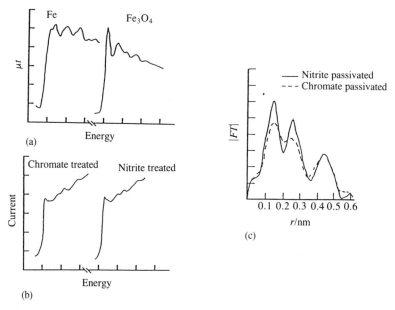

Fig. 12.8. Passivation of iron by chromate and nitrite. Raw X-ray absorption data of (a) Fe and Fe_3O_4 and (b) Fe films after treatment; (c) EXAFS data after Fourier transform (adapted from Ref. 31 with permission).

Other X-ray surface probe methods which have not as yet found widespread use but are applicable to electrode/solution interfacial studies are based on X-ray standing waves and glancing angle X-ray diffraction.

Second harmonic generation (SHG)[32,33]

In a non-centrosymmetric medium, two photons of frequency can be converted into one orientation-dependent photon with the second harmonic frequency 2ω. Such photons can also be created at the interface between two centrosymmetric media, for example between a face-centred cubic metal electrode and aqueous solution, within 1 nm of the interface. The fundamental or the second harmonic wavelength can be tuned to that of molecular electronic transitions, leading to resonance. SHG is thus an interfacial probe, enabling discrimination between surface adsorbed species and those in adjacent bulk media with sub-monolayer sensitivity as well as giving information on the orientation of the adsorbed species. Air/solid, liquid/air, liquid/liquid, and electrochemical interfaces have been studied. In the latter case this has been particularly at monocrystalline and polycrystalline platinum electrodes. So far, applications have been in the visible range using short light pulses from dye lasers. Application to vibrational transitions in the IR region have been hindered by the relatively low sensitivity of IR detectors, although in principle this problem can be circumvented by using IR-visible sum-frequency generation (SFG).

12.3 *Ex situ* spectroscopic techniques

These techniques[34] are used for studying solid electrode surfaces. The energy of bombardment of the electrode by photons, electrons, etc. is higher than for *in situ* techniques, and the experiments are carried out under vacuum conditions. Penetration of incident radiation is therefore greater. Electrodes are removed from the electrochemical cell with due care and mounted in a spectrometer in exactly the same way as any other solid sample. From recent studies, it appears that the double layer remains intact during the immersion process and introduction into the spectrometer, making the study of the double layer possible, and not only that of adsorbed species[35].

The most important techniques are described below.

Photoelectron spectroscopy (XPS)[36–39]

If a photon has sufficient energy, on bombarding atoms or ions it can cause electron emission. To a first approximation (*Koopman's theorem*)

$$h\nu = \tfrac{1}{2}m_e v^2 - E_i \tag{12.4}$$

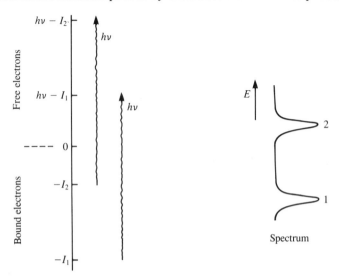

Fig. 12.9. A photoelectron spectrum of an isolated atom. A monochromatic source of electromagnetic radiation is used and the detector energy swept.

where $\frac{1}{2}m_e v^2$ is the kinetic energy of the emitted electron and E_i its ionization energy (Fig. 12.9). For studying surfaces usually X-ray photons are of interest as they interact with electrons close to the nucleus and the energy of these electrons is characteristic of the element, and not of the atomic environment. For this reason the technique was designated *electron spectroscopy for chemical analysis* (ESCA), but it is also known as XPS. (Use of photons in the UV region in UPS probes the ionic environment, emitting valence electrons.) The penetration of X-rays into solids is 5 nm, so XPS only gives information about surface structure. Fortunately, it is also possible to determine the oxidation state of the elements through XPS, since the ionization energy is higher the higher the oxidation state.

Applications have been principally to observe the surface of modified electrodes, anodic oxide films, and in electrodeposition. Figure 12.10 gives an example.

Auger electron spectroscopy (AES)[36,38]

In XPS a vacancy is created in an electronic level close to the nucleus by photon bombardment or, in certain cases, by electron bombardment. It is probable that this vacancy is filled by an electron coming from a higher electronic level further from the nucleus (Fig. 12.11). The excess energy

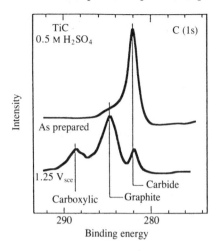

Fig. 12.10. C(1 s) photoelectron spectrum of a reactively sputtered TiC electrode on a glass substrate. (a) As prepared; (b) After 5 min at +1.25 V vs. SCE (passive region) in 0.5 M H_2SO_4 (from Ref. 39 with permission).

is then dissipated by

- photon emission in the X-ray region (*fluorescence*) and/or
- emission of a secondary electron (*Auger electron*).

As in XPS, Auger electrons are characteristic of the emitting species.

In AES applied to electrochemical studies, normally electron bombardment is employed as it furnishes higher intensities of incident radiation, increasing the number of Auger electrons and facilitating their

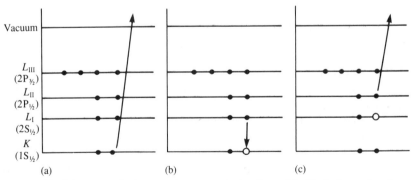

Fig. 12.11. Diagram of Auger electron production. (a) Emission of primary electron; (b) Decay of an electron from $L_1 \rightarrow K$; (c) Emission of Auger electron. Sometimes photons are emitted instead of electrons (X-ray fluorescence).

detection. It should be noted, however, that electrons have a more destructive effect than photons on surface structure. It is not a technique for studying the double layer.

AES is much used to make depth profiles of solid structure. The procedure consists in recording an Auger spectrum, bombarding the surface with high-energy ions (e.g. Ar^+) to remove a few atomic layers, taking another Auger spectrum, etc. This scheme is particularly useful in the investigation of porous and non-porous films formed anodically on metals such as nickel or aluminium.

Electron energy loss spectroscopy (EELS)

Electrons that bombard a surface are usually inelastically scattered. The energy can be transferred, in the same way as for photon bombardment, to cause secondary electron emission (Auger electrons) or to cause photon emission (X-ray fluorescence), which gives information on elemental composition as described above. *Electron energy loss spectroscopy* (EELS) consists of detecting the reflected or transmitted incident electron beam, which gives information on energy loss and momentum transfer, and can be related to elemental composition. Its particular use is in detecting light elements undetectable by X-ray fluorescence due to their low X-ray cross-sections. Thus the two techniques are complementary.

EELs is usually employed at high electron kinetic energies (up to 10^5 eV) with thin samples in the transmission mode. For the electrochemist, low kinetic energies (<1 eV) in the reflection mode are of greatest interest as they can be used to study the vibrational spectra of adsorbates[40].

Electrochemical mass spectrometry (ECMS)[41] and secondary ion mass spectrometry (SIMS)[42, 43]

Mass spectrometry (MS) is a gas-phase technique in which atoms or molecules present in the spectrometer chamber are ionized, and follow a trajectory through applied electric and magnetic fields which separates them according to their mass/charge ratio. A number of procedures have been developed to enable MS to be used for analysing species in the liquid and solid phases, and are based on species extraction into the gas phase. These include plasma desorption, ion bombardment, thermospray and electrospray ionization, and laser desorption. In this section we concentrate on techniques useful to electrochemistry.

To complement electrochemical studies two types of interface between

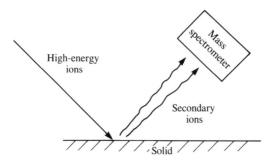

Fig. 12.12. Principle of secondary ion mass spectrometry (SIMS).

electrochemical (EC) cells and mass spectrometers (MS) have been devised:

• *porous interfaces* (porous electrodes or membranes) for the analysis of gaseous products of electrode reactions

• *thermospray interfaces* for determining polar and non-volatile compounds in solution

In the latter case, solution entering a vaporizer probe attached to the electrochemical cell is rapidly vaporized by resistive heating and sucked into the MS by the vacuum. Applications of ECMS have been in mechanistic analysis, and in the fields of electrocatalysis, batteries, sensors, and corrosion.

Adsorbates on solid electrode surfaces can also be examined by MS if they are removed from the surface by bombardment with a high-energy primary ion beam. The species removed are secondary ions derived from the surface constituents and are detected by the mass analyser (Fig. 12.12). Thus the technique is referred to as secondary ion mass spectrometry (SIMS). It turns out that sensitivity is greater than with AES or XPS. The obvious disadvantage is the destruction of the top layers of the solid and implantation in the solid of ions used in the bombardment.

Low-energy and reflection high-energy electron diffraction (LEED and RHEED)[44]

Whereas high-energy electron bombardment of a surface at high angles of incidence leads to X-ray and Auger electron emission, if the electrons are of lower energy, corresponding to between 10 and 500 eV, the interaction is different. In this case the wavelength of the electrons (0.05–0.4 nm) is of the same order of magnitude as interatomic distances and there is elastic scattering: low-energy electron diffraction (LEED)

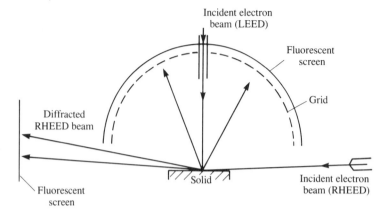

Fig. 12.13. Schematic view of electron diffraction apparatus for low-energy electron diffraction (LEED) and reflection high-energy electron diffraction (RHEED).

(Fig. 12.13). Constructive and destructive interferences give rise to points of diffracted light on a two-dimensional plot. Analysis of this data gives information about surface structure and adsorbates[45].

An electrode can be characterized by LEED before and after the electrochemical experiment. Differences give information on adsorption and eventual movements of atoms over the surface. The great practical difficulty is the necessity of locating the electrode in exactly the same position for the two diffraction experiments.

The diffraction pattern arising from reflection of a beam of high-energy electrons (RHEED) of energy 40 keV at grazing angles of incidence yields information concerning essentially only one crystallographic direction in a particular experiment. Experiments at different sample positions therefore need to be carried out. Advantages with respect to LEED arise principally from the lower interaction of the incident beam with the surface, so that multiple scattering effects can be neglected which can simplify data analysis.

12.4 *In situ* microscopic techniques

Recently a whole new family of microscopic techniques has been invented which enable the *in situ* imaging of solid surfaces using local probes. By local we mean that resolution can approach atomic dimensions, implying that probe size and accuracy of controlling its movement over the surface limit the resolution. The probe is usually located at atomic distances from the sample and is not influenced by the medium.

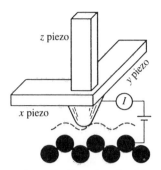

Fig. 12.14. Scheme of operation of piezoelectric drives for scanning local probe
microscopy.

Not only can experiments be performed *in situ,* but also the time
evolution of surface topography can be monitored. Accurate scanning of
the surface is accomplished by means of piezoelectric drives—the size of
a piezoelectric crystal changes linearly with the applied potential
difference. Drives are applied in *x, y* and *z* directions as shown
schematically in Fig. 12.14. Generally, successive scans are applied in the
x direction incrementing *y* between scans, so that a series of *x–z* profiles
are recorded. These can be converted by computer software into
gray-scale or coloured images if desired.

From a practical point of view, adequate vibration-free experimental
conditions must be assured given the resolution in question. Vibration-
free tables are available, but in many cases suspending a concrete block
on which the working part of the instrument is located from the ceiling
with elastic cords is perfectly adequate.

The most important of these techniques is *scanning tunnelling microscopy* (STM), the invention of Binning and Rohrer[45], for which they won
the Nobel Prize in Physics in 1986, followed by *atomic force microscopy*
(AFM)[47], and which are described in this section, indicating their
application to the study of electrode processes.

Other local probe techniques to be discussed, of an electrochemical
nature, which rely on much of the same instrumental technology, are
scanning electrochemical microscopy (SECM) and *scanning ion conductance microscopy* (SICM).

Scanning tunnelling microscopy[46,48,49]

This local probe technique relies on a tunnelling current being passed
between tip and substrate. For electron tunnelling to be possible the
distance between tip and substrate must be less than about 2 nm, and the

magnitude of the current increases markedly as the tip is brought closer. The tip itself must therefore also be of atomic dimensions. Materials used for the tips, which are usually etched, are tungsten and platinum/rhodium or platinum/iridium alloys. The phenomenon occurs essentially because of overlap between the electron clouds of substrate and tip. Strictly speaking, therefore, it is the electron density of the substrate that is mapped by scanning and not the atomic topography.

In general, a constant tunnelling current is imposed and the height of the tip adjusted accordingly, which leads to the topographic image. The tip/substrate voltage bias can be altered and topography registered for each bias. Since it is electron density that is recorded, changes with varying positive or negative voltage bias can give spectroscopic information. This is known as *current imaging tunnelling spectroscopy* (CITS)[50], although it has, as yet, been little explored.

Much of the initial work was done in vacuum and with semiconductor and metal monocrystals. These experiments showed evidence of surface reconstruction, and the presence of substances that adsorb on these substrates.

Studies can also be carried out in electrolyte solution, with the substrate acting as the working electrode. Thus real-time records can be made of nucleation and new phase formation, for example in under-potential deposition, surface corrosion, and adsorption in a general sense. The only limitation is that the substrate must be a conductor or a good semiconductor, such as silicon or germanium. Two examples of scanning tunnelling micrographs are shown in Fig. 12.15.

Atomic force microscopy (AFM)[47,53]

The principal limitation of STM is that it cannot be used with insulating substrates. However, at the sort of distances where tunnelling currents occur, there is an attractive or repulsive force between atoms in the tip and the substrate, which is independent of the conducting or non-conducting nature of the substrate. In order to measure this, the tip is mounted on the end of a soft cantilever spring, the deflection of which is monitored optically by interferometry or beam deflection. These cantilever springs are microfabricated photolithographically from silicon, silica, or silicon nitride and have lateral dimensions of around $100\,\mu\text{m}$ and thickness of $1\,\mu\text{m}$, to which tiny diamond tips are attached.

Resolution is generally not quite so high as with STM. However, applications to some forms of polymerization, protein adsorption on insulating surfaces, and corrosion of anodic oxide films are made possible by this arrangement.

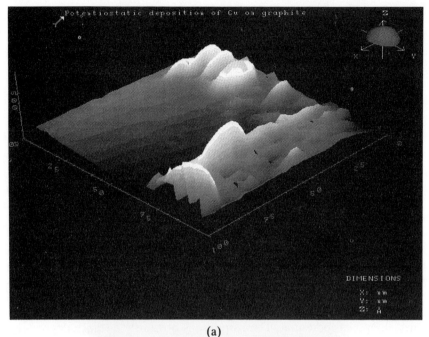

(a)

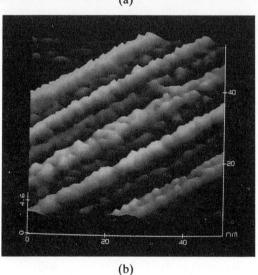

(b)

Fig. 12.15. Examples of scanning tunnelling micrographs. (a) Copper electrode-posited from $Cu(NO_3)_2$ solution on highly ordered pyrrolytic graphite and imaged at $-0.5\,V$ vs. $Ag\,|\,AgCl$ (tip bias potential $+0.6\,V$) (from Ref. 51 with permission); (b) Electropolymerized polypyrrole with poly(4-styrenesulphonate) anion on graphite substrate (from Ref. 52 with permission).

Scanning electrochemical microscopy (SECM)[54]

Scanning electrochemical microscopy seeks to overcome the lack of sensitivity and selectivity of the probe tip in STM and AFM to the substrate identity and chemical composition. It does this by using both tip and substrate as independent working electrodes in an electrochemical cell, which therefore also includes auxiliary and reference electrodes. The tip is a metal microelectrode with only the tip active (usually a metal wire in a glass sheath). At large distances from the substrate, in an electrolyte solution containing an electroactive species the mass-transport-limited current is therefore

$$I = 4nFaDc_\infty \tag{12.5}$$

as discussed in Section 5.5. As the tip is brought close to the substrate modifications in the current appear, as shown in Fig. 12.16. If the substrate is an insulator then diffusion of electroactive species from bulk solution is hindered and the current falls. If it is a conductor and held at a potential such that the electrode reaction occurring at the tip is reversed, then the current is enhanced—this may be due to the 'natural' potential of the substrate arising from open circuit charge transfer processes or through applied potential. The response is obviously distance dependent. Besides topographical imaging using constant current mode, chemical and electrochemical information can be obtained depending on electrolyte conditions, and potentials applied to tip and substrate.

There are two types of application of this microscopic technique;

● To monitor substances and electrode surfaces, from thin films to biological or polymer.

● To fabricate structures by electrochemical etching of metallic and semiconductor electrodes with high precision and accuracy.

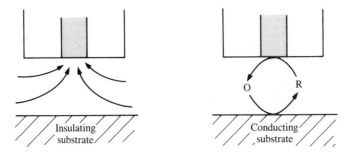

Fig. 12.16. Schematic effect of substrate type on microelectrode tip response: (a) insulating—current reduction; (b) conducting—current enhancement by feedback.

The resolution of the technique is limited by the size of the micro-electrode tip, at present 200 nm. In the future, reduction of tip size to tens of nanometres by use of novel microelectrode fabrication procedures should increase the applicability of SECM.

Scanning ion conductance microscopy (SICM)[55]

This technique is also done in electrolyte solution, but shows some important differences from the other *in situ* microscopic techniques. The tip is a tiny glass micropipette contained within which is a small electrode, whose function is to measure ion currents. A potential difference is applied between tip and substrate. The current which is registered diminishes as the tip approaches the substrate surface, disappearing altogether when they contact since the micropipette opening is blocked. Thus surface topography under constant current mode can be registered. However, the most exciting potential application is to detect ion channels in biomembranes, and determine what stimuli cause them to open and close, particularly useful if the technique can be made ion-specific. At present, resolution is about 200 nm.

12.5   *Ex situ* microscopic techniques: electron microscopy

Electron microscopy[56], most commonly employed as scanning electron microscopy (SEM), is now a widely used tool in examining the morphology of surfaces under vacuum conditions, and so in an electro-chemical context electrode surfaces and corroded surfaces. Instruments also permit chemical microanalysis to be carried out.

One of the reasons as to why we should use electron microscopy rather than optical microscopy to obtain topographical images is shown in Fig. 12.17: the resolution obtainable can reach almost atomic dimensions. The

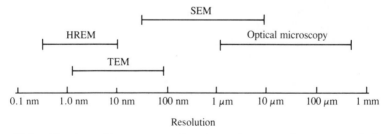

Fig. 12.17. Usual working ranges of various electron microscopy techniques: SEM, scanning electron microscopy; TEM, transmission electron microscopy; HREM, high-resolution electron microscopy.

other reason, in the region where electron and optical microscopy overlap, is the very high depth of focus of electron microscopy, unattainable with optical instruments.

In electron microscopy a sample is bombarded with a finely focused beam of monochromatic electrons. Products of the interaction of the incident electron beam with the sample are detected. If the sample is sufficiently thin—up to 200 nm thickness—the beam is transmitted after interacting with the sample, leading to the technique of transmission electron microscopy (TEM). TEM is used to probe the existence of defects in crystals and phase distributions. Scanning TEM instruments have been recently developed to obtain images over a wider area and to minimize sample degradation from the high-intensity beams.

If the sample is thicker, then interaction is with the surface of the sample and the products of this interaction follow trajectories away from the surface. In scanning electron microscopy (SEM), it is usually the secondary emitted electrons that are detected, the electron beam being scanned (rastered) across the sample surface. For reflection techniques non-metallic surfaces have to be coated with a thin metal film (about

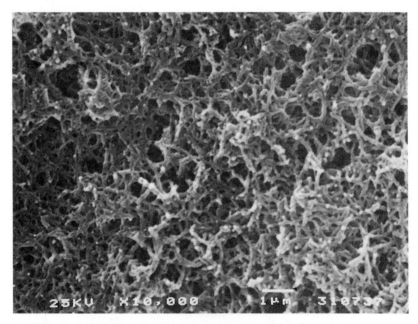

Fig. 12.18. Scanning electron micrograph of polyaniline, obtained by polymerization of 0.1 M aniline in 0.5 M H_2SO_4 for 10 min on a GC electrode, scanning the potential between -0.2 V and $+0.8$ V vs. SCE at 50 mV s^{-1}.

10 nm), for example by sputtering with gold, to prevent build-up of charge on the surface of the sample. An example of a scanning electron micrograph is shown in Fig. 12.18.

From the discussion of *ex situ* spectroscopic techniques earlier in the chapter it is clear that other products of the interaction between incident beam and the surface can be detected. One of these is backscattered electrons (BSE) which give an image in which heavy elements lead to high backscattering (white areas) and light elements lead to low backscattering (black areas). Thus a very qualitative form of elemental analysis can be performed by BSE detection.

Electron microprobes permit chemical microanalysis as well as SEM and BSE detection, often referred to as *analytical electron microscopy* (AEM), or electron probe microanalysis (EPMA)[56,57]. This is because another product of the surface interaction with an incident electron beam is X-ray photons which have wavelengths and energies dependent on element identity and on the electron shell causing the emission. Analysis of these photons can give a local chemical analysis of the surface. Resolution of 1 μm is attainable. Two types of X-ray spectrometer can be employed:

- *Wavelength dispersive spectrometer* (WDS) in which the wavelength is scanned, and the wavelengths corresponding to the ejected photons determined, by using diffraction from crystals mounted in the spectrometer and manipulating the angle between the crystal surface and the photon beam.

- *Energy dispersive spectrometer* (EDS),. in which a multichannel analyser gives the photon energy spectrum.

In both cases peaks can be assigned to particular elements and their areas to the percentage of the element present. WDS has a much better resolution than EDS, although until recently the former was applicable to significantly fewer elements. However, with state-of-the-art instruments and anticontamination devices for light elements, elements from carbon upwards can be quantitatively analysed by WDS.

TEM instruments can also perform microanalysis, EDS being the most used detection technique at present, together with EELS (see Section 12.3) for light elements.

One of the interesting applications is that, by fixing the wavelength (or energy) corresponding to a particular element, an image of the solid surface (for SEM) or solid sample (for STEM) showing the distribution of the element can be obtained. This can give very useful information on corrosion phenomena and on surface, in particular electrode surface, composition.

Complementary information to that obtainable by EPMA, particularly regarding *in situ* identification of adsorbed species may be by the newly-developed technique of scanning infra-red microscopy (SIRM).

12.6   Other *in situ* techniques

There are non-electrochemical techniques that are being used to characterize the electrode/electrolyte interface and that cannot be grouped into the categories of the previous sections. These are based on measurements of mass and heat change.

Measurement of mass change: the quartz crystal microbalance (QCM)[58,59]

Quartz crystals have a characteristic oscillation frequency which varies according to their mass. Although crystal wafers have been used as mass sensors in vacuum and gas-phase experiments for many years, it is only recently that they have been employed in contact with liquids or solutions. Quartz crystal wafers can be used as electrodes by depositing a thin film of electrode material on the exposed surface, and interfacial mass changes can then be monitored. It is then known as the electrochemical QCM or EQCM. It is a direct, but non-selective, probe of mass transport.

The design of the assembly is shown in Fig. 12.19. The quartz wafer is sandwiched between two electrodes that apply an oscillating electric field, resulting in a standing wave within the wafer and in mechanical oscillation at resonant frequencies, generally in the range from 2 MHz to 20 MHz. A wafer of thickness 320 μm oscillates at around 5 MHz. At this

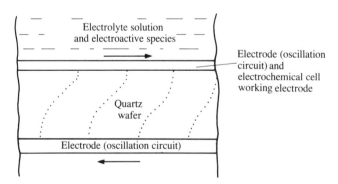

Fig. 12.19. Design of the quartz crystal microbalance for electrochemistry, showing the applied electric field and external contact.

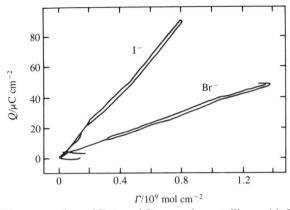

Fig. 12.20. Electrosorption of Br^- and I^- on polycrystalline gold. Plot of charge passed vs. coverage determined by EQCM. Slope is the electrosorption valency (from Ref. 60 with permission).

frequency a mass change of 18 ng cm^{-2} of an electrode causes a frequency change of 1 Hz, a resolution attainable with good frequency counters.

Examples have been in measuring monolayers obtained by underpotential deposition and sub-monolayer by electrosorption. The measurement of mass changes in thin films such as redox and conducting polymers has been studied, particularly when accompanying redox processes: QCM measurements can give information on both counter-ion and solvent incorporation. Figure 12.20 shows a plot of charge passed for adsorption of Br^- and I^- on polycrystalline gold electrodes vs. the coverage as measured by EQCM. Charge transfer from adsorbed species to the electrode reduces their ionic character—the charge transferred is called the *electrosorption valency*. The plot shows direct evidence of partial charge transfer from adsorbed bromide ion to the electrode (electrosorption valency -0.4), but total transfer in the case of iodide (electrosorption valency -1.0).

Measurement of absorbed radiation: thermal changes

If an electrode has an irregular surface, it is sometimes more profitable to detect absorbed radiation rather than transmitted or reflected radiation. Radiation absorption causes an increase in electrode temperature that can be detected directly with a thermocouple or thermistor (photothermal spectroscopy[61]).

Alternatively, a sharp change in temperature induced by an intermittent light beam of high intensity causes pressure fluctuations in the vicinity of the electrode. Detection of this pressure change is known as *photoacoustic spectroscopy*[62]. Detectors are conventionally piezoelectric or microphones. However, since the temperature change alters the refractive index of the fluid close to the electrode, the light beam (usually from a laser) is deflected from its original path—detection of this beam is known as *photothermal deflection* or *mirage effect detection*. At the solid/liquid interface this detection mode has been shown to be two orders of magnitude more sensitive than the other types of detector.

Electrochemical applications up the present have been few, but wider use can be predicted in the future.

12.7 Photoelectrochemistry

Photoelectrochemical reactions are those where incidence of photons on an electrode excites an electron within the electrode or excites a species in solution so that an electrochemical reaction occurs at the electrode. In both cases a current flows if the conditions are appropriate.

The first of these is particularly pertinent for semiconductor electrodes[63] (Section 3.7). This is because the energy of visible light is 1–3 eV, which corresponds to the energy bandgap in a number of semiconductors. These photochemical reactions are of extreme interest owing to possible technological applications using solar energy[64,65].

The mode of operation is as follows. Excitation of an electron in a semiconductor leaves a hole. Depending on the correspondence of energy levels, a species in solution could receive the excited electron or fill the hole with one of its own electrons, so that there are two possible types of reaction. Additionally electron–hole recombination can occur, the energy excess being released as thermal energy; this recombination often proceeds through surface states and it is desirable to avoid it. Measurement of the photocurrent gives information on semiconductor properties in the space–charge region[66].

Figure 12.21 shows the effect in an n-type semiconductor of promotion of electrons by incident photons and subsequent electrode reactions. This figure should be compared with Fig. 6.9 for an n-type semiconductor without incident radiation. Irradiation facilitates oxidation, a significant overpotential being unnecessary. Figure 12.22 compares schematically what is obtained at semiconductor electrodes with and without incident light. As is to be expected there is no photoeffect (except in rare cases) for potentials more negative than U_{fb}.

For a p-type semiconductor, there is no anodic photoeffect and irradiation promotes reduction and passage of a cathodic current.

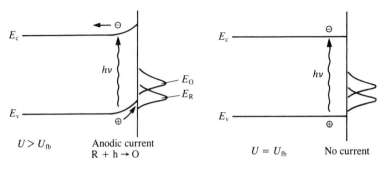

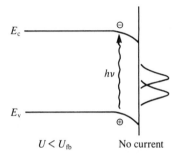

Fig. 12.21. The effect of incident light on an n-type semiconductor and on electron transfer. Electron-hole separation is promoted only in the space-charge region. The energy of the redox couple, E_{redox}, determines if there is oxidation or reduction.

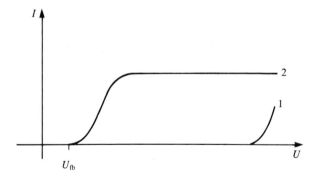

Fig. 12.22. Oxidation at an n-type semiconductor. Curve 1: no light; Curve 2: with light.

There are three types of photoelectrochemical cell. The cell, besides the semiconductor electrode, contains auxiliary and reference electrodes that are not light sensitive:

1. *Photovoltaic cells.* As the name suggests, these involve direct conversion of light into electric current. The reaction at the auxiliary electrode is the inverse of that at the semiconductor electrode. In principle, there is no change in electrolyte composition with time. An example would be

$$n\text{-}TiO_2 \,|\, NaOH, O_2 \,|\, Pt$$

with oxidation of OH^- at n-TiO_2 and reduction of oxygen at the platinum electrode.

2. *Photocatalytic cells.* As in (1) above the reaction functions in the sense $\Delta G < 0$ but the photons are used to overcome the activation energy barrier. These cells are used in converting substances. Probable applications are exemplified by the decomposition of acetic acid into ethane, carbon dioxide and hydrogen:

$$n\text{-}TiO_2 \,|\, CH_3COOH \,|\, Pt$$

3. *Photoelectrolytic and photogalvanic cells.* These cells involve the conversion of radiant energy in chemical energy for converting substances but unlike the situation in (2), $\Delta G > 0$. The potential applied to the electrode helps the conversion.

● *Photoelectrolytic cells.* Chemical compounds are converted irreversibly. Relevant examples of possible industrial importance are the decomposition of water or hydrogen sulphide into hydrogen and oxygen or hydrogen and sulphur respectively.

● *Photogalvanic cells*[67]. The conversion of substances is used as a way of storing photon energy. On connecting the external circuit the cell supplies current regenerating the initial chemical compounds: it is thus similar to a battery. Substances used are generally sulphonated dyes such as thionine. Unfortunately the efficiency is low, and thus their commercial application is unlikely.

The semiconductior electrode most studied in photoelectrolytic cells has been n-TiO_2, and in photogalvanic cells n-SnO_2. Because the bandgap energies are 3.0 eV and 3.5 eV respectively, they are not optimum semiconductors as they only make use of about 5 per cent of the solar energy. For this reason there has been research into other semiconductors, for example cadmium sulphide. In all cases the efficiency is fairly low.

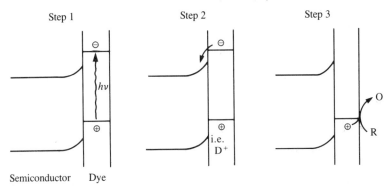

Fig. 12.23. Mediation of electron transfer at a semiconductor electrode by dye sensitization.

As a result of these difficulties the possibility of using light-sensitive mediators (i.e. dyes) that aid the semiconductor in using a larger percentage of incident light have been explored[66,69]. We illustrate this idea with n-TiO$_2$ covered by a thin film of a dye such as methylviologen. The dye, D (donor), is excited (sensitized) and injects an electron into the conduction band of the semiconductor, being transformed into D$^+$; following this D$^+$ reacts with R in solution regenerating D. This process is useful if the wavelength of the light necessary to excite the dye is sufficiently long. The process is shown in Fig. 12.23.

In recent years there has been interest in using semiconductor dispersions in the form of colloidal particles instead of macroscopic electrodes[70]. The area/volume ratio is clearly larger, which gives increased yields. Colloidal semiconductors investigated are principally n-TiO$_2$ and cadmium sulphide with adatoms (surface states) of platinum. The particles have to function simultaneously as cathode and anode. Figure 12.24 shows their mode of operation schematically for reduction of A, aided by oxidation of a dye, D.

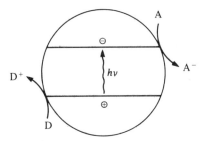

Fig. 12.24. Use of semiconductor particles for photoelectrolysis of A. D is a dye, for example methylviologen.

Recently a low-cost, high efficiency (about 12 per cent in diffuse daylight) photovoltaic cell based on dye-sensitized colloidal titanium oxide films with a conducting glass auxiliary electrode was described[71].

12.8   Electrochemiluminescence[72,73]

The study of the fate of radicals produced by electrode reactions is important in the investigation of species arising in the decomposition or homogeneous reaction of organic compounds. Sometimes radicals, on undergoing homogeneous reactions in solution, emit light. The study of this luminescence is called electrochemically generated chemilumines- cence or *electrochemiluminescence* (ECL).

Whenever there is light emission we can register a spectrum, which is very useful in the identification of the light-emitting species.

We exemplify with rubrene (R) and N,N,N',N',tetramethyl-*p*- phenlydiamine (TMPD) in acetonitrile. Rubrene can give cation or anion radicals, but TMPD gives only cation radicals. At a rotating ring-disc electrode, R^- is generated on the disc and $TMPD^+$ on the ring. If the reaction between the two radicals is fast then a very fine band of light coincident with the internal radius of the ring will appear; if it is slower than the band will be wider and shifted radially. These considerations have been quantified[74].

References

1. R. J. Gale (ed.), *Spectroelectrochemistry, theory and practice,* Plenum, New York, 1988.
2. R. G. Compton and A. Hamnett (ed.), *Comprehensive chemical kinetics,* Elsevier, Amsterdam, 1989, Vol. 29.
3. G. Gutiérrez and C. Melendres (ed.), *Spectroscopic and diffraction tech- niques in interfacial electrochemistry,* Proceedings of NATO ASI 1988, Kluwer, Dordrecht, 1990.
4. R. Varma and J. R. Selman (ed.), *Techniques for characterization of electrodes and electrochemical processes,* Wiley, New York, 1991.
5. H. D. Abruña (ed.), *Electrochemical interfaces: modern techniques for in-situ interface characterization,* VCH, New York, 1991.
6. T. Kuwana and N. Winograd, *Electroanalytical chemistry,* ed. A. J. Bard, Dekker, New York, Vol. 7, 1974, pp. 1–78; W. H. Heinemann, F. M. Hawkridge, and H. N. Blount, Vol. 13, 1984, pp. 1–113.
7. T. Kuwana and W. R. Heinemann, *Acc. Chem. Res.,* 1976, **9**, 241.
8. R. H. Müller (ed.), *Advances in electrochemistry and electrochemical engineering,* Wiley, New York, 1973, Vol. 9.
9. R. H. Müller, in Ref. 8, pp. 167–226.

10. J. D. E. McIntyre, in Ref. 8, pp. 61–166.
11. D. M. Kolb, in Ref. 1, Chapter 4.
12. A. Bewick, *Trends in interfacial electrochemistry,* ed. A. F. Silva, Proceedings of NATO ASI, 1984, Riedel, Dordrecht, 1985, pp. 331–358.
13. A. Bewick and S. Pons, *Advances in infrared and Raman spectroscopy,* ed. R. J. H. Clark and R. E. Hester, Wiley, Chichester, Vol. 12, 1985, pp. 1–63.
14. J. K. Foley, C. Korzeniewski, J. L. Daschbach, and S. Pons, *Electroanalytical chemistry,* ed. A. J. Bard, Dekker, New York, Vol. 14, 1986, pp. 309–440.
15. P. A. Christensen and A. Hamnett in Ref. 2, pp. 1–77.
16. J. Kruger, in Ref. 8, pp. 227–280.
17. R. Greef, in Ref. 2, pp. 427–452.
18. R. W. Collins and Y.-T. Kim, *Anal. Chem.,* 1990, **62,** 887A.
19. A. Hamnett and A. R. Hillman, *J. Electroanal. Chem.,* 1987, **233,** 125.
20. W. N. Hansen, in Ref. 8, pp. 1–60.
21. R. L. McCreery and R. T. Packard, *Anal. Chem.,* 1989, **61,** 775A.
22. R. P. Cooney, M. R. Mahoney, and J. R. McQuillan, *Advances in infrared Raman spectroscopy,* ed. R. J. H. Clark and R. E. Hester, Wiley, Chichester, Vol. 9, 1982, pp. 188–281.
23. R. E. Hester, in Ref. 2, pp. 79–104.
24. R. L. Garrell, *Anal. Chem.,* 1989, **61,** 410A.
25. M. Fleischmann, P. J. Hendra, and A. J. McQuillan, *J. Chem. Soc. Chem. Commun.,* 1973, 80.
26. T. M. McKinney, *Electroanalytical chemistry,* ed. A. J. Bard, Dekker, New York, Vol. 10, 1977, pp. 97–278.
27. A. M. Waller and R. G. Compton, in Ref. 2, pp. 297–352.
28. M. J. Fay, A. Proctor, D. P. Hoffmann, and D. M. Hercules, *Anal. Chem.,* 1988, **60,** 1225A.
29. H. D. Abruña, *Modern aspects of electrochemistry,* Plenum, New York, Vol. 20, 1989, ed. J. O'M. Bockris, R. E. White, and B. E. Conway, pp. 265–326.
30. H. D. Dewald, *Electroanalysis,* 1991, **3,** 145.
31. G. G. Long, J. Kruger, D. R. Black, and M. Kuriyama, *J. Electroanal. Chem.,* 1983, **150,** 603.
32. R. M. Corn, *Anal. Chem.,* 1991, **63,** 285A.
33. G. L. Richmond, *Electroanalytical chemistry,* ed. A. J. Bard, Vol. 17, 1991, pp. 87–180.
34. R. Parsons in Ref. 2, pp. 105–127.
35. D. M. Kolb, *Z. Phys. Chem. Neue Folge,* 1987, **154,** 179.
36. J. Augustynski and L. Balsenc, *Modern aspects of electrochemistry,* Plenum, New York, Vol. 13, 1979, ed. B. E. Conway and J. O'M. Bockris, pp. 251–360.
37. P. M. A. Sherwood, *Chem. Soc. Rev.,* 1985, **14,** 1.
38. K. W. Nebesny, B. L. Maschhoff, and N. R. Armstrong, *Anal. Chem.,* 1989, **61,** 469A.
39. R. Kötz, *Advances in electrochemical science and engineering,* ed. H. Gerischer and C. W. Tobias, VCH, Weinheim, Vol. 1, 1990, pp. 75–126.
40. M. A. Chesters and N. Sheppard, *Advances in infrared and Raman spectroscopy,* ed. R. J. H. Clark and R. E. Hester, Wiley, Chichester, Vol. 16, 1988, pp. 377–412.

41. B. Bittins Cattaneo, E. Cattaneo, P. Konigshoven, and W. Vielstich, *Electroanalytical chemistry*, ed. A. J. Bard, Vol. 17, 1991, pp. 181-220.
42. C. A. Evans Jr., *Anal. Chem.*, 1975, **47**, 818A, 855A.
43. J. C. Vickerman, A. E. Brown, and N. M. Reed, *Secondary ion mass spectrometry: principles and applications*, Oxford University Press, 1990.
44. G. Ertl and J. Küppers, *Low energy electrons and surface chemistry*, VCH, Weinheim, 1985.
45. P. N. Ross and F. T. Wagner, *Advances in electrochemistry and electrochemical engineering*, ed. H. Gerischer and C. W. Tobias, Vol. 13, 1983, pp. 69–112.
46. G. Binning, H. Rohrer, C. Gerber, and E. Weibel, *Phys. Rev. Lett.*, 1983, **50**, 120.
47. G. Binnig, C. F. Quate, and C. Gerber, *Phys. Rev. Lett.*, 1986, **56**, 930.
48. T. R. I. Cataldi, I. G. Blackham, A. D. Briggs, J. B. Pethica, and H. A. O. Hill, *J. Electroanal. Chem.*, 1990, **290**, 1.
49. R. Sonnenfeld, J. Schneir, and P. K. Hansma, *Modern aspects of electrochemistry*, Plenum, New York, Vol. 21, 1990, ed. R. E. White, J. O'M. Bockris, and B. E. Conway, pp. 1–28.
50. R. J. Hamers, *Annu. Rev. Phys. Chem.*, 1989, **40**, 531.
51. D. R. Yaniv and L. D. McCormick, *Electroanalysis*, 1991, **3**, 103.
52. R. Yang, K. Naoz, D. F. Evans, W. H. Smyrl, and W. A. Hendrickson, *Langmuir*, 1991, **7**, 556.
53. D. Rugar and P. Hansma, *Physics today*, Oct. 1990, 23.
54. A. J. Bard, G. Denuault, C. Lee, D. Mandler, and D. O. Wipf, *Acc. Chem. Res.*, 1990, **23**, 357.
55. P. K. Hansma, B. Drake, O. Marti, S. A. C. Gould, and C. B. Prater, *Science*, 1989, **243**, 641.
56. J. P. Eberhardt, *Structural and chemical analysis of materials*, Wiley, Chichester, 1991, Part 5.
57. D. E. Newbury, C. E. Fiori, R. B. Marinenko, R. L. Myklebust, C. R. Swyt, and D. S. Bright, *Anal. Chem.*, 1990, **62**, 1159A, 1245A.
58. M. R. Deakin and D. A. Buttry, *Anal. Chem.*, 1989, **61**, 1147A.
59. D. A. Buttry, *Electroanalytical chemistry*, ed. A. J. Bard, Vol. 17, 1991, pp. 1–86.
60. M. Deakin, T. Li, and O. Melroy, *J. Electroanal. Chem.*, 1988, **243**, 343.
61. G. H. Brlmyer, A. Fujishima, K. S. V. Santhanam, and A. J. Bard, *Anal. Chem.*, 1977, **49**, 2057.
62. D. S. Ballantine Jr. and H. Wohltjen, *Anal. Chem.*, 1989, **61**, 704A.
63. Yu. V. Pleskov and Yu. Ya. Farevich, *Semiconductor photoelectrochemistry*, Plenum, New York, 1986.
64. S. U. M. Khan and J. O'M. Bockris, *Modern aspects of electrochemistry*, Plenum, New York, Vol. 14, 1982, ed. J. O'M. Bockris, B. E. Conway, and R. E. White, pp. 151–193.
65. Yu. V. Pleskov, *Solar energy conversion. A photoelectrochemical approach*, Springer-Verlag, Berlin, 1990.
66. L. Peter, in Ref. 2, pp. 353–383.
67. W. J. Albery, *Acc. Chem. Res.*, 1982, **15**, 142.
68. M. Grätzel, *Acc. Chem. Res.*, 1981, **14**, 376.

69. M. Grätzel (ed.), *Energy resources through photochemistry and catalysis*, Academic Press, New York, 1983.
70. M. Grätzel, *Modern aspects of electrochemistry*, Plenum, New York, Vol. 15, 1983, ed. R. E. White, J. O'M. Bockris, and B. E. Conway, pp. 83–165.
71. B. O'Regan and M. Grätzel, *Nature,* 1991, **353,** 737.
72. L. R. Faulkner and A. J. Bard, *Electroanalytical chemistry,* ed. A. J. Bard, Dekker, New York, Vol. 10, 1977, pp. 1–95.
73. J. G. Velasco, *Electroanalysis,* 1991, **3,** 261.
74. L. R. Faulkner, *Meth. Enzymol.,* 1978, **57,** 494.

PART III

Applications

13

POTENTIOMETRIC SENSORS

13.1 Introduction

Electrochemical measurements for analytical purposes can be carried out under conditions of equilibrium (zero current) with potentiometric sensors or outside equilibrium (passage of current) with amperometric or voltammetric sensors. In this chapter potentiometric sensors currently in use will be described. The cell arrangement, with indicator and reference electrodes in solution and linked to a potentiometer, or high-impedance voltmeter, has already been described in Section 7.4. In principle any electrode can be used for a potentiometric measurement. The equilibrium potential, E_{eq}, results from the sum of the partial anodic and cathodic currents due to the various electrode reactions being equal to zero. In order to make the value of E_{eq} useful for quantitative interpretation, it is necessary to create conditions where secondary reactions can be neglected in relation to the reaction being studied. Selective electrodes were developed with this objective, selectivity being provided by the electrode material. The mode of operation varies according to the type of electrode material—glass, ionic salt, etc.—and its operation is different to that of a metal or semiconductor electrode.

After a description of the simple, but important, technique of potentiometric titrations, the rest of the chapter is devoted to the use and operation of potentiometric selective electrodes.

13.2   Potentiometric titrations[1]

Any titration involves the progressive change of the activities (or concentrations) of the titrated and titrating species and, in principle, can be done potentiometrically. However, for an accurate determination it is necessary that there is a fairly rapid variation in equilibrium potential in the region of the equivalence point. Useful applications are redox, complexation, precipation, acid–base titrations, etc. From the titration curve it is possible to calculate values of the formal potentials of the titrated and titrating species, as explained below.

An important question is whether we can use any indicator electrode. A 'redox' electrode, i.e. inert in the range of potential where measurements are being done, is a possibility, especially for redox titrations. In other cases, the use of electrodes selective to the ion being titrated is better, such as pH electrodes in acid–base titrations. The method of analysis of the data obtained is, naturally, the same in all cases and independent of electrode material.

We illustrate with the general case of a simple redox titration:

$$\underset{\text{titrated}}{n_2 O_1} + \underset{\text{titrant}}{n_1 R_2} \rightarrow n_1 O_2 + n_2 R_1$$

corresponding to the half-reactions

$$O_1 + n_1 e^- \rightleftarrows R_1$$

$$O_2 + n_2 e^- \rightleftarrows R_2$$

An example is $O_1 = Ce(IV)$ and $R_2 = Fe(II)$. The appropriate Nernst equations are

$$E_1 = E_1^{\ominus'} + \frac{RT}{n_1 F} \ln \frac{[O_1]}{[R_1]} \tag{13.1}$$

$$E_2 = E_2^{\ominus'} + \frac{RT}{n_2 F} \ln \frac{[O_2]}{[R_1]} \tag{13.2}$$

Before the equivalence point the couple O_1/R_1 is in excess and determines the potential, and after the equivalence point the couple in excess is O_2/R_2 and this determines the potential. Therefore, by use of expressions (13.1) and (13.2) it is possible to construct the theoretical titration curve if the values of $E_1^{\ominus'}$ and $E_2^{\ominus'}$ are known (Fig. 13.1).

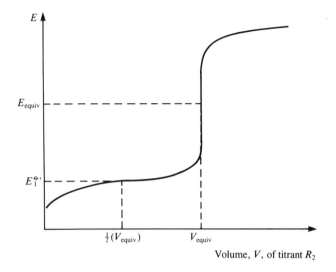

Fig. 13.1. Potentiometric titration between O_1 (titrated) and R_2 (titrant). V_{equiv} is the volume of titrant added at the equivalence point at potential E_{equiv}.

At the equivalence point the potential, $E_{equiv} (= E_1 = E_2)$, can be expressed as the sum of (13.1) and (13.2) after multiplying by n_1 and n_2 respectively:

$$(n_1 + n_2)E_{equiv} = n_1 E_1^{\ominus'} + n_2 E_2^{\ominus'} + \frac{RT}{F} \ln \frac{[O_1][O_2]}{[R_1][R_2]} \qquad (13.3)$$

Since, at the equivalence point, $[O_1] = [R_2]$ and $[O_2] = [R_1]$, (13.3) reduces to

$$E_{equiv} = \frac{n_1 E_1^{\ominus'} + n_2 E_2^{\ominus'}}{n_1 + n_2} \qquad (13.4)$$

$E_1^{\ominus'}$ is the potential when half of the volume necessary to reach the equivalence point has been added—see (13.1); $E_2^{\ominus'}$ is the potential when twice the titrant volume needed to reach E_{equiv} has been added—it can also be calculated from (13.4), given a knowledge of E_{equiv} and $E_1^{\ominus'}$.

As can be seen, an accurate determination of the formal potentials and concentration of the species being titrated depends heavily on a correct measurement of the equivalence point. For this reason the following plots of various functions of the titration curve are often done (Fig. 13.2):

- The first derivative (Fig. 13.2a). The equivalence point corresponds to the top of the peak.

● The second derivative (Fig. 13.2*b*). The equivalence point is where the curve crosses the *V*-axis.

● The Gran plot (Fig. 13.2*c*[2]). This method consists of the mathematical transformation of the titration curve into straight lines via rearranged Nernst equations. Using a selective electrode that responds only to a

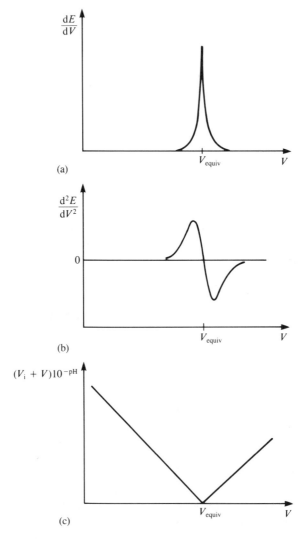

Fig. 13.2. Methods for determining the equivalence point of a potentiometric titration curve (including acid–base titrations). (a) First derivative; (b) Second derivative; (c) Gran plot for titration of a strong acid with a strong base; V_i is the initial volume of acid and V the volume of base added.

cation A we have

$$\lg [A] = \frac{nF}{2.3RT}(E - E') = \frac{n}{S}(E - E') \qquad (13.5)$$

where E' is the sum of the constant terms in the measured potential and $S = 2.3RT/F$. Thus

$$[A] = K \times 10^{nE/0.059} = K \times 10^{E/S} \qquad (13.6)$$

Taking into account the volume change during the titration, we can deduce that a plot of $(V_i + V)10^{E/S}$ vs. V is linear and should be zero at $V = V_{equiv}$. After the equivalence point another redox couple is in excess and a different line is obtained with slope of opposite sign, as shown in Fig. 13.2c. Using this type of plot, points on the curve throughout the titration can be employed, thus reducing the error in determining V_{equiv} which can be measured by extrapolation. The equivalence point is the minimum where the two straight lines intersect.

The accuracy of any one of these methods depends on the size of the increments of titrant volume: the smaller these are the better. Calculations with few experimental points and assuming a linear variation between them, i.e. calculating $\Delta E/\Delta V$, leads to obviously incorrect results.

More complex redox titrations show a dependence on solution pH, the expression for E_{equiv} being slightly more complicated, for example the reduction of MnO_4^- to Mn^{2+} in acid medium:

$$MnO_4^- + 8H^+ + 5e^- \rightarrow Mn^{2+} + 4H_2O$$

with Nernst equation

$$E = E^{\ominus\prime} + \frac{RT}{5F} \ln \frac{[MnO_4^-][H^+]^8}{[Mn^{2+}]} \qquad (13.7)$$

In acid–base titrations of weak acids or bases the expressions are also more complicated. Nevertheless, the introduction of concentration dependence as a function of the composition of the medium in the Nernst equations always leads to the correct result.

Bipotentiometric titrations, that is potentiometric titrations with a constant imposed passage of current of the order of $5-10\,\mu A$, usually between two platinum electrodes, should also be mentioned here. These are not strictly speaking potentiometric titrations, since $I \neq 0$, but they involve a reading of potential. The current flow provokes the occurrence of a half-reaction. Where there is a dominant redox couple before and after the endpoint, the potential difference registered is more or less constant, but in the zone of the equivalence point there is generally a sharp change in the potential difference, ΔE, which can be monitored. The form of the plot of potential difference/titrant volume depends to a great extent on the species involved. In the case of the Fe(II)/Ce(IV) titration, before and after the equivalence point where the iron and cerium redox couples dominate respectively, ΔE is small, and near the equivalence point there is a sudden increase in ΔE. The technique is restricted to polarizable electrodes, i.e. those that can pass current easily, since nearly all selective electrodes, such as the pH electrode, are unable to pass current. Its most popular application is in the determination of water by the Karl–Fischer method[3].

13.3 Functioning of ion-selective electrodes

The functioning of an ion-selective electrode (ISE)[4-6] is based on the selectivity of passage of charged species from one phase to another leading to the creation of a potential difference. The fundamental theoretical formulation is the same as that developed for liquid junction potentials (Section 2.11). In the case of ISEs one phase is the solution and the other a membrane (solid or liquid in a support matrix). The membrane potential, E_m, for an ion, i, of charge z_i is

$$E_m = -\frac{RT}{z_i F} \ln \frac{\alpha_i^{\beta}}{\alpha_i^{\alpha}} \tag{13.8}$$

where α and β are the two phases: α is the solution and β the membrane. If a_i^{β} remains constant then E_m varies in a Nernstian fashion with the activity of species i in solution. We can write

$$E_m = \text{constant} + \frac{RT}{z_i F} \ln a_i^{\alpha} \tag{13.9}$$

which shows a variation of $59/n$ mV per decade of variation in activity at 298 K. Real electrodes do not exhibit exactly this value, but the reading should be close—the range of activities that it is possible to measure is usually from 10^{-6}–10^{-1}.

There are interferences to this simple functioning of the ISE according to (13.9). These are due to the fact that membranes are not perfectly selective and respond to some extent to species other than the desired ion. If we consider a linear concentration gradient within the membrane then the Henderson equation (equation (2.60)) can be applied, writing it in the form

$$E = \text{constant} + \frac{RT}{z_i F} \ln \left(a_i + \sum_j k_{i,j}^{\text{pot}} a^{z_i/z_j} \right) \tag{13.10}$$

where j are the interfering species, $k_{i,j}^{pot}$ is called the *potentiometric selectivity coefficient* and should have the lowest value possible for all interferents. It should be remembered, nevertheless, that it is the product $k_{i,j}^{pot} a^{z_i/z_j}$ that determines the extent of interferences.

There are three basic types of selective electrode: those based on glass membranes, on inorganic salt solid membranes, and on ion exchange. Other more complex electrodes are sensitive to dissolved gases and enzymes. These various types are now described.

13.4   Glass electrodes and pH sensors

Glass is an amorphous solid containing complex forms of silicates and various other ions, whose presence or absence affects the physical properties of the glass. It is permeable to H^+ and to Group IA cations, Na^+ and K^+. For this reason glass membranes were developed that, owing to their permeability to these ions, create a potential difference across the membrane, the magnitude of which depends on ion activity. By altering the composition of the glass, the membrane can be made sensitive to pH, Na^+, or K^+, but there will always be some mutual interferences.

It is important to see how the potential difference across the membrane, usually of thickness $50~\mu m$, arises. The glass is hydrated to about $50~nm$ depth on each side, regions m' and m'', the interior of the membrane, region m, remaining dry. There is adsorption and desorption of ions on the hydrated layers. It appears that transport through the membrane is entirely by cations, e.g. Na^+ or Li^+.

Consideration of Fig. 13.3 shows that the potential across the membrane is given by

$$E_m = (\phi^\beta - \phi^{m''}) + (\phi^{m''} - \phi^m) + (\phi^m - \phi^{m'}) + (\phi^{m'} - \phi^\alpha) \quad (13.11)$$

The first and last terms are differences of interfacial potential originating

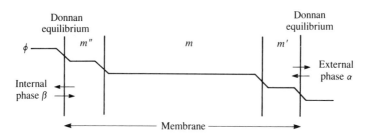

Fig. 13.3. Model of transport through a glass membrane: m' and m'' are hydrated layers, and m the interior of the membrane.

from an equilibrium known as *Donnan equilibrium*. The other two terms are due to diffusion within the membrane.

The 'alkaline error', often found in pH electrodes, arises because in very alkaline solution $[Na^+]$ or $[K^+]$ is normally very high, making a significant extra contribution to the potential as expressed through the Henderson equation (13.10). Minimization of this error is done by using glass of special composition and with very low selectivity coefficients for Na^+ and K^+.

We now turn to the experimental method of measurement of E_m. The potential on the membrane exterior is measured by an $Ag \mid AgCl$ or SCE reference electrode. The interior potential is very difficult to measure through a direct metal contact (only in some solid state and in hybrid sensors, Section 13.10) and one opts for another reference electrode called the internal reference. Thus a typical cell would be

$$\underbrace{Ag \mid AgCl \mid KCl(3\ \text{M})}_{\text{external reference}} \parallel \text{test solution} \mid \text{membrane} \mid \underbrace{HCl(0.1\ \text{M}) \mid AgCl \mid Ag}_{\text{internal reference}}$$

where the solution to be tested affects the exterior membrane potential. One hopes that any change in potential between the solution and external reference electrode can be neglected. A schematic design of a pH electrode is shown in Fig. 13.4. The external reference electrode can be placed in the same package as the pH electrode—forming a combined electrode—which has some advantages for routine analysis for for small volume samples.

In some experimental situations it is not desirable to use a glass electrode to measure pH. In these cases solid-state oxide-based pH sensors can be used. Obviously the pH range in which they can be

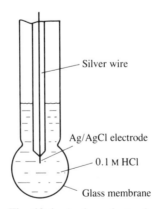

Silver wire

Ag/AgCl electrode

0.1 M HCl

Glass membrane

Fig. 13.4. A glass electrode.

employed is limited by acidic reaction and complex formation, and they must have a small solubility product. They are of two types:

• Those based on the pH-dependent redox reaction between the metal and a stable oxide, i.e. antimony, iridium and palladium.

• Those based on the pH dependent change between oxides of different oxidation states: titanium, ruthenium, rhodium, tantalum, platinum, and zirconium.

The development of these sensors has recently been summarized[7].

13.5 Electrodes with solid state membranes

In this type of selective electrode, the membrane is an ionic solid which must have a small solubility product in order to avoid dissolution of the membrane and to ensure a response that is stable with time. Conduction through the membrane is principally ionic and is due to point defects in the crystal lattice, relying on the fact that no crystal is perfect.

Point defects in crystal lattices can be classified into two essential types (Fig. 13.5[8]):

• In a *Frenkel defect,* an ion leaves its lattice position for an interstitial site, producing a vacancy. Since in well-packed structures the cation is normally smaller than the anion, the Frenkel defect is more probable for the cation. Crystal volume remains almost unaltered by defect formation.

• In *Schottky defects,* an ion leaves its lattice position and migrates to the surface; in this case the probability of cation or anion movement is equal. The creation of this type of defect results in a volume increase and density decrease of the crystal.

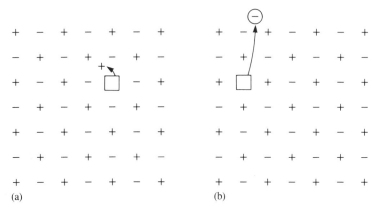

Fig. 13.5. Defects in crystal lattices: (a) Frenkel defect; (b) Schottky defect.

Other defects are combinations of those mentioned above, the most common being the formation of ion-pair vacancies.

These defects are natural, or intrinsic, defects, the total charge of the solid remaining unaltered. In certain cases we can alter the structure of the solids and introduce defects externally by doping, in interstitial positions or by substitution of ions in the lattice by others with a different charge. This latter procedure can increase the electronic conductivity and turn an insulator into a semiconductor. There is also the possibility of creating defects by electromagnetic radiation.

A solid-membrane selective electrode that demonstrates well the simultaneous importance of defects and doping is the lanthanum fluoride electrode. Doping with Eu^{2+} or with Ca^{2+} creates anion vacancies, improving ionic conduction of fluoride ion. Figure 13.6 shows the migration of Frenkel and Schottky defects, as would occur in this electrode. Thus, the electrode is sensitive to fluoride ions in solution, the magnitude of ionic conduction depending on ionic activity in solution through adsorption and desorption of fluoride ions at the electrode surface.

Other electrodes are based on silver salts or metal sulphides, and are prepared by pressing the salts into a disc together with a polymeric support matrix made of rubber, silicone or PVC, for example. Silver salts conduct via Ag^+ ions, and silver sulphide is added to the metal sulphides to improve conductivity. Examples are given in Table 13.1.

Due to the fact that the membranes are formed from sparingly soluble salts, adsorption or desorption on the surface can be of cation or anion, the electrode being sensitive to both species. It is also sensitive to any other ion that forms a sparingly soluble precipitate on the membrane surface. For example, the silver sulphide electrode responds to Hg^{2+}.

The variation of potential with activity is more complicated than in the

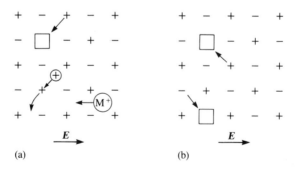

Fig. 13.6. Migration of ionic defects in an electric field: (a) Frenkel defects; (b) Schottky defects.

Table 13.1. Examples of electrodes with solid state membranes

Ion	Material	Detection limit	Ion	Material	Detection limit
S^{2-}	Ag_2S	10^{-6}	F^-	LaF_3	10^{-7}
Ag^+	Ag_2S	10^{-7}	Cl^-	$AgCl$	10^{-5}
Pb^{2+}	$PbS(Ag_2S)$	10^{-7}	Br^-	$AgBr$	10^{-6}
Cd^{2+}	$CdS(Ag_2S)$	10^{-7}	I^-	AgI	10^{-8}
Cu^{2+}	$CuS(Ag_2S)$	10^{-8}			

case of the glass electrode, as the electrode responds to both cations and anions. We exemplify with a silver halide membrane, AgX, where the internal contact can be an internal reference electrode, as in glass electrodes, but a metallic contact is more common, often called *ohmic contact* (Fig. 13.7). (In the case of ohmic contact the migration of ions through the contact should strictly speaking be considered—normally this contribution to the total potential is neglected). The cell can be schematized as

Ag | AgCl | KCl(0.1 M) ‖ test solution containing | AgX membrane | Ag
$\qquad\qquad\qquad\qquad\qquad$ Ag$^+$ but not X$^-$ $\qquad\qquad\qquad\qquad$ |
$\qquad\qquad\qquad\qquad\qquad\qquad\qquad\qquad\qquad\qquad\qquad\qquad\qquad$ ohmic
$\qquad\qquad\qquad\qquad\qquad\qquad\qquad\qquad\qquad\qquad\qquad\qquad\qquad$ contact

The cell e.m.f. is given by

$$E_{cell} = E^{\ominus}_{Ag^+/Ag} + \frac{RT}{F} \ln a_{Ag^+} - E^{\ominus}_{AgCl/Ag} \qquad (13.12)$$

$$= E_{ISE} - E^{\ominus}_{AgCl/Ag} \qquad (13.13)$$

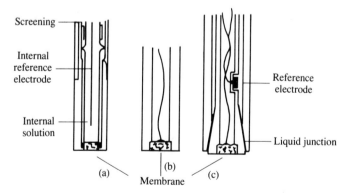

Screening

Internal reference electrode

Internal solution

Reference electrode

Liquid junction

(a)

(b)

(c)

Membrane

Fig. 13.7. Forms of ion-selective electrodes with solid state membranes: (a) with internal reference electrode; (b) with ohmic contact; (c) with ohmic contact and combined reference electrode.

or E_{ISE} is given by

$$E_{ISE} = E^{\ominus}_{Ag^+/Ag} + \frac{RT}{F} \ln a_{Ag^+} \qquad (13.14)$$

$$= E^{\ominus}_{Ag^+/Ag} + \frac{RT}{F} \ln K_{sp} - \frac{RT}{F} \ln a_{X^-} \qquad (13.15)$$

We can also write

$$E_{ISE} = E^{\ominus}_{X_2/X^-} - \frac{RT}{F} \ln a_{X^-} \qquad (13.16)$$

As demonstrated in (13.15) the detection limit (Section 13.9) is affected by the solubility product of the membrane. Following simple equilibrium reasoning $(a_{Ag^+})_{tot}$ is always equal to the sum of a term due to membrane dissolution and one due to Ag^+ activity in solution, $_{sol}a_{Ag^+}$, that is

$$(a_{Ag^+})_{tot} = {}_{sol}a_{Ag^+} + \frac{K_{sp}}{(a_{Ag^+})_{tot}} \qquad (13.17)$$

Substituting in (13.14),

$$E = E^{\ominus}_{Ag^+/Ag} + \frac{RT}{F} \ln \left\{ \frac{{}_{sol}a_{Ag^+} + [{}_{sol}a^2_{Ag^+} + 4\,K_{sp}]^{1/2}}{2} \right\} \qquad (13.18)$$

If $_{sol}a^2_{Ag^+} \gg (4\,K_{sp})^{1/2}$, there will be Nernstian response. Rigorously, the activity of Ag^+ from lattice defects should also be considered: if this is much larger than $(4\,K_{sp})^{1/2}$ then the detection limit is given by the concentration of lattice defects.

It is not difficult to show that the equation corresponding to (13.18) for the anion X^- is

$$E = E_{X_2/X^-} - \frac{RT}{F} \ln \left\{ \frac{{}_{sol}a_{X^-} + [{}_{sol}a^2_{X^-} + 4\,K_{sp}]^{1/2}}{2} \right\} \qquad (13.19)$$

The detection limit for a silver chloride electrode is shown in Fig. 13.8, which is far above the theoretical limit predicted from the solubility product. It has been shown in solutions of buffered silver ion activity that the limit can be reached. Note that the sensitivity is linked to resolution of the potential measurement and not with the detection limit.

The presence of complexing agents alters membrane solubility, and the detection limit is also affected. Additionally, there is a shift in potential to more negative values owing to the lower concentration of free cations.

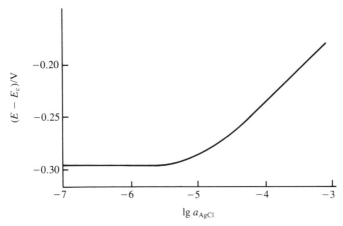

Fig. 13.8 Detection limit for an AgCl ion-selective electrode $(K_{sp} \sim 10^{-10})$, following the equation $E = E_c + (RT/2F) \ln K_{sp}$ for the horizontal straight line (see (13.18)).

As seen in Table 13.1 and already stated, silver sulphide is added to electrodes based on sulphides of copper, cadmium, or lead to improve conductivity. In fact the functioning is not altered as the solubility product of silver sulphide is significantly lower than that of the other sulphides. The expression for the potential is

$$E = E^{\ominus}_{Ag^+/Ag} + \frac{RT}{2F} \ln \frac{K_{sp}(Ag_2S)}{K_{sp}(MS)} + \frac{RT}{2F} \ln a_{M^{2+}} \tag{13.20}$$

which describes the behaviour fairly accurately.

13.6 Ion-exchange membranes and neutral carrier membrane electrodes

Instead of using membranes for selective electrodes that are permeable to ions via adsorption, porous membranes can be employed, where the species to be measured traverses from one side of the membrane to the other. These are of two types: ion exchange[9] and neutral carrier[10]. Examples are given in Fig. 13.9 and Table 13.2. These membranes have a polymeric matrix of PVC, of silicone rubber, etc., and contain a solvent and chelating agents which are selective to the species of interest. The agents are usually macrocyclic and transport is by exchange of the species between adjacent macrocycles.

The most important electrode of the ion-exchange type is the calcium electrode. A water-immiscible hydrophobic solvent is used—for example

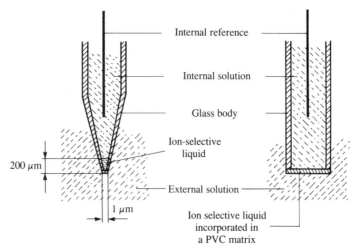

Fig. 13.9. Scheme of an ion-selective electrode with porous membrane.

dioctylphenyl-phosphonate. The function of the solvent is simultaneously to exclude ions of the opposite charge and to let selectivity of the chelating agent actuate. Using the less polar decanol as solvent instead of dioctyl-phenylphosphonate, makes the electrode sensitive to all divalent ions. A chelating agent of the type $(RO)_2PO_2^-NA^+$ is the most usual.

In the case of a neutral non-ionic chelating agent we have neutral carrier-selective electrodes: transport is achieved by selective complexation of certain ions. The best-known electrode of this kind is the potassium-selective electrode, whose membrane consists of a valinomycin macrocycle immobilized in phenylether. The important criterion appears to be the size of the cavity in the centre of the macrocycle and interferences are from cations with similar hydrated ionic radius, such as Rb^+ and Cs^+.

Table 13.2 Electrodes based on ion exchange and neutral carrier membranes

---

Ion exchange
    Ca^{2+} calcium di-(n-octylphenyl) phosphate (active material)
    NO_3^- ammonium tetradecylammonium nitrate (active material)
    ClO_4^-, BF_4^-, CO_3^{2-}, Cl^-

Neutral carriers
    K^+ valinomycin (active material)
    NH_4^+ nonactin (active material)
    Ba^{2+}, Ca^{2+}, UO_2^{2+}

Table 13.3 Potentiometric sensors
selective to dissolved gases

Gas	ISE
NH_3	pH
CO_2	pH
NO_x	pH
SO_2	pH
H_2S	S^{2-}

13.7 Sensors selective to dissolved gases[11]

The principle of these electrodes is a little different, and is normally based on the measurement of the pH of a solution of electrolyte placed between a membrane and a glass electrode, the membrane being porous to the species it is desired to determine (Table 13.3). The dissolved gas conditions the pH of the solution behind the membrane. Membranes can be microporous (for example PTFE) or homogeneous (for example silicone rubber).

It should be noted that some gas-sensitive sensors are amperometric, as in the Clark electrode for oxygen (Section 14.3).

Another type of sensor is a high-temperature solid-state potentiometric sensor for oxygen ($>400°C$) in industrial processes. These are based on the measurement of the e.m.f. of a concentration cell of the type

$$O_2 \text{ (test)}, M' \,|\, \text{zirconia solid electrolyte } (ZrO_2) \,|\, M'', O_2 \text{ (ref.)}$$

where M' and M'' are metallic contacts[12].

13.8 Enzyme-selective electrodes

Enzymes are substances that react very selectively with a substrate in a very specific reaction. Their immobilization on a membrane which is then placed over an electrode in a solution together with the substrate to be determined leads to reaction products that can be detected at the electrode covered by the membrane. An example is the degradation of urea by urease with an internal sensor element (i.e. ion-selective electrode) sensitive to ammonium ion:

$$CO(NH_2)_2 + H_2O \xrightarrow{\text{urease}} CO_3^{2-} + 2NH_4^+$$

In other cases the enzyme reaction alters the pH, and the internal sensor element is a glass electrode.

These electrodes are described in Chapter 17.

13.9   Some practical aspects

On opting to use potentiometric sensors there are aspects of a practical nature that have to be taken into account:

1. What is the detection limit? Given that the variation of potential with activity is not linear below a certain activity, eventually becoming constant, it is important to define parameters for determining the limit. The present criterion for the detection limit, according to IUPAC, is shown in Fig. 13.10. Other criteria are based on the precision of potential readings and their resolution at specified confidence levels.

2. During how long does the electrode response remain Nernstian—often even with a new electrode the response is sub- or supra-Nernstian (i.e. slopes of potential vs. activity are less or greater than $59/n$ mV at 298 K)? Or, at the very least, does the electrode give a variation of potential with activity that is stable and reproducible? This condition is related to the electrode lifetime and varies with the type of utilization (contact with which solutions and for how long) of the selective electrode. In the case of solid-state membrane electrodes the potential and its reproducibility depend on the preconditioning—polishing with abrasive etc.—since, by polishing, a fresh electrode surface is produced each time.

3. What is the response time of the electrode, that is how long does it take to reach equilibrium after dipping the electrode in solution, or after changing the concentration of solution (an activity step)? Clearly this must be as short as possible: optimized times would be of the order of

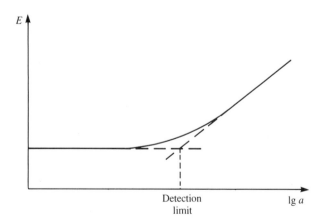

Fig. 13.10. Determination of the detection limit of an ion-selective electrode, following IUPAC[13].

30 s. How should this be defined[14]?

4. What is the electrode selectivity in relation to other species in solution? Remember that this is given by $(k_{i,j}^{pot} a^{z_i/z_j})$ and not only by the selectivity coefficient $k_{i,j}^{pot}$.

All these factors have to be considered. The effect of the variation of potential with temperature and alteration of the slope of the potential vs. activity profile can be minimized by periodic calibration. This period depends on the type of analysis being carried out, but calibration cannot be dispensed with.

13.10   Recent developments: miniaturization

There has been wide application of ion-selective electrodes. Some of these applications require small electrodes, for example measurements *in vivo*. Miniaturization of glass electrodes in the form of micropipettes has gone some way to enabling electrophysiological measurements to be carried out. Other applications require good reproducibility at low cost. Two types of ISE have been developed with both these criteria in mind: ISFETs and coated wire electrodes, besides so-called hybrid sensors. The first of these has the extra advantage that signal processing is done *in situ*, resulting in a low impedance signal (a conventional electrode gives a high impedance signal) that improves the signal/noise ratio.

ISFETs

The field-effect transistor (FET) that is ion selective (ISFET) was developed in the 1970s and 1980s[15,16]. It was developed bearing in mind that miniaturizing a conventional configuration ISE would give a fairly bad signal/noise ratio, affecting detection limit and sensitivity. In an ISFET, signal amplification is done directly in the membrane in contact with the solution, instead of in the measuring instrument at the end of a cable which is susceptible to the electric and magnetic fields of the local environment.

A diagram of a typical ISFET is shown in Fig. 13.11. Instead of using a metallic gate as in a normal FET, a thin film of a material sensitive to an ion is used (ISM). The difference in potential between the ISM and the solution is a function of ion activity. The potential developed alters the concentration of carriers in the region marked 'channel' which in turn alters the $I-V$ characteristics between source and drain. The current is a low-impedance signal that can be related directly to the activity of the ions in solution.

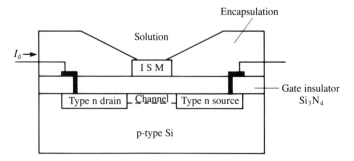

Fig. 13.11. An ion-selective field effect transistor (ISFET).

This sensor is extremely small. The construction of many ISFETs on the same chip is possible, either identical or sensitive to different ions. A chip of 1×2 mm can have five or six ISFETs.

One of the important applications possible with these sensors is for *in vivo* studies[17].

Coated wire electrodes[18]

A thin film of a membrane-forming material is deposited on a metal wire (silver, platinum, nickel, etc.) or on carbon, the membrane being, for example, a liquid exchanger immobilized in PVC (Fig. 13.12). This electrode is thus similar to an ion-exchange selective electrode, but without solution and without internal reference. The detailed mechanism of its operation is not clear but it is certain, and logical, that the interface

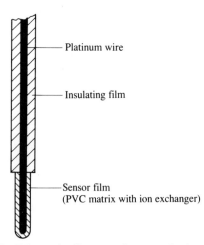

Fig. 13.12. Schematic diagram of a coated wire electrode.

between membrane and wire is not totally blocked (the wire is not totally covered) so that effectively there is an internal reference electrode. If this were not the case there could be no passage of current to establish equilibrium.

The great advantage of these sensors is that they are cheap and disposable; the disadvantage is their poor reproducibility. However, there have been many applications, especially in the quantitative detection of drugs of abuse.

Hybrid sensors

In these sensors the technology developed for ISFET construction is used in conventional electrodes. Links between the membrane and internal reference are metallic (ohmic contact), by deposition of the metal on the membrane (solid state membranes), or by deposition of an ion-selective membrane on a metal. This latter is an *integrated sensor*.

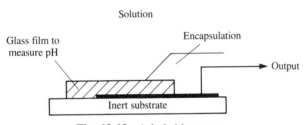

Fig. 13.13. A hybrid sensor.

An example being researched for possible use is a pH electrode (Fig. 13.13): a metal contact is deposited on an inert substrate followed by a glass film (in an oven) that totally encapsulates the electrical contact[19]. This would be a considerably more robust pH electrode than those presently used, and easier to maintain.

13.11 Potentiometric sensors in flow systems

Flow methods[20] permit the placing of electrochemical sensors at essential control points. Normally a branching of the principal flow is aranged that passes by the sensors any reagents, etc., being added after branching and before the sensor. An on-line response can be extremely important for efficient control.

Any potentiometric sensor or combination of potentiometric sensors can, in principle, be used in a flow system. Certain precautions have to be taken, some particularly important owing to solution movement:

1. Positioning of the external reference electrode. The signal can be affected by local electric fields between indicator and external reference electrodes, an effect that is increased by solution flow, making it important to reduce the distance between the electrodes as much as possible.

2. The electrode response time is increased relative to stationary solution.

3. Solution movement has the tendency to remove species from the membrane surface, i.e. membrane deterioration is faster than in stationary solution, and there is a consequent shortening of electrode lifetime. Periodic calibration becomes extremely important.

4. Removal of solid residues by filtering, and the adjusting of solution conditions to optimize the response, e.g. control of pH, ionic strength, elimination of ionic interferences, should be done frequently.

13.12 Electroanalysis with potentiometric sensors

Owing to their specificity, sensitivity, and range of measurable concentrations, potentiometric sensors based on ionic or enzymatic selectivity have wide application in analytical determinations. Other potentiometric sensors using electrodes of the first kind ($M^{n+} \mid M$), used in precipation titrations, are not easy to manipulate and redox electrodes have a reduced application owing to their lack of selectivity, reacting to any oxidizable or reducible species.

Since ISEs can be used in continuous flow systems or in flow systems with sample injection (*flow injection analysis,* FIA)[21] their application is wide, not limited to discrete samples. Analysis time becomes shorter, with faster recycling. Additionally, in flow systems the experimental assembly and data analysis can be controlled automatically by microcomputer, including periodic calibration. Another development is the use of sensors for the detection of eluents of chromatographic columns in high-pressure liquid chromatography (HPLC). Miniaturization has permitted an increase in the use of sensors in foods, biological tissues, and clinical analyses in general.

It remains an objective to develop potentiometric sensors with longer lifetimes, greater reproducibility and greater stability. The importance of an appropriate statistical treatment of the results in order to determine their precision is stressed. Frequent calibration is necessary, at least at the beginning of each measurement session and in a medium as similar as possible to that where the sensors are to be employed, in order to ensure the accuracy of the analytical determinations.

References

1. D. T. Sawyer and J. L. Roberts, *Experimental electrochemistry for chemists*, Wiley, New York, 1974, Chapter 9.
2. M. Mascini, *Ion Selective Electrode Rev.*, 1980, **2,** 17.
3. S. K. Macleod, *Anal. Chem.*, 1991, **63,** 557A.
4. P. L. Bailey, *Analysis with ion selective electrodes*, Heyden, London, 1976.
5. J. Koryta and K. Stulik, *Ion selective electrodes*, Cambridge University Press, 1983.
6. H. Freiser (ed.), *Ion selective electrodes in analytical chemistry*, Plenum, New York, 1981, 2 vols.
7. S. Glab, A. Hulanicki, G. Edwall, and F. Ingrun, *CRC Crit. Rev. Anal. Chem.*, 1989, **21,** 29.
8. N. N. Greenwood, *Ionic crystals, lattice defects and non-stoichiometry*, Butterworth, London, 1968.
9. G. J. Moody, B. B. Saad, and J. D. R. Thomas, *Ion Selective Electrode Rev.*, 1988, **10,** 71.
10. D. Amman, W. E. Morf, P. C. Meier, E. Pretsch, and W. Simon, *Ion Selective Electrode Rev.*, 1983, **5,** 3.
11. S. Bruckenstein and J. S. Symanski, *J. Chem. Soc. Faraday Trans*, I, 1986, **82,** 1105.
12. J. Fouletier and E. Siebert, *Ion Selective Electrode Rev.*, 1986, **8,** 133.
13. *Pure Appl. Chem.*, 1976, **48,** 127.
14. E. Lindner, K. Toth, and E. Pungor, *Pure Appl. Chem.*, 1986, **58,** 469.
15. J. Janata, *Chem. Rev.*, 1990, **90,** 691.
16. P. Bergveld, *IEEE Trans.*, 1970, **BME-17,** 70; 1972, **BME-19,** 342.
17. B. A. McKinley, B. A. Houtchens, and J. Janata, *Ion Selective Electrode Rev.*, 1984, **6,** 173.
18. R. W. Cattrall and I. C. Hamilton, *Ion Selective Electrode Rev.*, 1984, **6,** 125.
19. R. G. Kelly and A. E. Owen, *J. Chem. Soc. Faraday Trans*, I, 1986, **82,** 1195.
20. K. Stulik and V. Pacakova, *Electroanalytical measurements in flowing liquids*, Ellis Horwood, Chichester, 1987.
21. J. Ruzicka and E. H. Hansen, *Flow injection analysis*, 2nd edn, Wiley, New York, 1988.

14

AMPEROMETRIC AND VOLTAMMETRIC SENSORS

14.1 Introduction

An amperometric sensor measures a current at a fixed applied potential, that is at one point on the current–voltage curve. A voltammetric sensor records a number of points on, or a chosen region of, the current–voltage profile. Thus an amperometric sensor is a fixed-potential voltammetric sensor.

The larger part of this book has been devoted to studies related to electrochemical measurements away from equilibrium. As demonstrated, these permit the determination of kinetic and thermodynamic parameters of the electrode processes, whereas measurements at equilibrium furnish only thermodynamic data. So, whilst potentiometric analysis is a powerful tool in the determination of activities or concentrations, the specificity arising from the electrode material, amperometric or voltammetric analysis permits other parameters besides these to be obtined.

The most important selectivity parameter of electrodes for voltammetric sensors is the *applied potential*. Ideally, the electrode potentials of the redox couples would be sufficiently far apart for there to be no interference between different species. Unfortunately this is not the case, and it is necessary to look for greater selectivity. We can discriminate better between the different species present in solution through a correct choice of conditions for the study of the electrode reaction: electrode material (Chapter 7) in some cases through surface modification; use of hydrodynamic electrodes (Chapter 8); application of potential sweep

(Chapter 9) or potential pulses (Chapter 10); alternating current voltammetry (Chapter 11) and so on. Additionally, kinetic and mechanistic information can sometimes be obtained from the same set of experiments.

Recent developments in amperometric and voltammetric sensors have been in improving selectivity, increasing sensitivity, and lowering detection limits. Hydrodynamic electrodes, besides permitting greater sensitivity, also bring greater reproducibility and the possibility of use in flow systems for process control, quality control, and as chromatographic detectors. The widespread use of pulse techniques, particularly *differential pulse voltammetry* and more recently *square wave voltammetry,* has enabled lower detection limits to be reached. At the same time, there has been much progress in the automation of these sensors, not only to control experiments but also to analyse the data obtained.

In this chapter, after describing the useful technique of amperometric titrations, recent developments in amperometric and voltammetric sensors are summarized. Their application as biosensors to the study of biological compounds and *in vivo,* is described in Chapter 17.

14.2 Amperometric titrations

Not all redox titrations have a well-defined equivalence point, and amperometric titrations[1], in which a potential corresponding normally to that necessary to attain the mass-transport-limited current is applied to the working (indicator) electrode, permit the calculation of the titration endpoint through measurements done far from the equivalence point. Titrations can be done in flow systems, and in this sense it is possible to alter the quantity of added titrant so as to obtain greater accuracy in the determination of the equivalence point[2].

Simple amperometric titrations

The electrical circuit consists of two electrodes: a redox indicator electrode and a reference electrode that also passes current. A fixed potential difference is applied and the equivalence point is calculated from the intersection of the two straight lines that show the variation of current before and after the endpoint in a plot of current as a function of added titrant volume. The plots can have various forms, depending on whether the titrated species or titrant are or are not electroactive. Figure 14.1 shows the four possible cases. Sometimes the potential difference applied is less than that necessary to reach the mass-transport-limited current, but sufficient to give good results.

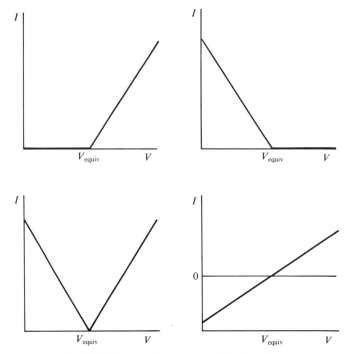

Fig. 14.1 Forms of amperometric titration.

As this is an indirect method of determining the equivalence point it has some advantages. Dilute solutions can be titrated, allowing titration of sparingly soluble precipitates with no interference from supporting electrolyte. Non-electroactive compounds can be titrated so long as the titrant is electroactive, or vice versa. Titrations are fast, only three points before and three points after the equivalent point being necessary.

Some examples are titration of Pb^{2+} with CrO_4^{2-}, Ni^{2+} with diacetyl-dioxime, and barbituric acids with $Hg_2(ClO_4)_2$.

Biamperometric titrations

Biamperometric titrations involve the use of two redox electrodes in solution, and are applicable only to titrations where there is a reversible system before or after the endpoint; there is no reference electrode. The application of a potential difference causes one electrode to be anode and the other cathode. A current passes due to oxidation or reduction, respectively, of a species present in solution, decreasing to $I = 0$ at the equivalence point; alternatively it may be that $I = 0$ until the equivalence

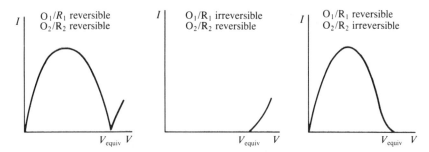

Fig. 14.2 Forms of curve obtained in biamperometric titrations.

point and that the current begins to rise after this. Since $I = 0$ in all cases at the equivalence point, this method is sometimes called the dead-stop endpoint method. The forms of the curves that can be obtained are shown in Fig. 14.2. Applications of this technique are in titrations of iodine, bromine, titanium(III) and cerium(IV) and in the determination of water in non-aqueous solvents with the Karl-Fischer reagent.

$$R_1 + \underset{\text{titrant}}{O_2} \rightarrow O_1 + R_2 \qquad \Delta E \sim 0.1 \rightarrow 0.2 \text{ V}$$

Amperometric titrations with double hydrodynamic electrodes

Titrations of non-electroactive compounds can be carried out by homogeneous reaction with titrants electrogenerated *in situ*. This can be done using double hydrodynamic electrodes in the diffusion layer microtitration technique described in Section 8.7.

In this technique the upstream electrode (generator) is galvanostatically controlled to generate a species that reacts with species X in solution in a second-order homogeneous reaction, the detector electrode being used to quantify the fraction of the electrogenerated species that did not react:

generator $A + n_1 e^- \rightarrow B$

solution $B + X \rightarrow \text{products}$

detector $B + n_2 e^- \rightarrow C \text{ (or A)}$

Figure 8.11 shows the type of curve that is obtained, which allows the determination of the concentration of X. The rotating ring-disc electrode[3] and the wall-jet ring-disc electrode in continuous flow[4] (Fig. 14.3) and with sample injection into potassium bromide solution[5] have been used—this last procedure reduces the amount of sample necessary and

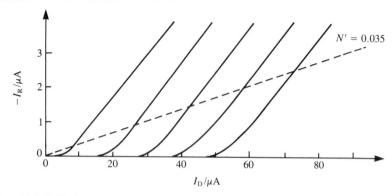

Fig. 14.3 Diffusion-layer microtitration curves at the wall-jet ring disc electrode for titration of As(III) (X) with bromine (B) generated at the disc electrode from bromide. Solution 10^{-2} M KBr + 0.5 M H_2SO_4. Analysis of the curve leads to [X] (from Ref. 4 with permission).

makes quantification faster. An important application is in the quantitative determination of amino-acids and proteins by reaction with Br_2 or OBr^- electrogenerated from Br^-, which has been automated[6].

14.3 Membrane and membrane-covered electrodes

In order to make an electrode more selective we can arrange for only certain species to reach its surface, with the additional advantage of reducing electrode poisoning. This can be done by modifying the electrode surface (see next section), by a porous membrane touching the electrode or separated from it by a thin film of electrolyte, or by using a metallized membrane as indicator electrode.

Direct coverage of electrodes with porous membranes[7] can diminish poisoning problems, avoiding reduction in response with time. This is a particularly pertinent problem in solutions containing reasonable amounts of organic compounds. An example is the successful use of cellulose acetate to impede the irreversible adsorption of proteins on glassy carbon electrodes[8].

The best-known example of a membrane-covered electrode is the Clark electrode for determination of dissolved oxygen (Fig. 14.4[9,10]). A membrane (usually PTFE) with pores of a size that lets only oxygen diffuse through is placed over a thin film of electrolyte on top of a platinum or gold electrode, the potential of this being controlled so as to reduce oxygen. The anode is usually a silver disc that acts simultaneously as reference electrode. It can be used for oxygen determination in gas or liquid phases.

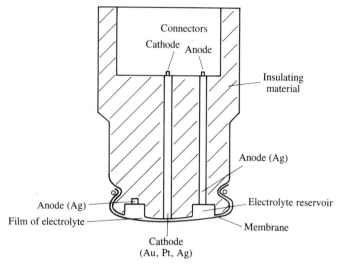

Fig. 14.4 The Clark electrode for determination of dissolved oxygen.

Other electrodes functioning in a similar way have been developed for other dissolved gases, with important clinical applications (Table 14.1). Since the porous membrane does not let past species that can poison the electrode, these electrodes are ideal for measurements in biological fluids.

Metallized membrane electrodes[11] have the inner, non-exposed surface of the membrane metallized: this is the indicator electrode, and contacts with an internal electrolyte solution. Typical metals are gold and platinum and typical membrane materials PTFE and polyethylene; ion-exchange membranes such as Nafion can be used to improve selectivity; membranes can be porous or non-porous—in the latter case transport through the membrane is by activated diffusion. The external medium can be liquid or gas phase and so this type of detector can be used in gas chromatography. Gases such as oxygen, carbon monoxide, hydrogen, chlorine and nitrous oxide can be determined, and also anions

Table 14.1. Amperometric sensors for dissolved gases

Gases dissolved in aqueous phase
O_2, NO, halothane, CO_2
Gas phase
H_2S, HCN, CO, NO, NO_2, Cl_2

and cations and some alcohols after conversion to volatile gas-phase products which are introduced into a carrier stream.

14.4 Modified electrodes

At a modified electrode[12-15] the electrode surface is deliberately altered by adsorption, by physical coverage, or by bonding of specific species. The result is to block direct access to the electrode, inhibiting some electrode processes and promoting others. Modification can therefore be an important aid in obtaining greater selectivity, and thence its importance in analysis[16]. This modification can be done to microelectrode arrays to produce tiny specific chemical sensors[17]. Normally the modifier layer is electroactive, acting as a mediator between the solution and the electrode-substrate in electron transfer (Fig. 14.5).

There are several ways of preparing different types of modified electrodes:

1. *Chemical modification* (*chemical bonding*). An electroactive species is immobilized on the electrode surface by chemical reaction. Normally the fact that the electrode is covered by hydroxyl groups owing to the oxygen in the atmosphere is used. For example, the silanization process is

$$\text{—OH} + \text{X—Si—R} \longrightarrow \text{—O—Si—R} + \text{HX}$$

where $X = OR$ or Cl; the silane group then reacts with the species of interest. These methods tend to give monolayers, with the exception of chemical bonding of polymers, e.g. polymerized ferrocene.

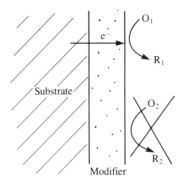

Fig. 14.5 Functioning of a surface-modified electrode. Reduction of O_2 to R_2 is inhibited.

2. *Adsorption.* Adsorption can be reversible or irreversible. This method has been used particularly for the preparation of polymer-modified electrodes. A solution of polymer is either painted on the electrode and the solvent evaporated, or the electrode is immersed in a solution of the polymer. Relevant examples are polymers that let charge pass through the film: polyvinylpyridine (PVP), polyvinylferrocene (PVF), porphyrins, and phthalocyanines. Direct deposition in the gas phase or sputtering are also possible.

3. *Electroadsorption*—adsorption carried out with an applied electrode potential. The quantity deposited is a function of deposition time, multilayer formation being possible, as is the case with thionine. On the other hand, application of a potential, in the correct conditions, in the presence of a molecule susceptible to polymerization, can produce radicals, initiating polymerization and subsequent electrode modification. Examples of these conducting polymer monomers are pyrrole, *N*-phenylpyrrole and *N*-methylpyrrole, aniline, and thiophene.

4. *Plasma.* A plasma is used to clean the electrode surface, leaving unbonded surface atoms and, thus, an activated surface. Carbon is much used for this: subsequent exposure to amines or ethenes, for example, results in chemical bond formation. Plasma discharge in the presence of radical monomers in solution, leading to polymer formation on the surface, is equivalent to chemical activation. The use of lasers in this area may be interesting, but has been little exploited as yet.

Characterization of modified electrodes can be carried out by electrochemical, spectroscopic, and microscopic methods. Of the electrochemical methods we stress cyclic voltammetry, chronocoulometry, and impedance, which combined together measure the number of redox centres, film conductivity, kinetics of the electrode processes, etc. Almost all the non-electrochemical techniques described in Chapter 12 have been employed for the characterization of modified electrodes.

Modified electrodes often give rise to currents that are higher than in the absence of the modifier. Sometimes, on placing the modified electrode in a solution that contains supporting electrolyte only, the voltammetric characteristics of the immobilized species are observed. This is extremely useful for diverse applications such as, for example, in electroanalysis[18].

Applications are varied, from catalysis of organic and inorganic reactions to electron transfer to and from molecules of biological interest. For example, it has been shown that ruthenium(IV) immobilized inside PVP catalyses organic oxidations such as that of propan-2-ol to acetone[19]. The electroreduction of oxygen (important in fuel cells, Section 15.10) is catalysed by metalloporphyrins and metallophthalocyanines[20]. The development of the electrochemistry of proteins and enzymes was hindered by the strong attraction of these compounds to the electrode surface causing poisoning, and, additionally in the case of enzymes, their degradation by the electrode material. A mediator facilitates electron transfer and minimizes attraction and repulsion effects between the biological molecule and the electrode. However, a careful choice of substrate may lead to the development of methods that do not need mediator and allow direct immobilization (see Chapter 17).

In technology, conducting polymers will probably have an important role, as they can be successively oxidized or reduced[21,22]

$$\text{insulating polymer} \rightleftarrows \text{oxidized polymer}$$
$$\quad\text{neutral} \qquad\qquad\qquad \text{conducting}$$

by alteration of applied potential in a switching action; often this is accompanied by a colour change. For this reason conducting polymers are being investigated for electrochromic displays, energy stores, etc. As they are conducting they are also being considered as protective coatings of metal surfaces against corrosion, especially to protect against photocorrosion.

In specific cases, electrode *bulk* as opposed to surface modification can be employed, as with carbon paste electrodes[23]. The modifier, a substance that reacts preferentially in some way with a species to be determined, is mixed directly with the carbon paste. The mode of action is either by catalyzing the analyte reaction or pre-concentrating the analyte on the surface before determination.

14.5 Increase in sensitivity: pre-concentration techniques

A sensitivity increase and lower detection limit can be achieved in various ways with the use of voltammetric detectors rather than amperometry at fixed potential or with slow sweep. The principle of some of these methods was already mentioned: application of a pulse waveform (Chapter 10) and a.c. voltammetry (Chapter 11). There is, nevertheless, another possibility—the utilization of a pre-concentration step that accumulates the electroactive species on the electrode surface before its quantitative determination, a determination that can be carried out by control of applied current, of applied potential or at open circuit. These pre-concentration (or stripping) techniques[24–26] have been used for cations and some anions and complexing neutral species, the detection limit being of the order of 10^{-10} M. They are thus excellent techniques for the determination of chemical species at trace levels, and also for speciation studies. At these levels the purity of the water and of the reagents used in the preparation of the supporting electrolyte are extremely important.

The process effectively consists of two, or sometimes three steps:

• Deposition or adsorption of the species on the electrode during a time t (preconcentration step). This step occurs under potential control or at open circuit;

• Change to an inert electrolyte medium. This step can sometimes be unnecessary;

• Reduction/oxidation of the species that was accumulated at the electrode. This can be achieved by varying the applied potential, registering a current peak (or its integral) proportional to concentration. Alternatively, a current can be applied, or its equivalent in terms of oxidation/reduction by another chemical species (reducing or oxidizing agent) in solution—the variation of potential with time is recorded as a chronopotentiogram, the transition time being proportional to concentration.

The four variations of this technique are to be found in Table 14.2. The schemes of operation are shown in Fig. 14.6. Important applications for trace metals are the use of anodic stripping voltammetry (ASV) to determine trace quantities of copper, cadmium, lead and zinc, and adsorptive stripping voltammetry (AdSV) of trace quantities of nickel and cobalt—pre-concentration by adsorption accumulation of the oxime complexes followed by reduction to the metal is employed, as reoxidation of these metals in ASV is kinetically slow and does not lead to well-defined stripping peaks.

Table 14.2 Principles of pre-concentration techniques (adapted from Ref. 27)

Method	Preconcentration step	Determination step	Measurement
A Stripping voltammetry	Potential control	Potential control	I vs. t (or I vs. E)
B Adsorptive stripping voltammetry	Adsorption (with or without applied potential)	Potential control	I vs. t (or I vs. E)
C Potentiometric stripping analysis	Potential control	Reaction with oxidant or reductant in solution	E vs. t
D Stripping chronopotentiometry	Potential control	Current control	E vs. t

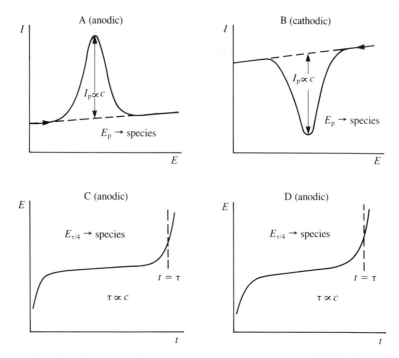

Fig. 14.6 Determination step in stripping techniques of Table 14.2.

The use of hydrodynamic electrodes in these experiments has been very important in that they increase sensitivity because of higher mass transport, ensure good reproducibility, and sometimes gives better resolution in solutions of mixtures. The use of cells such as the wall-jet in flow systems is particularly useful, as response is fast and it is easy to introduce them at any point in the flow system.

Another important factor is the electrode material. Many applications of pre-concentration techniques are for heavy metals that are reduced to the corresponding metal at potentials that are reasonably negative. At these potentials the background current for a lot of electrode materials is appreciable, which reduces the usefulness of the experiment. Mercury is the best electrode material for this purpose, having a very low background current because of its large hydrogen overpotential. It can be used in the form of the hanging or static mercury drop in stationary solution, or in forced convection systems as a thin film electrodeposited on an appropriate substrate. Since it is important that the mercury does not dissolve in the substrate (as happens with gold for example) and that it adheres well, the choice of substrate is limited[28]. A substrate that works reasonably well and which is much used is glassy carbon. This has the additional advantage that in the zone of positive potentials glassy carbon itself can be used as electrode for anion determinations, etc. A wide range of potentials is thus possible. It has been suggested that mercury is electrodeposited on glassy carbon at $-1.0\,V$ vs. SCE so that the tendency of mercury to form small droplets on the surface is minimized. Other substrates suggested are iridium and silver.

In anodic stripping voltammetry the mercury film and the metal ion to be determined are often co-deposited (called *in situ* mercury deposition). The thin mercury film has characteristics similar to a thin-layer cell, described in Section 9.10. Additionally, it can be easily used in hydrodynamic systems[29].

Use of mercury as electrode material resolves the problem of the negative potential zone being too small but brings others, as it is a liquid. When a metal ion is reduced on the mercury surface to the metal, this can diffuse to within the mercury film, forming Hg–M bonds, or, if there is more than one dissolved metal, intermetallic compounds can be formed within the mercury, as is the case of Cu–Zn. Reoxidation of intermetallic

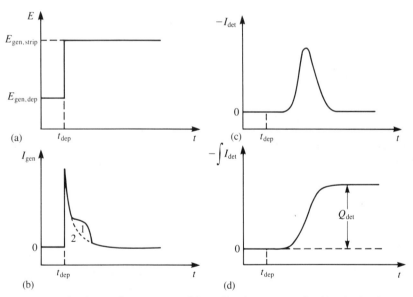

Fig. 14.7 Stripping voltammetry with collection at a double hydrodynamic electrode, with accumulation at the generator electrode: (a) Potential step at $t = t_{dep}$ at the generator electrode to remove the deposit; (b) Chronoamperogram at the generator electrode with faradaic (1) and capacitive (2) components; (c) Chronoamperogram at the detector electrode with only faradaic component (potential unaltered throughout the experiment); (d) Integral of the curve in (c), the height being proportional to the concentration of species to determine through $Q_{det} = -N_0 I_{gen,dep} t_{dep}$.

compounds occurs at a potential different from the individual metals, which can cause difficulties in the resolution of adjacent peaks. Naturally, there are ways of reducing this problem, and there is even the possibility of using other pre-concentration techniques. The study of the formation and properties of Cu–Zn is very important, because in nature the presence of one of these elements implies the presence of the other.

Figure 14.7 shows the technique of anodic stripping voltammetry with collection at a double hydrodynamic electrode[30]. The fact that it is possible to control the potentials of generator and detector electrodes independently is used to increase sensitivity. This procedure has been used with success at rotating and wall-jet electrodes[4,27]

14.6 Amperometric and voltammetric electroanalysis

The great possibilities of amperometric and voltammetric electroanalysis both in stationary solution and in continuous flow are evident from the previous sections. However, the choice of which technique and experimental protocol to use depends on various factors, such as:

- The concentration range of the species to be determined
- Possible interferences to its exact determination, i.e. matrix composition
- The accuracy and precision necessary
- The quantity of sample
- The required speed with which an answer is required

In the case of trace quantities, or any determination close to detection limits, accuracy is much more important than precision, as the former is the primary consideration as to whether a given trace level is acceptable or not. Theoretical detection limits are often defined in the literature as three times the standard deviation of the measurements, but practical detection limits are usually higher and are determined by levels of contamination that can be introduced through reagent and sample manipulation.

It is probable that there will be much development in new solid electrode detectors for flow systems in the near future which, owing to their hydrodynamic character, give fast on-line voltammetric responses[31–33]. When the amount of sample is small, a flow injection system[34] is indicated; applications of stripping analysis[35] and modified electrodes[36] in this mode have been reviewed. Chromatographic detection is similar in many ways to flow injection, and there are many applications of

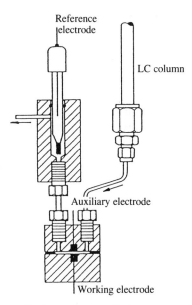

Fig. 14.8 A thin-layer cell for use as a high pressure liquid chromatography electrochemical detector (courtesy of Bioanalytical Systems).

electrochemical detectors[36,37] as there are in capillary electrophoresis[38]. A thin-layer cell for coupling to an HPLC column is shown in Fig. 14.8.

The use of pulse techniques for electroanalytical determinations has been much publicized, and is applicable to both solid electrodes and the HMDE/SMDE. The development in recent years of square wave voltammetry (SWV)[39] widens the possibilities beause of its rapidity (Section 10.9); it is especially useful because the time necessary to do an experiment is only 2 s, which means that a SMDE drop in the dropping mode can also be used for micromolar determinations. Progress obtained with pulse techniques[40,41] has meant that applications of a.c. voltammetry have been few, but there is no theoretical reason for this. The very low detection limits achieved in stripping voltammetry result not only from the pre-concentration step but also from the use of pulse waveforms in the determination step.

Another important future direction is in the use of microelectrodes and microelectrode arrays. They are often easier to manipulate by the inexperienced, and instrumentation is simpler. They can be used in highly resistive 'dirty' media where conventional electrodes may be unuseable and are able to probe localized concentrations. Composite electrodes[42], of which carbon paste is an example, if conveniently prepared, can act as microelectrode assemblies. In a more general sense, lithographic and screen-printing processes can be used to fabricate microelectrode assemblies, which either all do the same determination to increase the analytical signal, or are independently controlled to determine different species.

Finally, the possibilities of automation of amperometric and voltammetric electroanalysis should be stressed, as well as the use of solvents other than water[43]. Pulse techniques are semi-automated by nature; the responses can be transmitted directly to a microcomputer for immediate analysis. Fast on-line analysis in flow systems with automated calibration is one of the great advantages, which will be much exploited in the future.

References

1. D. T. Sawyer and J. L. Roberts, *Experimental electrochemistry for chemists*, Wiley, New York, 1974, Chapter 9.
2. K. Toth, G. Nagy, Zs, Feher, G. Horvai, and E. Pungor, *Anal. Chim. Acta*, 1980, **114**, 45.
3. W. J. Albery, S. Bruckenstein, and D. C. Johnson, *Trans. Faraday Soc.*, 1966, **62**, 1938.
4. W. J. Albery and C. M. A. Brett, *J. Electroanal. Chem.*, 1983, **144**, 211.
5. A. M. Oliveira Brett and L. Santos, unpublished results.
6. W. J. Albery, L. R. Svanberg and P. Wood, *J. Electroanal. Chem.*, 1984, **162**, 45.
7. K. Doblhofer and R. D. Armstrong, *Electrochim. Acta*, 1988, **33**, 453.
8. J. Wang, *Electroanalysis*, 1991, **3**, 255.
9. L. C. Clark Jr., *Trans. Am. Soc. Artif. Intern. Organs*, 1956, **2**, 41.
10. M. L. Hitchman, *Measurement of dissolved oxygen*, Wiley-Interscience, New York, 1978.
11. F. Opekar, *Electroanalysis*, 1989, **1**, 287.
12. R. W. Murray, *Electroanalytical chemistry*, ed. A. J. Bard, Dekker, New York, Vol. 13, 1984, pp. 191–368.
13. W. J. Albery and A. R. Hillman, *Ann. Rep. Prog. Chem. Sect. C*, 1981, **78**, 377.
14. A. J. Bard, *J. Chem. Ed.*, 1983, **60**, 302.
15. M. Kaneko and D. Wöhrle, *Adv. Polymer. Sci.*, 1988, **84**, 143.
16. S. Dong and Y. Wang, *Electroanalysis*, 1989, **1**, 99.
17. M. J. Natan and M. S. Wrighton, *Progress in inorganic chemistry*, ed. S. J. Lippard, Wiley-Interscience, New York, Vol. 37, 1990, pp. 391–494.
18. R. Guadalupe and H. D. Abruña, *Anal. Chem.*, 1985, **57**, 142.
19. B. A. Moyer, M. S. Thompson and T. J. Meyer, *J. Am. Chem. Soc.*, 1980, **102**, 2310.
20. e.g. J. P. Collman, P. Denisevich, Y. Jonai, M. Marrocco, C. Koval, and F. C. Anson, *J. Am. Chem. Soc.*, 1980, **102**, 6027.
21. G. K. Chandler and D. Pletcher, *RSC Spec. Per. Reports, Electrochemisty*, Vol. 10, 1984, p. 117.

22. A. F. Diaz, J. F. Rubinson, and H. B. Mark Jr., *Adv. Polymer Sci.*, 1988, **84**, 113.
23. K. Kalcher, *Electroanalysis*, 1990, **2**, 419.
24. V. Vydra, K. Stulik, and E. Julakova, *Electrochemical stripping analysis*, Ellis Horwood, Chichester, 1976.
25. J. Wang, *Stripping analysis. Principles, instrumentation and applications*, VCH Publishers, Deerfield Beach, Florida, 1985.
26. J. Wang, *Electroanalytical chemistry*, ed. A. J. Bard, Dekker, New York, Vol. 16, 1989, pp. 1–87.
27. C. M. A. Brett and M. M. P. M. Neto, *J. Chem. Soc. Faraday Trans. 1*, 1986, **82**, 1071.
28. S. P. Kounaves and J. Buffle, *J. Electroanal. Chem,*, 1987, **216**, 53.
29. C. M. A. Brett and A. M. C. F. Oliveira Brett, *J. Electroanal. Chem.*, 1989, **262**, 83.
30. D. C. Johnson and R. E. Allen, *Talanta*, 1973, **20**, 305.
31. K. Stulik and V. Pacakova, *Electroanalytical measurements in flowing liquids*, Ellis Horwood, Chichester, 1987.
32. H. Gunasingham and B. Fleet, *Electroanalytical chemistry*, ed. A. J. Bard, Dekker, New York, Vol. 16, 1989, pp. 89–179.
33. C. M. A. Brett and A. M. C. F. Oliveira Brett, *Port. Electrochim. Acta*, 1989, **7**, 657.
34. J. Ruzicka and E. H. Hansen, *Flow injection analysis*, Wiley, New York, 1988.
35. M. D. Luque de Castro and A. Izquierdo, *Electroanalysis*, 1991, **3**, 457.
36. E. Wang, H. Ji and W. Hou, *Electroanalysis*, 1991, **3**, 1.
37. G. Horvai and E. Pungor, *CRC Crit. Rev. Anal. Chem.*, 1989, **21**, 1.
38. P. D. Curry Jr., C. E. Engstrom-Silverman, and A. G. Ewing, *Electroanalysis*, 1991, **3**, 587.
39. J. G. Osteryoung and J. J. O'Dea, *Electroanalytical chemistry*, ed. A. J. Bard, Dekker, New York, Vol. 14, 1986, pp. 209–308.
40. J. G. Osteryoung and M. M. Schreiner, *CRC Crit. Rev. Anal. Chem.*, 1988, **19**, 81.
41. G. N. Eccles, *CRC Crit. Rev. Anal. Chem.*, 1991, **22**, 345.
42. D. E. Tallman and S. L. Petersen, *Electroanalysis*, 1990, **2**, 499.
43. E. A. M. F. Dahmen, *Electroanalysis: theory and applications in aqueous and non-aqueous media and in automated chemical control*, Elsevier, Amsterdam, 1986.

15

ELECTROCHEMISTRY IN INDUSTRY

15.1 Introduction

Electrochemistry has a very important role in industry in various types of application including electrolysis (conversion of substances), metal processing and finishing, batteries and fuel cells, and water and effluent treatment[1-3]. Extensive, often fairly specialized, books on various aspects exist in the literature. Reference 1 provides a modern, up-to-date, general view with many examples. In this chapter a brief survey of these applications is provided. At industrial level primary considerations are naturally economic, the product or process yield in space and in time and the energy consumption being very important. In the case of batteries and fuel cells their efficiency, lifetime, stability, the current supplied are all of importance. Corrosion, unwanted in all but a very few cases such as electrochemical machining, but important because of its economic and social consequences, is discussed in the next chapter.

The use of electrochemistry in industry is affected by the price of electricity and its ease of supply, principally in cases where there is an alternative production method. For this reason large-scale energy-intensive electrolysis processes such as metal extraction have developed where electricity can be generated at low cost. This criterion is more important than the cost of transporting the ores, owing to the huge thermal losses in transmission of electrical energy. Now that superconductors with critical temperatures significantly higher than the boiling point of liquid nitrogen are being discovered and developed, the transmission of electricity along superconductor cables without energy losses could and probably will become reality within the next few decades, a development that will very much benefit the electrochemical industry.

15.2 Electrolysis: fundamental considerations

Electrolysis is the conversion of electrical energy into chemical energy in order to convert substances by oxidation or reduction, so that products are formed as the element or an appropriate compund. Also included is the generation of charged intermediates that link to other species, as in electrosynthesis. The design of the cells where these reactions take place, together with associated operations, is *electrochemical engineering*[4].

As has been explained throughout this book, electrochemists have under their control parameters such as the solvent, supporting electrolyte, concentration of electroactive species, solution movement, shape and material of the electrodes, the electrochemical cell, applied potential or current, and temperature. All these factors affect the kinetics and mechanism of electrode processes, and the electrochemical engineer must be aware of all of them. In industry one attempts to increase the rate of the required mechanism whilst minimizing the rate of any other mechanism or parallel process that can occur simultaneously, in order to maximize the yield in space and time. To achieve a good yield it is necessary to maximize the contact between electrode and electrolyte and sometimes to apply fairly large overpotentials to overcome solution resistance. In order to optimize these factors, a good knowledge of the fundamental principles of electrochemisty is required. In the past this was often achieved by trial and error, but at present there is a lot of effort towards optimizing industrial processes through scientific research into process development.

In certain cases of metal extraction, electrolysis has to be employed of necessity. Metals are, with rare exceptions such as gold, found in nature in positive oxidation states. In principle, reduction to the metal can be carried out by a chemical reducing agent. Chemical reducing agents that are used on a large scale are hydrogen, carbon, carbon monoxide, or mixtures of these. However, for the very electropositive metals, the reducing power of these compounds is not sufficient to extract the metal except under very extreme conditions of high temperature and pressure, so that electrolysis has to be employed. Examples are the elements of Group IA, magnesium, and aluminium.

In other cases electrochemistry is used because of the control of the selectivity of the reaction that is achieved through applied potential, when an ore or other sample contains various species with similar chemical but different electrochemical characteristics. The substances converted are obtained with a high degree of purity, but often at high energy cost.

Cell voltages for industrial electrochemical cells are expressed in the following way:

$$E_{cell} = E_c - E_a - \sum |\eta| - IR_{cell}$$

where E_c and E_a are the thermodynamic potentials for cathodic and anodic reactions respectively, η are the overpotentials, and (IR_{cell}) is the contribution from the resistance of electrolyte and electrodes, dependent on cell design. Thus electrolysis cell voltages are always negative. It is easy to see that the cell's energy efficiency is given by

$$\text{percentage energy efficiency} = \frac{(E_c - E_a)}{E_{cell}} \times 100$$

15.3 Electrochemical reactors

There are various types of electrochemical reactor[5,6]; the classification is similar to that used for other chemical processes. The three basic types of electrochemical reactor are shown in Fig. 15.1:

1. *Batch reactor.* The process requires total conversion of the reagents, and therefore includes a time for discharging and recharging. It is not easily adaptable to industrial situations, because of:

• low conversion rate towards the end of the electrolysis step

• time lost with discharging and recharging.

This reactor is best for laboratory use in investigations of electrolysis kinetics and mechanism.

2. *Plug-flow reactor.* Reagents are continuously introduced and the electrolyte (reagent + product) continuously removed. In general, conversion is less than 100 per cent but there are not the disadvantages of the batch reactor. Use of porous electrodes or electrodes with small anode–cathode spacing significantly improves the yield. Because of this it has much industrial application.

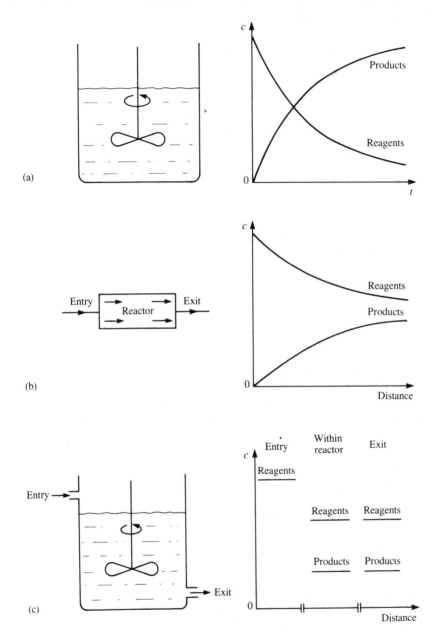

Fig. 15.1. Types of electrochemical reactor; (a) Batch reactor; (b) Plug flow reactor; (c) Backmix flow reactor.

3. *Backmix flow reactor or continuously stirred tank reactor.* The conversion rate is lower than for plug-flow reactors because the reagent is immediately diluted on being introduced into the reactor. Many flow reactors, e.g. tubular reactors, and especially in the turbulent regime are in this class.

Plug-flow and backmix flow reactors can be used as single-pass, with recirculation or in cascade, leading to many possible configurations, but always with the aim of optimizing product yield in space and time.

In practice, there exist much more complex reactors. An extremely important factor is the flow regime. Reactor design has to be such that the current density is uniform at all the electrodes (except in rare cases) and, preferably, high.

It is possible to design reactors so that the anode and cathode are positioned close to each other and with the cells containing more than one pair of electrodes. In this way conversion can be almost 100 per cent (compare with thin-layer cells, Section 9.10).

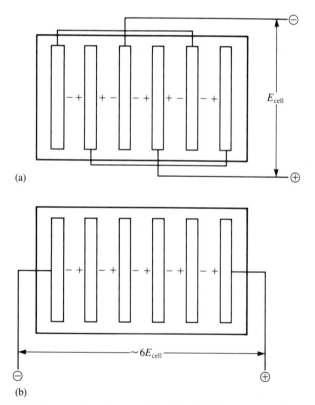

Fig. 15.2. Scheme of multielectrode cells: (a) Monopolar; (b) Bipolar.

There are two types of multielectrode reactor: monopolar and bipolar cells, as shown in Fig. 15.2. The bipolar configuration has the advantage that the electrical circuit has only to be linked at the ends of the electrode pile; the disadvantage is limitation to certain electrode materials: when the anode and cathode are of the same material or when they can be easily glued to each other.

Another way to obtain high degrees of conversion, besides the close positioning of anode and cathode, is through the use of porous and packed bed electrodes. Since these electrodes have great importance in the laboratory as well as at an industrial level, they are described separately.

15.4   Porous and packed-bed electrodes

Porous electrodes are fabricated from a piece of material, such as reticulated vitreous carbon (RVC), that contains interlinked pores which let solution pass through. *Packed-bed electrodes* consist of small contacting conducting particles which fill the volume of a reactor leaving holes between the particles for solution to pass: they can be of, for example, metal fragments or carbon fibres. There are also fluidized-bed electrodes in which the reactor is not completely filled with particles and solution passes vertically from the bottom to the top, forcing movement of the particles and consequent fluid behaviour. Whilst in the porous and packed-bed electrodes there is a large resistance to solution movement, in fluidized-bed systems high flow rates are possible. With all these types of electrode, potential gradients along the electrode assembly are created. The disadvantage of the fluidized-bed electrode is that there is less control over the potential of the particles, and for this reason its future application is not clear. Porous-bed and packed-bed electrodes will now be described in greater detail, for which, because of their structural similarities there exists, with some approximations, a common theoretical description[7,8].

It is possible to show that, if the size and distribution of the pores is uniform, then the fraction, R, of species electrolysed is

$$R = 1 - \exp\left(bU^{\alpha-1}sL\right) \qquad (15.1)$$

where U is the linear flow velocity, b is a proportionality factor defined by

$$k_{\mathrm{d}} = bU^{\alpha} \qquad (15.2)$$

with k_{d} the mass transfer coefficient and α is a constant that has a value between 0.33 and 0.50 for laminar flow, and between 0.50 and 1.00 for turbulent flow. s is the *specific area* of the electrode:

$$s = \frac{a}{LA} \qquad (15.3)$$

where A is the transverse section of the electrode, L its length, and a the internal electrode area, taking in the sum of the areas of all the pores. As can be seen, $R = 1$ when

$$bU^{\alpha-1}sL \gg 0 \qquad (15.4)$$

and as $\alpha < 1$, (15.4) predicts that efficiency decreases with increase in U, as would be expected. The maximum possible value of U for total conversion is

$$U_c = \frac{2LD\epsilon}{r^2} = \frac{LDs^2}{2\epsilon} \qquad (15.5)$$

where r is the average pore radius and ϵ the material porosity (97 per cent for RVC).

These expressions assume mass transport control; kinetic limitations complicate the analysis. However, it is reasonable to assume that a potential corresponding to the mass-transport-limited current can almost always be chosen.

Principal applications are in electrosynthesis, metal recovery and, sometimes, electroanalysis.

15.5 Examples of industrial electrolysis and electrosynthesis

In this section some examples of electrolysis and electrosynthesis of great industrial relevance will be described. Nevertheless, it should be made clear that there are many other important processes which are not described here. A summary is given in Table 15.1.

The chlor-alkali industry[8–12]

This is one of the largest electrochemical industries in the world. It consists in the electrolysis of sodium chloride as brine to give chlorine and caustic soda. Chlorine is used in the preparation of vinyl chloride for PVC, as a bleaching agent for paper and paper pulp, as a disinfectant, besides other chloration applications. Caustic soda is important in mineral processing, and in the paper, textile, and glass industries. Table 15.2 shows recent data for industrial consumption of chlorine and caustic soda in the USA.

Table 15.1. Industrial electrolysis and electrosynthesis

Chlor-alkali industry[9-13]	Extraction of chlorine and sodium hydroxide from NaCl
Metal extraction	Aluminium (Hall–Heroult process)[14-16] Sodium, magnesium, lithium (electrolysis of the fused salts) Electrolysis in aqueous solution (principally copper and zinc)
Electrolysis in the preparation of inorganic compounds[17]	Strong oxidizing agents: $KMnO_4$, $K_2Cr_2O_7$, $Na_2S_2O_8$, F_2, $NaClO_3$. Active metal oxides: MnO_2, Cu_2O Hydrogen and oxygen by water electrolysis[18-20]
Electro-organic synthesis[21-24]	Hydrodimerization of acrylonitrile (Monsanto process) Direct processes e.g. reduction $Me_2CO \rightarrow i\text{-PrOH}$ Indirect processes—an inorganic reagent is used as catalyst, being oxidized or reduced at the electrode to give a species that reacts with the organic compound e.g. Electrode(Pb): $Cr^{3+} \rightarrow Cr_2O_7^{2-}$ Solution: $Cr_2O_7^{2-}$ + anthracene $\rightarrow$ anthraquinone + Cr^{3+}

Table 15.2. Chlorine and caustic soda consumption in the USA (1990 data)[25]

Chlorine	Per cent	Caustic soda	Per cent
Vinyl chloride	24	Paper pulp, paper	24
Propylene oxide	8	Organics	22
Chlorinated methanes	7	Soaps and detergents	8
Chlorinated ethanes	7	Petroleum	8
Epichlorohydrin	5	Textiles	5
Other organics	12	Inorganics	12
Paper pulp, paper	14	Alumina	3
Inorganic compounds	8	Water treatment	8
Water treatment	5	Misc.	10
Titanium dioxide	3		
Export	4		
Misc.	3		
Consumption/Mtons (1989) 11.1			11.2

The electrode reactions are

$$\text{anode:} \qquad 2Cl^- \rightarrow Cl_2 + 2e^- \qquad E^{\ominus} = +1.36 \text{ V}$$

$$\text{cathode:} \quad 2H_2O + 2e^- \rightarrow H_2 + 2OH^- \qquad E^{\ominus} = -0.84 \text{ V}$$

and in a mercury cell (see below) at the anode

$$2Na^+ + 2e^- + 2Hg \rightarrow 2NaHg \qquad E^{\ominus} = -1.89 \text{ V}$$

followed by washing with water:

$$2NaHg + 2H_2O \rightarrow 2Hg + 2Na^+ + H_2 + 2OH^-$$

At the anode there is the possibility of the parallel unwanted reactions

$$2H_2O \rightarrow O_2 + 4H^+ + 4e^-$$

$$6OCl^- + 3H_2O \rightarrow 2ClO_3^- + 4Cl^- + 6H^+ + 3/2O_2 + 6e^-$$

and if the anode is of carbon, there is even the reaction

$$C + 2H_2O \rightarrow CO_2 + 2H_2O$$

There are three types of cell: mercury, diaphragm, and membrane, as shown in Fig. 15.3. Owing to the corrosive power of chlorine and oxidation of the electrode itself, conventional anodes of carbon or graphite, which have a high overpotential of 0.5 V, have been replaced by materials based on titanium covered with RuO_2 containing other transition metal oxides such as Co_3O_4. These *dimensionally stable anodes* (DSA) hardly corrode and their overpotential is 5–40 mV. They have yet another advantage: the unwanted oxygen evolution side reaction is reduced to a very low percentage (1–3 per cent).

1. *Mercury cell* (Fig. 15.3a). The cathode is mercury so that the sodium metal produced at the cathode reacts immediately with the mercury to form an amalgam, NaHg, thus being separated from the other products. Posterior treatment with water converts NaHg into caustic soda, hydrogen, and mercury, the mercury then being reused. The sum of the thermodynamic potentials in a commercial cell is −3.1 V; however, even with DSAs −4.5 V is necessary, the difference of 1.4 V being to overcome solution, electrode resistance, etc. The yield is good, but there is the huge problem of mercury toxicity. For this reason this type of cell is being gradually withdrawn as an industrial process.

2. *Diaphragm cell* (Fig. 15.3b). In this cell there is a physical barrier between the anode (DSA) and the cathode (steel) which is an asbestos diaphragm supported by a steel net. Sometimes separator and cathode are joined. Caustic soda is generated directly at the cathode with the

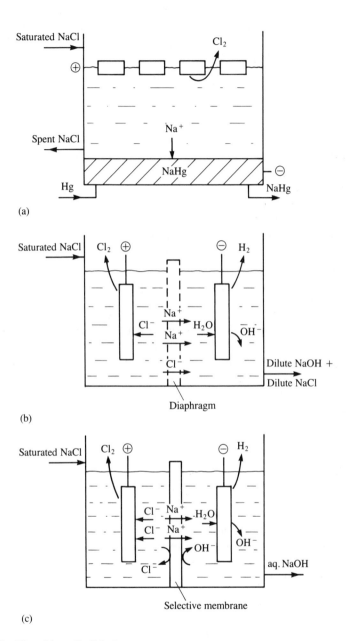

Fig. 15.3. The chlor-alkali industry. Types of cell (schematic): (a) Mercury cell; (b) Diaphragm cell; (c) Membrane cell.

corresponding release of hydrogen; unfortunately its concentration cannot go above 10 per cent or there is significant diffusion of OH^- to the anode compartment to produce chlorate, thus reducing cell efficiency. The caustic soda has to be concentrated by evaporation afterwards. Disadvantages of the cell separator are its short lifetime, its resistance, and the fact that it lets past all species. The big advantage with respect to the mercury cell is that the cell potential is -3.45 V (reversible potential -2.20 V).

3. *Membrane cell* (Fig. 15.3c). This is similar to the diaphragm cell, except that a selective membrane, which lets only certain ions through, is employed instead of a physical separator. In this way higher concentrations of caustic soda can be produced than in the diaphragm cell. Membranes used are of Nafion® (a tetrafluoroethylene copolymer), Flemion®, and other similar materials, as well as bilayer membranes which improve selectivity and impede hydroxide diffusion, so that up to 40 per cent sodium hydroxide solution can be produced. The cell potential is -2.95 V (reversible potential -2.20 V). Energy consumption is the lowest of the three processes, and product purity is high.

Other steps in these processes are: purification of the sodium chloride brine before electrolysis, evaporation of water to further concentrate the caustic soda, removal of oxygen from the chlorine. The hydrogen obtained in partitioned cells is very pure and can be used, for example, in the food industry. There is now a general changeover to membrane cells worldwide.

Metal extraction: aluminium[14-16]

Aluminium is one of the most abundant elements in the earth's crust, but, under feasible industrial conditions, can only be extracted by electrolysis. The process used is the electrolysis of aluminium hydroxide in molten cryolite (Na_3AlF_6) at 1030°C, pure aluminium hydroxide having been prepared from the mineral bauxite (hydrated aluminium oxide containing silica and some metal oxides such as iron) by the Bayer process. The cathode is carbon covered by molten aluminium metal and the anode is carbon, the total reaction being

$$2Al_2O_3 + 3C \rightarrow 4Al + 3CO_2$$

which consumes the anode. The mechanism is obviously more complex and involves the electrolyte. This is the Hall–Heroult process, represented in Fig. 15.4.

The molten cryolite is ionized. The reactions that can follow are shown

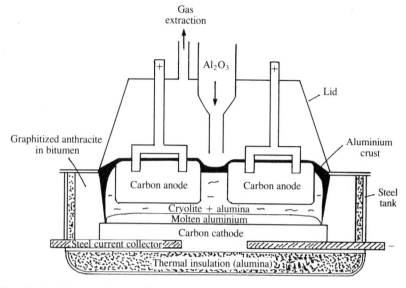

Fig. 15.4. Scheme of a cell for aluminium extraction by the Hall–Heroult process.

below, and involve complex ion formation:

$$Na_3AlF_6 \rightarrow 3Na^+ + AlF_6^{3-}$$

$$AlF_6^{3-} \rightleftharpoons AlF_4^- + 2F^-$$

$$4AlF_6^{3-} + Al_2O_3 \rightarrow 3Al_2OF_6^{2-} + 6F^-$$

$$2AlF_6^{3-} + 2Al_2O_3 \rightarrow 3Al_2O_2F_4^{2-}$$

At the cathode, there is evidence that the complex ions AlF_6^{3-} and AlF_4^- dismute, then following reduction of Al^{3+} to Al,

$$AlF_6^{3-} \rightleftharpoons Al^{3+} + 6F^-$$

$$AlF_4^- \rightleftharpoons Al^{3+} + 4F^-$$

$$Al^{3+} \rightarrow Al + 3e^-$$

Na^+ is not reduced and F^- produced by the cathode reactions neutralizes the charge of the sodium ions.

At the anode the electrode reactions involve the aluminium oxyfluoride complexion:

$$AlO_2F_4^{2-} + 2AlF_6^{3-} + C \rightarrow 4AlF_4^- + CO_2 + 4e^-$$

$$2AlOF_6^{2-} + 2AlF_6^{3-} + C \rightarrow 6AlF_4^- + CO_2 + 4e^-$$

Finally we need to know the cell potential. Under typical conditions it is

reversible potential	-1.2 V
anode overpotential	-0.5 V
IR drops: at electrodes	-1.1 V
in electrolyte	-1.5 V
	-4.3 V

Thus only 30 per cent of the total potential is used for the electrode reactions. It would clearly be very advantageous if there were another more energy-efficient process, preferably at a lower temperature.

The only reasonably successful advance in this sense is the Alcoa process, based on the electrolysis of aluminium trichloride in a 2–15 per cent concentration at 700°C in a 3:2 mixture of molten sodium chloride and potassium chloride using carbon electrodes. Aluminium oxide is previously converted into aluminium chloride using chlorine from electrolysis. The reactions are thus

$$2Al_2O_3 + 3C + 6Cl_2 \rightarrow 4AlCl_3 + 3CO_2$$
$$2AlCl_3 \rightarrow 2Al + 3Cl_2$$

Energy efficiency is stated to be about 10 per cent better than the Hall–Heroult process.

Water electrolysis[17–19]

Hydrogen gas is very important for hydrogenation of inorganic molecules, in semiconductor fabrication and in ammonia synthesis. It is normally produced by separating carbon monoxide and hydrogen resulting from high-temperature gasification of coal, or from petroleum products. However, this hydrogen is not very pure.

High-purity hydrogen is necessary in the food industry to produce margarines etc. Pure oxygen, generated *in situ*, is important in life support systems in submarines, space vehicles, and so on. Electrolysis generally leads to high-purity products, and only water electrolysis permits hydrogen and oxygen to be obtained with sufficient purity for the applications described. (It should be noted however, that the hydrogen produced in the electrolysis of sodium chloride in diaphragm or membrane cells is also of high purity—see above.)

The basic configuration of a cell for water electrolysis is fairly simple (Fig. 15.5). Anode and cathode are separated by a physical barrier, usually asbestos. The electrodes can be of various materials, such as carbon, and are often doped with an electrocatalytic material so as to

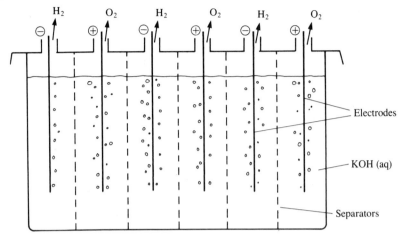

Fig. 15.5. Scheme of a cell for water electrolysis. Note the separators between the H_2 and O_2 produced. The electrical link is monopolar (but in other designs can be bipolar).

reduce the energy consumption. The electrolyte is potassium or sodium hydroxide, as corrosion problems are less in alkaline media. High-pressure electrolysers are often used, since they are more compact than those operating at atmospheric pressure, but cell engineering can be difficult.

Conventional high-pressure electrolysers are being gradually replaced by solid polymer electrolysers which are light and compact and operate on a different principle. Only the anode is exposed to water, where oxidation of water gives oxygen and $H^+(aq)$, the latter diffusing through a solid electrolyte membrane to the cathode where gaseous hydrogen is evolved. An important advantage is that pure water, as opposed to corrosive electrolyte, can be used and product gases are separated.

There has recently been interest in the photoelectrolysis of water (Section 12.4). In this process a large part of the energy necessary for electrolysis is provided photochemically by solar radiation, promoting the reactions by exciting the valence electrons in semiconductor electrodes. This electronic energy is then transferred to the water molecules, helping to break the O–H bonds. Efficiencies achieved so far are still not high, and it is not clear at present what future this will have.

Organic electrosynthesis: the Monsanto process[23]

The industrial organic electrosynthesis reaction of greatest impact must be the Monsanto process for the hydrodimerization of acrylonitrile to adiponitrile, one of the steps in the fabrication of nylon 66. Chemical synthesis routes, such as gas-phase catalysis of butene, exist and are used but feedstocks are more expensive than for the electrochemical route.

The electrode reactions are:

anode: $2H_2O \rightarrow O_2 + 4H^+ + 4e^-$

cathode: $2CH_2\!\!=\!\!CHCN + 2H_2O + 2e^- \rightarrow \begin{array}{c} CHCH_2CN \\ | \\ CHCH_2CN \end{array} + 2OH^-$

via $[CH_2\!\!=\!\!CHCN]^{\cdot -}$

At the cathode, other reactions are possible such as reaction of acrylonitrile with OH^- or direct protonation of the radical anion $[CH_2\!\!=\!\!CHCN]^{\cdot -}$ to give propionitrile. The existence of several possible pathways is general in the synthesis of organic compounds, and optimized experimental conditions, such as choice of electrode potentials that minimize unwanted lateral reactions, must be sought. In this particular case pH control is also of obvious importance in minimizing the unwanted reactions.

The present process is shown schematically in Fig. 15.6. The 'solution' is an emulsion of acrylonitrile and 10–15 per cent disodium hydrogen-phosphate in water containing a quaternary ammonium salt, hexamethylene-bis(ethyldibutylammonium) phosphate, to conduct the current. The anode is carbon steel and the cathode is cadmium (a sheet fixed on to the anode); this cell is a good example of the use of bipolar electrodes (Section 15.3). EDTA and borax are added in small quantities

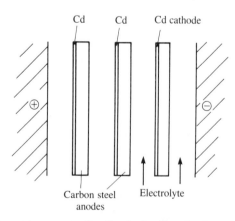

Fig. 15.6. The Monsanto process for the hydrodimerization of acetonitrile. The cell has bipolar electrodes.

to minimize anode corrosion. The cell potential is $-3.8\,\text{V}$ and the selectivity is 88 per cent.

15.6 Electrodeposition and metal finishing

Metals are used for many purposes, but they are often susceptible to corrosion (Chapter 16). Protection against corrosion[26] brings huge economic benefits. Often protection is done by electrodeposition of a layer of another metal, more inert (and more expensive) on the substrate[27], for example on iron and steels. Because of the importance of efficient protection there has been much laboratory investigation into electrodeposition mechanisms, but there are still empirical factors to be explained satisfactorily.

The mechanism of electrodeposition or electrocrystallization[28,29] involves, as a first step, the reduction of a cation on the substrate surface (aided by an applied potential or current) to form an *adatom,* and its migration over the surface to an energetically favourable site. Other atoms of the electrodeposit aggregate with the first, forming the nucleus of a new phase. The nucleus grows parallel and/or perpendicular to the surface. Clearly, a number of nuclei can form and grow on the surface. When all the electrode surface is covered with at least a monolayer, deposition is on the same metal rather than on a different metal substrate. As is to be expected, the formation of the first layers determines the structure and adhesion of the electrodeposit.

Qualitatively the process is very similar to the formation of a precipitate in homogeneous solution. The difference is in the structure of the precipitate[30], as well as its formation, in homogeneous solution being affected by the degree of supersaturation and in electrodeposition by the overpotential.

We now consider briefly how nucleus formation on an electrode surface, *electrocrystallization,* can be studied. Nucleation normally follows a first-order rate law

$$N = N_0(1 - \exp{(-At)}) \tag{15.6}$$

where N_0 is the number of nucleation sites and A the nucleation constant. There are two limiting cases of (15.6):

• *instantaneous nucleation*: $N = N_0$, $At \gg 1$, which is probable on applying a high overpotential;

• *continuous nucleation*: $N = AN_0t$, $At \ll 1$.

We are assuming an equal nucleation energy for all nucleation sites. In reality, the energy is less where there are breaks in structure such as grain boundaries, dislocations, etc.

In the growth phase, the nucleus can grow parallel to and/or perpendicular to the surface. If the growth probability is equal in all three directions hemispheres are formed of surface area $2\pi r^2$, where r is the radius of the sphere.

For a kinetically controlled process the current per nucleus is

$$I_i = nFk(2\pi r^2) \tag{15.7}$$

$$= \frac{2\pi nFM^2k^3}{\rho^2}t^2 \tag{15.8}$$

where we introduce the dependence of r on t; M is the molecular mass of the electrodeposit, ρ its density, and k the rate constant. When there are many nuclei then, for instaneous nucleation,

$$I = \frac{2\pi nFM^2k^3N_0}{\rho^2}t^2 \tag{15.9}$$

and for progressive nucleation,

$$I = \frac{2\pi nFM^2k^3AN_0}{\rho^3}t^3 \tag{15.10}$$

If the process is diffusion controlled then, for small t,

$$I = \frac{\pi nF(2Dc_\infty)^{3/2}M^{1/2}N_0}{\rho^{1/2}}t^{1/2} \tag{15.11}$$

for instantaneous nucleation, and

$$I = \frac{4\pi nF(Dc_\infty)^{3/2}M^{1/2}AN_0}{3\rho^{1/2}}t^{3/2} \tag{15.12}$$

for progressive nucleation. The dependence of I on t for the four cases is summarized in Table 15.3.

Table 15.3 Dependence of current with time for the growth of hemispherical nuclei: values of n where $I \propto t^n$

	Nucleation	
	Instantaneous	Progressive
Diffusion control	1/2	3/2
Kinetic control	2	3

Thus, a study of the variation of I with t can give information on the mechanism of electrocrystallization. However, experimental observation gives a weighted average of the various types of growth (not necessarily hemispherical) and its deconvolution is difficult. In this sense electron microscopy and scanning tunnelling microscopy are valuable tools, as are reflection techniques (Chapter 12). In these growth studies, initial nucleation is provoked by application of a short pulse at a high, negative potential.

As in any electrode process, the potential applied to the electrode determines the reaction rate. In electrodeposition, we expect that it affects the rate of deposition and thence the structure of the deposit: a low overpotential signifies more time available to form an electrodeposit of perfectly crystalline structure. This can be observed experimentally (Fig. 15.7). Another factor arises from differences in current density between different parts of the electrode owing to electrode shape, which affects mass transport and thus accessibility to the cations to be deposited. Generally, it is best to apply a potential corresponding to the formation of polycrystalline deposits. A more perfect crystalline structure would be desirable, but the low rate of electrodeposition does not compensate for using such low overpotentials.

In an electrodeposition bath there exist, besides the cation for electrodeposition and an inert electrolyte for good conduction, various additives to improve the quality of the deposit—their mode of functioning is often not well understood. Organic compounds and surfactants can make the deposit smoother and brighter and modify its structure, probably in the initial nucleation step. The addition of complexing agents is useful for altering the deposition potential and avoiding spontaneous

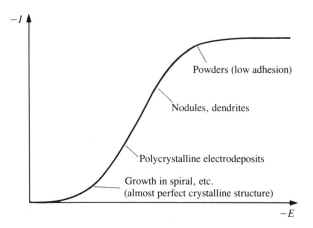

Fig. 15.7. Variation of electrodeposit structure with applied potential.

Table 15.4. Examples of industrial electrodeposition and its applications

Metal electrodeposited	Applications
Nickel	Domestic use and engineering
	Undercoat for chromium electrodeposition
Chromium	Domestic use, car components, surfaces with reduced friction in tools and valves in motors (applied normally over an undercoat of nickel or copper)
Copper	Contacts in electronic industry. Decoration. Undercoat for nickel and chromium
Tin	Protection of cans, electrical contacts for soldering
Silver and gold	Decorating, mirrors, electrical contacts
Cadmium and zinc	Protection of steel alloys

chemical reactions, for example between iron substrates and Cu^{2+}. Some examples of industrial electrodeposition are given in Table 15.4.

Other important metal finishings to protect against corrosion are conversion coatings such as anodization (especially for aluminium), electroless plating, and electrophoretic painting. The first is done to form a passive layer, and is described in greater detail in Section 16.4.

Electroless plating does not use electrodes, and is an autocatalytic reaction on the metal surface after nucleation has begun. A reducing agent R reduces metal ions to the corresponding metal on the substrate surface by undergoing surface-induced oxidation. The most important examples are nickel on steel substrates and copper on printed circuit boards and plastics, for both functional and decorative applications. Electrolytes are buffer solutions, usually containing hypophosphite as reducing agent, the reaction for nickel being

$$Ni^{2+} + H_2PO_2^- + H_2O \rightarrow Ni + H_2PO_3^- + 2H^+$$

Additives are often used to reduce the rate of plating, and easily adsorbable molecules such as thiourea to minimize the inherent instability of the plating solution in the presence of foreign particles on which plating would otherwise occur. Particular advantages of electroless plating relative to electroplating are that, other than in exceptionally unfavourable circumstances, coverage is uniform, and that coatings are harder and of lower porosity leading to increased resistance to corrosion and wear (the latter possibly due to the presence of other elements such as phosphorus in the coating). These advantages can compensate the higher cost of electroless as compared to conventional electroplating. The harder coatings do lead to lower ductility and higher residual stress, so that the type of application must be carefully studied.

Electrophoretic painting[31] is much used in the automobile industry. It consists of the electrodeposition of a polymeric film on a metal from a polymer solution, which totally covers and protects it. The word 'electrophoretic' is not very correct in describing the process, and recently the term 'electrodeposited paint' has been used. The polymer contains acidic or basic groups which form micelles; the solution also contains pigments and other organic solids in suspension. The mechanism involves migration of the micelles to the electrode, neutralization of their charge, precipitation of the neutralized polymer with occlusion of pigment molecules, and, finally, removal of the water solvent from the polymer layer by electroosmosis. The advantage in using water is clear; the coating is very good, even within crevices. It is a pity that it is only possible to apply one coat since, after the surface is coated, it no longer conducts. An important application is in the priming of steel car bodies after phosphate conversion coating (see Chapter 16).

15.7 Metal processing

Electrodeposition and electrodissolution of metals has an important role in the fabrication of metal articles with shapes that are difficult to make by conventional methods[32], We exemplify with two types of processing: electroformation and electrochemical machining.

In *electroformation*, a metal is electrodeposited in the shape and form of the desired object on top of a substrate (the mould) that is afterwards removed. An example is the manufacture of foils for electrical shavers: the mould (cathode) is a cylinder with a design containing insulated parts where there is no electrodeposition and that turns slowly; the anode is also cylindrical and concentric with the cathode, and the foil is separated from the mould with a knife (Fig. 15.8). The most commonly used metals

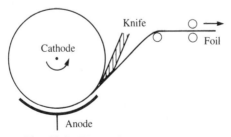

Fig. 15.8. Electroformation of foils.

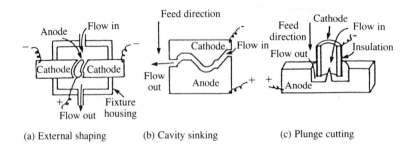

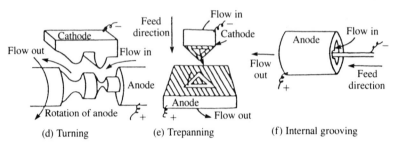

Fig. 15.9. Electrochemical machining: principle of operation for various types of shaping (from Ref. 32 with permission).

are nickel and copper, for economic reasons. Apart from foils, bulk materials of complex shape can be formed, such as tools and forms for plastics, high-frequency waveguides, and so on.

Metal objects with complex shapes can be formed by *electrochemical machining* (electroerosion), especially important when mechanical machining is not possible. The object is the anode, where dissolution occurs, and the tool is the cathode, having the form of a mould for the object. The cathode has small holes from which jets of electrolyte exit so that there is a layer of electrolyte between anode and cathode (Fig. 15.9). An extremely important example is the manufacture of components such as blades for turbines.

15.8 Batteries

Batteries are stores of chemical energy which, on applying an external load, can convert this energy into electrical energy. Besides economic factors, there are certain requirements for a battery to be useful:

● its voltage should be constant and with a minimum overpotential

Table 15.5 Batteries in current use

Battery	Discharge reactions		Electrolyte	Electrode materials		E_{cell}/V	Applications
	Cathode (+ve)	Anode (−ve)		Cathode	Anode		
Secondary							
Lead/acid	$PbO_2 + 4H^+ + SO_4^{2-} + 2e^-$ $\rightarrow 2H_2O + PbSO_4$	$Pb + SO_4^{2-}$ $\rightarrow PbSO_4 + 2e^-$	Sulphuric acid (aq)	Pb	Pb	2.05	Automobiles, traction, industry
Nickel/cadimum	$NiO(OH) + H_2O + e^-$ $\rightarrow Ni(OH)_2 + OH^-$	$Cd + 2OH^-$ $\rightarrow Cd(OH)_2 + 2e^-$	KOH(aq)	Ni	Cd	1.48	Industry, starters for aeroplane engines, railway lighting
Primary							
Zinc/carbon (Leclanché)	$2MnO_2 + H_2O + 2e^-$ $\rightarrow Mn_2O_3 + 2OH^-$	$Zn \rightarrow Zn^{2+} + 2e^-$	$NH_4Cl/ZnCl_2/$ $MnO_2/$damp C powder	Graphite	Zn	1.55	Portable voltage sources (dry batteries)
Alkaline	$2MnO_2 + 2H_2O + 2e^-$ $\rightarrow 2MnO(OH) + 2OH^-$	$Zn + 2OH^-$ $\rightarrow ZnO + H_2O + 2e$	$NH_4Cl/ZnCl_2/$ MnO_2/C powder/ $NaOH$ (aq)	$MnO_2/$ graphite	Zn	1.55	High-quality dry batteries
Silver oxide/zinc	$Ag_2O + H_2O + 2e^-$ $\rightarrow 2Ag + 2OH^-$	$Zn + 2OH^-$ $\rightarrow ZnO + H_2O + 2e^-$	KOH (aq)	$Ag_2O/$ graphite	Zn	1.5	Watches, cameras
Mercury oxide/zinc	$HgO + H_2O + 2e^-$ $\rightarrow Hg + 2OH^-$	$Zn + 2OH^-$ $\rightarrow ZnO + H_2O + 2e^-$	KOH (aq)	$HgO/$ graphite	Zn	1.5	Watches, cameras

- on discharging the battery, the voltage remains constant
- it has a sufficient, normally high, capacity to supply current
- storage without self-discharge

Real batteries satisfy these criteria up to the point judged necessary for the application in mind. Batteries are classified into two types:

- *primary*, which discharge only once, e.g. dry batteries;

- *secondary*, which are rechargeable, e.g. lead–acid accumulators. Another criterion to consider in this category is the number of possible recharge cycles.

Some commonly used batteries are shown in Table 15.5, and two are drawn schematically in Fig. 15.10. From these it can be seen that important components are the container, the anode/cathode compartment separators, current collectors to transport current from the electrode material (usually a porous, particulate paste), the electrode material itself, and the electrolyte. It should be noted that the electrode reactions can be significantly more complex than those indicated in Table 15.4, and there will probably be parallel reactions. By stacking the batteries in series, any multiple of the cell potential can be obtained.

Because of the great utility of batteries, the search for systems that offer higher energy densities and voltages than present systems continues[33,34]. Lithium batteries in non-aqueous media, which are now

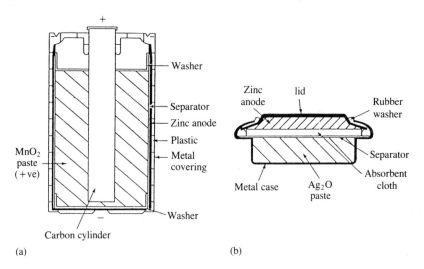

Fig. 15.10. Two batteries in common use: (a) The cylindrical Leclanché battery; (b) The silver oxide/zinc button cell.

reaching the market for medical and consumer purposes, have lithium negative electrodes and can in principle exhibit very high battery voltages, because of the very negative electrode potential of the Li^+/Li couple, high storage density and high discharge current density—they are usually primary batteries. Use of reduction of organic polymers, such as $(CF)_n$, or of oxidized conducting polymers such as polyacetylene, polyaniline, or polypyrrole, is being investigated as positive electrode reaction. The high chemical reactivity of lithium metal causes difficulties in battery design.

Amongst other new systems under study are the sodium/sulphur battery with sodium β-alumina solid electrolyte operating at 300–375°C and Li–FeS batteries operating at about 450°C. Long-term battery research is directed towards batteries that can operate at room temperature with aqueous electrolyte, such as zinc–halogen, aluminium–air, and iron–air.

15.9 Fuel cells

Fuel cells[33,34] are, in effect, batteries in which the reactants are fed externally. A fuel undergoes oxidation through controlled half-reactions, in order to convert chemical energy into electrical energy. Direct electrochemical oxidation of fuels appears very difficult to achieve in the case of hydrocarbons. For example, methane would follow the half-reactions

$$\text{anode} \qquad CH_4 + 2H_2O \rightarrow CO_2 + 8H^+ + 8e^-$$
$$\text{cathode} \qquad O_2 + 4H^+ + 4e^- \rightarrow 2H_2O$$

It seems to be necessary to convert the primary fuel into hydrogen or carbon monoxide first; after this the cell functions well. The best-known application is in the combustion of hydrogen in aqueous potassium hydroxide electrolyte with nickel electrodes at 200°C, as used in the Apollo series space flights.

Present research is mainly into four types of cell:

1. *Phosphoric acid* (PAFC). The electrolyte is concentrated H_3PO_4 absorbed on to a solid matrix, and operates at 200°C. The electrodes are carbon loaded with platinum particles; the anode fuel is hydrogen and the cathode fuel is air. The cell voltage is usually around 0.67 V. This type of cell has been tested commercially, producing 4.8 MW for several months at 40 per cent efficiency.

2. *Molten carbonate* (MCFC). The cell operates at 650°C and uses hydrogen or carbon monoxide as anode fuel, which reacts with carbonate

in the electrolyte (40 per cent $LiAlO_2$, 28 per cent K_2CO_3, 32 per cent Li_2CO_3) to give carbon dioxide, whilst oxygen is reduced at the cathode by carbon dioxide to carbonate. Cell voltages are around 0.9 V. Problems in functioning can arise with contamination by sulphur or chlorine.

3. *Solid oxide* (SOFC). The electrolyte is a ceramic oxide and operates at 1000°C and can consume hydrogen or hydrogen/carbon monoxide mixtures. A high electrical efficiency of over 50 per cent is reported.

4. *Alkaline.* Fuels are hydrogen and oxygen in a concentrated solution of potassium hydroxide at room temperature. The possible advantage is the use of non-platinum catalysts such as Raney nickel and silver on carbon supports. This is at an earlier stage of development than the other cells.

Possible future applications, besides small-scale power generation at remote sites such as in space, are large-scale power generation, vehicle traction, and burning side-products such as hydrogen from industrial chemical processes in order to recover electrical energy. Other reasons for continuing investigations are:

• the possibility of obtaining an efficiency of electrical energy production greater than that possible by classical combustion (which is about 35 per cent), thus saving natural fuel resources

• the amount of carbon dioxide released into the atmosphere is less per MW of electricity than in other electricity generation processes, which is very important for environmental reasons.

15.10 Electrochemistry in water and effluent treatment

Electrochemistry can be used for a number of purposes linked to water and effluent treatment. The most obvious of these involve the removal of ionic components from waters by application of an appropriate potential. This is employed to remove metal ions from process streams and often leads to recovery of the metal, which can be reused. Clearly, cell designs which favour high electrode surface area/catholyte volume ratios are to be recommended.

Cleaning of organic contamination in effluents and in sewage can be aided by electrochemistry in the following ways:

• Electrochemically *in situ* generated oxidants: hydrogen peroxide, ozone, and hypochlorite or chlorine

• Generating gas bubbles electrolytically at the base of tanks through which the effluent enters slowly at one end at the top and is continuously removed from the bottom at the other end. The gas bubbles take suspended matter with them to the surface from where it is scraped off (*electroflotation*). A flocculating agent can be added by controlled electrochemical anode dissolution, using anodes such as aluminium or iron, the process then being known as *electroflocculation*.

Ionic species in waters can be concentrated, and the water purified at the same time, by *electrodialysis*. Ion-exchange membranes are employed with an applied electric field in order to force ionic salts to pass from dilute into concentrated solutions. Examples of the use of electrodialysis are in the concentrating of Ni^{2+} in used nickel plating solutions for recirculation, and in desalination plants in the purification of sea water or well water to acceptable levels to make it fit for drinking.

References

1. D. Pletcher and F. C. Walsh, *Industrial electrochemistry*, 2nd edn., Chapman and Hall, London, 1990.
2. A. T. Kuhn (ed.), *Industrial electrochemistry*, Elsevier, Amsterdam, 1971.
3. J. O'M. Bockris, B. E. Conway, E. Yeager, and R. E. White (ed.), *Comprehensive treatise of electrochemistry*, Plenum, New York, Vol. 2, 1981.
4. E. Heitz and G. Kreysa, *Principles of electrochemical engineering*, VCH, Weinheim, 1986.
5. R. C. Alkire, *J. Chem. Ed.*, 1983, **60,** 274.
6. M. I. Ismail (ed.), *Electrochemical reactors*; *their science and technology*, Elsevier, Amsterdam, 1989.
7. R. E. Sioda, *Electrochim. Acta*, 1970, **15,** 783; *J. Appl. Electrochem.*, 1975, **5,** 221; 1978, **8,** 297.
8. J. Newman and W. Tiedemann, *Advances in electrochemistry and electrochemical engineering*, ed. H. Gerischer and C. W. Tobias, Wiley, New York, Vol. 11, 1978.
9. D. L. Caldwell, in Ref. 3, pp. 105–166.
10. D. M. Novak, B. V. Tilak, and B. E. Conway, *Modern aspects of electrochemistry*, Plenum, New York, Vol. 14, 1982, ed. B. E. Conway and J. O'M. Bockris, pp. 195–318.
11. W. N. Brooks, *Chem. Brit.*, 1986, **22,** 1095.
12. S. Venkatesh and B. V. Tilak, *J. Chem. Ed.*, 1983, **60,** 276.
13. F. Hine, B. V. Tilak, and K. Viswanathan, *Modern aspects of electrochemistry*, Vol. 18, 1986, ed. R. E. White, J. O'M, Bockris, and B. E. Conway, pp. 249–302.
14. W. E. Haupin and W. B. Frank, in Ref. 3, pp. 301–325.
15. W. E. Haupin, *J. Chem. Ed.*, 1983, **60,** 279.
16. A. R. Burkin (ed.), *Production of aluminum and alumina, critical reports on applied chemistry*, Wiley, Chichester, Vol. 20, 1987.
17. N. Ibl and H. Vogt, in Ref. 3, pp. 167–250.

18. B. V. Tilak, P. W. T. Lu, J. E. Colman, and S. Srinivasan, in Ref. 3, pp. 1–104.
19. F. Gutmann and O. J. Murphy, *Modern aspects of electrochemistry*, Plenum, New York, Vol. 15, 1983, ed. R. E. White, J. O'M. Bockris, and B. E. Conway, pp. 1–82.
20. M. Grätzel, *Modern aspects of electrochemistry*, Plenum, New York, Vol. 15, 1983, ed. R. E. White, J. O'M. Bockris, and B. E. Conway, pp. 83–165.
21. K. Köster and H. Wendt, in Ref. 3, pp. 251–299.
22. J. H. Wagenknecht, *J. Chem. Ed.*, 1983, **60**, 271.
23. M. M. Baizer (ed.), *Organic electrochemistry*, Dekker, New York, 1973.
24. D. K. Kyriacou and D. A. Jannakoudakis, *Electrocatalysis for organic synthesis*, Wiley-Interscience, New York, 1986.
25. H. S. Burney and J. B. Talbot, *J. Electrochem. Soc.*, 1991, **138**, 3140.
26. A. T. Kuhn (ed.), *Techniques in electrochemistry, corrosion and metal finishing—a handbook*, Wiley, London, 1987.
27. C. J. Rands, in Ref. 3, pp. 381–397.
28. E. D. Budevski, *Comprehensive treatise of electrochemistry*, Plenum, New York, Vol. 7, 1983, ed. B. E. Conway, J. O'M. Bockris, E. Yeager, and R. E. White, pp. 399–450.
29. M. Sluyters–Rehbach, J. H. O. G. Wigenberg, E. Bosco, and J. H. Sluyters, *J. Electroanal. Chem.*, 1987, **236**, 1.
30. A. R. Despic, *Comprehensive treatise of electrochemistry*, Plenum, New York, Vol. 7, 1983, ed. B. E. Conway, J. O'M. Bockris, E. Yeager, and R. E. White, pp. 451–528.
31. F. Beck, in Ref. 3, pp. 537–569.
32. J. P. Hoare and M. L. LaBoda, in Ref. 3, pp. 399–520.
33. A. F. Sammells, *J. Chem. Ed.*, 1983, **60**, 320.
34. M. Hayes, *Chem. Brit.*, 1986, **22**, 1101.

16

CORROSION

16.1 Introduction

Corrosion refers to the loss or conversion into another insoluble compound of the surface layers of a solid in contact with a fluid. The solid is normally a metal, but the term corrosion is also used to refer to the dissolution of ionic crystals or semiconductors. In the majority of cases the fluid is water, but an important exception is the reaction of metallic surfaces with air at high temperature, leading to oxide formation, or, in industrial environments, to sulphides, etc. In the context of this book, corrosion of metals or semiconductors in contact with aqueous solution or humid air at normal temperatures is of predominant interest.

Owing to the tremendous economic damage it can cause, corrosion has and continues to be the subject of extensive study especially with a view to its minimization at acceptable expense—economic and environmental. We attempt to give an idea of the forms of corrosion, how to investigate it by electrochemistry, and how it can be minimized, or at least reduced and controlled. As will be seen, given the complexity of corrosion processes, the mechanism of which can alter significantly depending on the local environment, the more specialized literature should be consulted for details on specific cases, for example Refs. 1–6.

16.2 Fundamentals

The corrosion of a metal in contact with an aqueous solution can be represented by the generic half-reactions

$$M \rightarrow M^{n+}(aq) + ne^-$$

with

- *in acid environment*

$$O_2 + 4H^+(aq) + 4e^- \rightarrow 2H_2O$$

and/or

$$2H^+ + 2e^- \rightarrow H_2$$

- *in alkaline environment*

$$O_2 + 2H_2O + 4e^- \rightarrow 4OH^-$$

and/or

$$2H_2O + 2e^- \rightarrow H_2 + 2OH^-$$

The metal ions can react immediately with OH^- to form insoluble oxides/hydroxides that cover the metal surface, or the metal ion can be released to bulk solution. The reactions that occur depend on pH, and we note that reduction half-reactions all alter the pH in the vicinity of the metal surface.

Thus, factors that affect the rate of corrosion are essentially pH, partial pressure of oxygen, and solution conductivity; there are also other less general factors that will be specified in the following sections. The half-reactions have to occur at different sites on the interface in order to form an electrical circuit, and thence the importance of solution conductivity. In particular cases other cathodic reactions can take place due to, for example, reduction of a species already present in solution such as Fe^{3+} reduced to Fe^{2+}.

As in any chemical reaction, thermodynamic and kinetic aspects have to be considered.

Thermodynamic aspects

The thermodynamic information is normally summarized in a *Pourbaix diagram*[7]. These diagrams are constructed from the relevant standard electrode potential values and equilibrium constants and show, for a given metal and as a function of pH, which is the most stable species at a particular potential and pH value. The ionic activity in solution affects the position of the boundaries between immunity, corrosion, and passivation zones. Normally ionic activity values of 10^{-6} are employed for boundary definition; above this value corrosion is assumed to occur. Pourbaix diagrams for many metals are to be found in Ref. 7.

The Pourbaix diagram for iron under these standard conditions is shown schematically in Fig. 16.1a. The two dotted lines on the diagram correspond to

$$O_2 + 4H^+ + 4e^- \rightarrow 2H_2O \qquad E^{\ominus} = (1.23 - 0.059\,pH) \quad V$$
$$2H^+ + 2e^- \rightarrow H_2 \qquad E^{\ominus} = -0.059\,pH \quad V$$

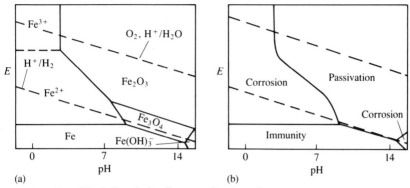

(a)   (b)

Fig. 16.1. Simplified Pourbaix diagram for iron in pure water—ionic activities 10^{-6}: (a) The diagram in terms of most stable species; (b) The diagram in terms of the type of reaction that can occur.

Whether a particular corrosion process is possible is determined by whether the line for hydrogen evolution or for oxygen reduction lies above (at a more positive potential than) the boundary for the oxidation half-reaction. This corresponds to a total negative free energy change. Nevertheless passivation often occurs, blocking further corrosion. In Fig. 16.1a an example would be the zone where Fe_2O_3 is formed.

Figure 16.1b is a representation of Fig. 16.1a in terms of the type of reaction: it shows, as a function of pH, the zones where corrosion by dissolution occurs, where initial corrosion forms insoluble oxides on the surface that impede further reaction (passive zone) and the region where the metal is stable (immune zone).

It is obvious that Fig. 16.1 corresponds to an unreal situation, in the sense that ionic activity in ordinary water is always greater than 10^{-6},

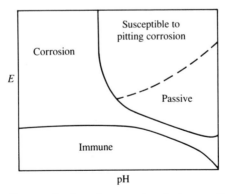

Fig. 16.2. Pourbaix diagram for iron in seawater, in terms of zones showing the type of reaction that occurs.

there may be foreign ions present and the local pH at the surface or at parts of the surface may differ from the bulk. A real situation is the corrosion of iron in seawater ($[Cl^-] = 0.7$ M; pH ≈ 7.5) which is shown in Fig. 16.2. The zone where pitting corrosion, caused by Cl^-, occurs should be noted; this involves rupture of the passive hydroxide film.

Kinetic aspects

Fortunately, kinetics makes corrosion more difficult, so that it is much less prejudicial than predicted thermodynamically. In the electrochemistry laboratory corrosion can be studied by voltammetry and kinetic parameters can be predicted from Tafel plots and from impedance data.

An important measurement is the corrosion potential, E_{cor}. This is the open circuit potential, whose value can change with time. E_{cor} is a mixed potential, since the anodic and cathodic reactions are different. The partial anodic or cathodic current that flows at this potential is called the corrosion current, I_{cor}, and is directly related to the rate constant of the electrode reaction.

A representation of E vs. I is called the *Evans diagram* (Fig. 16.3), a diagram much used by metallurgists. This diagram is useful for taking qualitative conclusions, especially when the kinetics of H^+ reduction is strongly dependent on the metal. For quantitative conclusions, it is clear to the electrochemist that a Tafel plot of E vs. $\lg |I|$ contains the same information and gives the possibility of determining I_{cor} with greater accuracy, as in Fig. 16.4. Besides this, it is possible to correct the data in

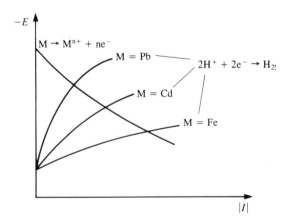

Fig. 16.3. Evans diagram for metallic corrosion in acid medium. The concentrations are adjusted for E_{eq} to be equal for the three metals.

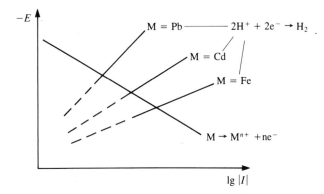

Fig. 16.4. A Tafel plot for the situation represented in Fig. 16.3 (in normal representation rotated 90° anticlockwise).

a Tafel plot for transport effects. Note that it is $\lg |I|$ (current) and not $\lg |j|$ (current density) that is plotted, owing to differences of geometric area where anodic and cathodic reactions occur.

The Tafel plot permits the calculation of the rate constants of the reactions from the intersections, and the charge transfer coefficient from the slope. Corrosion researchers use the parameter b (the inverse of the slope of the Tafel plot) extensively in their studies, the reason being the representation of the Tafel plot in the way illustrated in Fig. 16.4. In fact,

$$b_a = \frac{2.3RT}{\alpha_a n'F} \qquad b_c = \frac{-2.3RT}{\alpha_c n'F} \tag{16.1}$$

where n' is the number of electrons in the rate-determining step.

For potentials close to E_{cor}, we can obtain a relation that allows us to calculate the corrosion current. Considering the anodic half-reaction we know that

$$I_{cor} = I_{0,a} \exp\left[\frac{\alpha_a n'F(E_{cor} - E_{eq,a})}{RT}\right] \tag{16.2}$$

$$= I_{0,a} \exp\left[\frac{2.3(E_{cor} - E_{eq,a})}{b_a}\right] \tag{16.3}$$

where $E_{eq,a}$ is the equilibrium potential for the redox couple of the anodic half-reaction and $I_{0,a}$ the respective exchange current. In a similar way for a cathodic half-reaction,

$$I_{cor} = I_{0,c} \exp\left[\frac{-2.3(E_{cor} - E_{eq,c})}{|b_c|}\right] \tag{16.4}$$

in which $E_{eq,c}$ and $I_{0,c}$ have the corresponding meaning for the cathodic reaction. Therefore, for an applied potential different from E_{cor}, we have

$$I = I_{cor}\left\{\exp\left[\frac{2.3(E - E_{cor})}{b_a}\right] - \exp\left[\frac{2.3(E_{cor} - E)}{|b_c|}\right]\right\} \qquad (16.5)$$

If $(E - E_{cor}) = \Delta E$ is small, leading to a current ΔI, then we can make the approximation $\exp(x) = 1 + x$ and obtain

$$\Delta I = 2.3 I_{cor}\left(\frac{\Delta E}{b_a} + \frac{\Delta E}{|b_c|}\right) \qquad (16.6)$$

On rearranging,

$$I_{cor} = \frac{1}{2.3}\frac{b_a \cdot |b_c|}{b_a + |b_c|}\frac{\Delta I}{\Delta E} \qquad (16.7)$$

which is the *Stern–Geary relation*[8]. In this expression $\Delta I / \Delta E$ is called the *polarization conductance*, K_{cor}, and its inverse $\Delta E / \Delta I$ the *polarization resistance*, R_P. The utility of this relation is because, knowing K_{cor} or R_P and b_a and b_c, we can determine I_{cor}.

There are two basic methods for determining R_P in the laboratory:

1. Through an impedance experiment

$$R_P = Z_{\omega \to 0} - Z_{\omega \to \infty} \qquad (16.8)$$

2. With small-amplitude cyclic voltammetry[9], at a sufficiently slow scan rate (about $0.1\,\text{mV s}^{-1}$) and with an amplitude of $E_{cor} \pm 10\,\text{mV}$. A straight line of slope $1/R_P$ should be obtained.

Of these two methods, the second is the safer, owing to difficulties in the determination of $Z_{\omega \to 0}$, it often being necessary to use frequencies less than $10^{-3}\,\text{Hz}$.

Alternatively, for fieldwork, but less exactly, a two-electrode probe placed on the specimen with a $20\,\text{mV}$ potential difference applied will settle around E_{cor}, the current flow can be measured and R_P calculated.

Another aspect to consider is the presence or absence of oxygen in solution. The effect of its presence in acidic solution is demonstrated in Fig. 16.5, the result being greater corrosion. Sometimes it catalyses corrosion by oxidizing ionic species in solution, for example

$$4Fe^{2+} + O_2 + 4H^+ \longrightarrow 4Fe^{3+} + 2H_2O$$

$$2Fe^{3+} + 2e^- \xrightarrow{\text{surface}} 2Fe^{2+}$$

$$Fe \longrightarrow Fe^{2+} + 2e^-$$

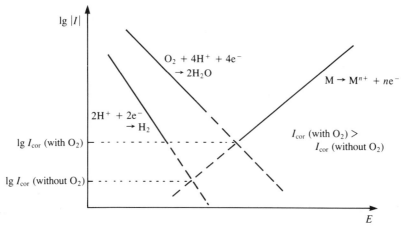

Fig. 16.5. Tafel plot in acidic medium for metallic corrosion in the absence and in the presence of oxygen.

Solution flow affects the reaction rate, given that it removes or supplies species in contact with the metal, such as oxygen.

The formation of passive films of metal oxides was already mentioned, the metal becoming immune to corrosion, even when corrosion is thermodynamically very favourable. On sweeping the potential in the positive direction, we first observe active corrosion, but after reaching a certain value of applied potential (the *Fladé potential*) there is passivation. Passivation is due to two factors: the solubility product of a hydroxide is reached or there is a structural change in a hydroxide film which already exists in a porous form and which changes to a non-porous form. Eventually, for very high positive potentials, there is film rupture and release of oxygen. Figure 16.6 shows a typical voltammetric profile for this type of metal, e.g. nickel. It should be noted that the presence of ions that cause pitting can make the passive region non-existent.

Utilization of impedance methods and simulation of the spectra obtained from analogue electrical circuits or transmission lines (Chapter 11) can lead to a better comprehension of the complex corrosion process and its inhibition[10,11]. It can also be used as a practical diagnostic of corrosion. An example of an impedance spectrum in the complex plane format and also in the *Bode format* (the modulus impedance and phase angle vs. $\lg(f/\text{Hz})$, which is extensively employed in impedance studies of corrosion, is shown in Fig. 16.7. The advantage of the Bode plot is that it shows high-frequency features which are hidden in the complex plane plot because in the latter the high-frequency points are very close together.

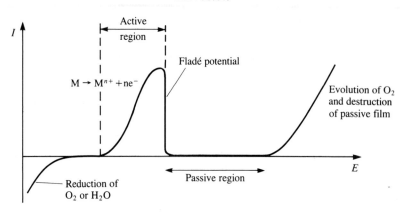

Fig. 16.6. Voltammetric curve for a metal that forms a passive film, e.g. nickel.

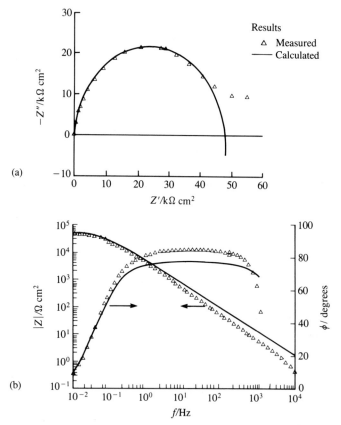

Fig. 16.7. Impedance of AISI316 stainless steel in 3 per cent (wt) NaOH at 80°C; (a) Complex plane plot; (b) Bode plot (from Ref. 10 with permission).

16.3   Types of metallic corrosion

There are many forms of corrosion: this section summarizes the various types. Besides this, knowing the corrosion rate (with or without protection against corrosion) and how it occurs, the engineer can orient calculations in the design of tubing, walls, etc. so that they have sufficient thickness to minimize rupture. In many of these cases where materials have reached their final form through application of mechanical forces (stress) or are subjected to mechanical forces periodically (resulting in fatigue), corrosion can be accelerated owing to the weakening of the material in certain places where cracks appear.

Uniform attack on a metal results in uniform corrosion. This is exploited in the processing and finishing of metals (Chapter 15). However, metallic structures are rarely homogeneous and surfaces are rough: corrosion occurs, preferentially within fissures in the surface (crevice corrosion), making corrosion faster.

In general, metals or alloys that are used are covered with oxide or hydroxide films. Formation of cracks and fissures can destroy the passivation. The depth of crevices increases rapidly because it is only there that the metal is not covered with a protective layer of oxide/hydroxide (see Fig. 16.8). The result is an increase in surface roughness and possible problems due to reduction in mechanical strength.

In the case of scratches in paint films that cover the metal the principle is the same (Fig. 16.9). There is oxide formation at the scratch site, the corrosion continuing, therefore, in a direction parallel to the metal surface.

Many other types of corrosion can occur. An important example is bimetallic corrosion which occurs at the junction between two different

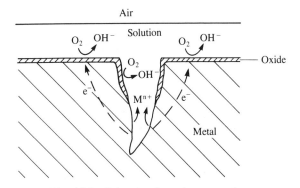

Fig. 16.8. Scheme of crevice corrosion.

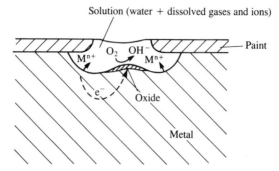

Fig. 16.9. Corrosion due to a paint scratch.

metals, a thin film of solution being necessary to make the electrical circuit. Contact can be between two metal sheets or pieces (Fig. 16.10), or between a metallic substrate and a metallic electrodeposit, through a scratch (Fig. 16.11). The more reactive metal is dissolved in either of the situations, as predicted thermodynamically by the order of electrode potentials.

Another form of corrosion is due in part to the mechanical forces applied to metals, *stress corrosion*. When the corrosion reaction occurs with hydrogen evolution, hydrogen atoms, owing to their small size, can enter the metallic lattice and thus reduce the strength of the interatomic bonds. This is known as *hydrogen embrittlement*. If afterwards we apply a mechanical stress to the metal there is a greater possibility that it will rupture. Corrosion fatigue can have similar effects. This has been held responsible for some aeroplane crashes.

Finally the formation of small, often semi-spherical holes or pits in the metal surface at certain sites in the presence of aggressive species should

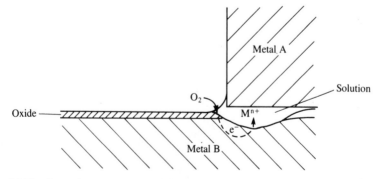

Fig. 16.10. Corrosion of metal B (more active than metal A) in a fissure formed at the junction between the two metals.

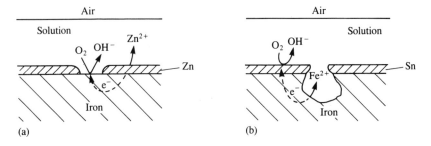

Fig. 16.11. Corrosion at the contact between electrodeposits and metal substrates; (a) Corrosion of the electrodeposit, e.g. zinc on iron; (b) Corrosion of the metal substrate, e.g. tin on iron.

be mentioned: this is *pitting corrosion*. Pitting corrosion is often caused by the presence of chloride ions that manage to pass through the passive film and initiate corrosion, resulting in rupture of the passive film. It is one of the most destructive forms of corrosion because it progresses rapidly, principally in maritime environments.

16.4   Electrochemical methods of avoiding corrosion

In a given situation, various aspects which contribute to corrosion can be altered in order to avoid it:

● selection of a bulk material with higher corrosion resistance

● coating of the metal with a suitable, sufficiently thick and homogeneous protective film: oxide, paint, etc.

● application of an external voltage or current or of use of a sacrificial anode to set the voltage of the material in the passive zone (anodic protection) or at a sufficiently negative potential such that the rate of corrosion is very low (cathodic protection)

● removal of the reducible and aggressive species in solution, e.g. increase of the pH, removal of oxygen (for example by chemical reaction with hydrazine to give nitrogen and water) or, in humid atmospheres, reduction of humidity

● avoidance of mechanical stress

● avoidance of bimetallic contacts where this can lead to enhanced corrosion.

Electrochemical methods that achieve these objectives are discussed in the rest of this section.

Electrochemically produced protective barriers

Much of metal electrodeposition is carried out with the aim of minimizing corrosion, the most common electrodeposits being tin, zinc, nickel and chromium on a cheaper metal substrate, such as iron. Since there is chemical bonding between substrate and electrodeposit, this is better than covering with paint (except electrophoretic painting, see Chapter 15) and additionally the surface generally becomes harder, as it does in nickel electroless plating.

Another process of physical protection is the formation of an oxide layer that makes the metal passive. This procedure is used for aluminium. Aluminium is normally anodized in 10 per cent sulphuric acid with steel or copper cathodes until an oxide thickness of 10–100 μm is obtained. As the more superficial part of the oxide layer has a fairly open structure it is possible to deposit metals (cobalt, nickel, etc.) or organic pigments in the pores and seal with boiling water or with an alkaline solution. The colours after metallic deposition are due to interference effects. Chromic and oxalic acids are also used significantly as electrolyte.

Anodization is also important for titanium, copper, and steel and in the fabrication of electrolytic and non-electrolytic capacitors from aluminium, niobium, and tantalum.

Phosphating provides a corrosion-resistant undercoat for paint finishes on steel (particularly automotive bodies), and to a lesser extent on zinc and aluminium. The usual process consists of immersion in a bath containing phosphoric acid, a metal phosphate (usually iron or zinc), and an accelerator, the pH varying between 1.8 and 3.2, at 60–90°C.

The disadvantage of physically protective barriers is the rapid and localized corrosion that occurs when the protective layer is scratched or removed locally (Figs. 16.10 and 16.11). Thus, in many cases, the utilization of methods involving continuous electrochemical protection is necessary.

Sacrificial anodes

Electrochemical protection can be achieved by forming an electrolytic cell in which the anode material is more easily corroded than the metal it is desired to protect. This is the case of zinc in contact with iron (Fig. 16.11): in this example there is a sort of cathodic protection. Protection of ship hulls, of subterranean pipeline tubings, of oil rigs, etc. is often done using sacrificial anodes that are substituted as necessary. The requisites for a good sacrificial anode are, besides its preferential corrosion, slow corrosion kinetics and non-passivation. Sacrificial anodes in use are, for this reason, normally of zinc, magnesium, or aluminium alloys, their surfaces partially active and partially passivated, so that corrosion is not very fast. A problem that can arise is a decrease in the degree of protection far from the anodes.

Methods of impressed current/potential

In this type of protection the potential of the metal surface is maintained constant by application of an appropriate potential or current (this latter is electrically easier) in a zone where the oxidation current is very low. Referring to Fig. 16.6 one sees that there are two possibilities:

1. *Cathodic protection* in the negative potential zone where reduction of oxygen or water commences, and where the rate of metal oxidation is low. In this case there has to be an inert auxiliary electrode close to the surface to be protected. The protection process consumes current, the quantity depending on solution resistance between the surface to be protected and the anode. This protection can be expensive in terms of energy consumption, and even more if there is hydrogen release and, consequently, hydrogen embrittlement.

2. *Anodic protection*, normally done potentiostatically, by application of a potential within the passive region. Given the form of Fig. 16.6, it is not so easy to control the potential by impressed current. The advantages are that there is no release of hydrogen and that often the current, and thus the energy consumed, is low.

Examples of cathodic protection with impressed current are, at the present time, protection of steel pipelines in maritime environments or in subsoil. An important example of anodic protection is in the storage of acids in steel tanks—the anodic current passivates the steel (see Fig. 16.1*a*).

Corrosion inhibitors

Corrosion inhibitors are organic or inorganic species added to the solution in low concentration and that reduce the rate of corrosion. Inhibition can function in three different ways:

1. A reagent that promotes the appearance of a precipitate on the metal surface, possibly catalysing the formation of a passive layer, for example hydroxyl ion, phosphate, carbonate, and silicate.

2. Oxidants such as nitrite and chromate which function by shifting the the surface potential of the metal in the positive direction until the passive zone in Fig. 16.6 (note that if these components are present in insufficient quantity the metal stays in the active zone, with potentially disastrous consequences).

3. A reagent that is adsorbed on the metal surface, diminishing either metal dissolution or the reduction of $H_2O/O_2/H^+$. In either of these cases, corrosion is reduced. Substances that inhibit metallic dissolution are organic and include aromatic and aliphatic amines, sulphur compounds, and those containing carbonyl groups; the release of hydrogen is inhibited by compounds containing phosphorus, arsenic, and antimony.

The great impact of the social and economic consequences of corrosion, with many tons of materials being corroded each day, and also from a safety point of view, means that research into these electrode processes and the search for new methods to reduce and control corrosion must continue.

References

1. U. R. Evans, *The corrosion and oxidation of metals*, Arnold, London, 1960.
2. A. T. Kuhn (ed.), *Techniques in electrochemistry, corrosion and metal finishing*, Wiley, London, 1987.
3. J. M. West, *Basic corrosion and oxidation*, 2nd edn, Ellis Horwood, Chichester, 1986.
4. A. J. B. Cutler, C. D. S. Tuck, S. P. Tyfield, and D. E. Williams, *Chem. Brit.*, 1986, **22**, 1109.
5. D. Pletcher and F. C. Walsh, *Industrial electrochemistry*, 2nd edn., Chapman and Hall, London, 1989, Chapter 10.
6. N. Sato, *Corrosion*, 1989, **45**, 354.
7. M. Pourbaix, *Atlas of electrochemical equilibria in aqueous solutions*, Pergamon Press, Oxford, 1966.
8. M. Stern and A. L. Geary, *J. Electrochem. Soc.*, 1957, **104**, 56.
9. D. D. Macdonald, *J. Electrochem. Soc.*, 1978, **125**, 1443.
10. D. C. Silverman and J. E. Carrico, *Corrosion*, 1988, **44**, 280.
11. K. Jüttner, *Electrochim. Acta*, 1990, **35**, 1501; F. Mansfeld, *Electrochim. Acta*, 1990, **35**, 1533; R. Oltra and M. Keddam, *Electrochim. Acta*, 1990, **35**, 1619.

17

BIOELECTROCHEMISTRY

17.1 Introduction

Faraday, Galvani and others showed in their experiments, many years ago, the existence of processes occurring in biological systems, which we now know have an electrochemical foundation. Despite the lack of research in this area during the past century, nobody today doubts the great importance of studying this kind of reactions for clarifying and comprehending the more relevant biological processes[1-5]. This is the objective of a relatively new branch of interdisciplinary research, bioelectrochemistry.

Electron transfer occurs in various ways in the biological sphere. New theories suggest that molecular electron transfer is crucial in the biological regulation of organisms' defence as well as in the growth of cancerous cells[6]. It has been proposed that many enzymatic reactions are of an electrochemical nature[7], the enzymes being not only catalysts but also conductors between active sites. Other processes linked with electrochemical reactions are photosynthesis, nerve excitation, blood coagulation, vision, smell, functioning of the thyroid gland, the origin of the biological electric potential, etc.

The complexity of biological systems means that it is very important that their study should be based on firm foundations. Many pieces in the biological puzzle are still missing, but each piece represents a valuable contribution towards attaining the goal.

At the present time, with the development of new electrochemical methods and new electrode materials, a large amount of research has been carried out in the electrochemistry of proteins, enzymes, and cellular components. Nevertheless, much remains to be done. Electrochemical experiments, in conjunction with other techniques, such as spectroscopy, may give a better answer to these questions.

In this chapter the intention is to give a view of present developments and research in bioelectrochemistry. It is not possible to describe the electrochemical aspects of all the kinds of biological events and processes occurring in living systems, but some examples will be presented and discussed to give an idea of the extent of bioelectrochemistry.

17.2 The electrochemical interface between biomolecules: cellular membranes, transmembrane potentials, bilayer lipid membranes, electroporation

A large fraction of biological molecules has dissociable groups in aqueous solution, giving rise to $H^+(aq)$ or other cations or anions, the extent of ionization depending on pH.

Examples of these molecules are amino acids, which exist in the zwitterionic form in acid solution, and in which the molecule contains an equal quantity of positively and negatively charged groups. Thus, zwitterionic molecules are dipolar ions that can have a total positive or negative charge or be neutral, according to solution pH.

Membrane lipids are amphiphatic molecules: they contain both a hydrophilic and a hydrophobic moiety. Molecules of this kind can lead to various types of interface. A natural example is the cellular membrane, a bilayer arrangement of such molecules, that marks the frontier between cells. The principal constituents of this membrane are lipids and proteins (Fig. 17.1).

Lipids are characterized by a predominantly hydrocarbon structure, are very soluble in organic solvents and sparingly soluble in water, and have physical properties that are in agreement with their hydrophobicity. Lipids are divided into several classes or families, some having a polar part; they have important biological functions besides being essential membrane components. Links with other molecules are via covalent bonding or van der Waals forces.

Proteins are made of a large number of L-amino acids united by peptide links (—CONH—), formed from the carboxyl group of one amino acid and an amine group of another with release of a water molecule. The name protein comes from a Greek word πρωτεΐνη meaning 'of the first importance'. Proteins are constituents of the cell and

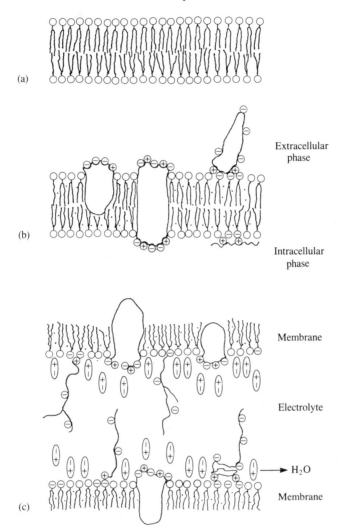

(a)

Extracellular
phase

(b)

Intracellular
phase

Membrane

Electrolyte

H$_2$O

Membrane

(c)

Fig. 17.1. Models of biomembranes: (a) Bilayer lipid membrane (BLM); (b) Lipid-proteic bilayer membrane; (c) Two opposed cell membrane surfaces.

the cell membrane. There is a very large number of them, corresponding to different genetically determined amino acid sequences. Proteins often have a very specific biological function. Table 17.1 lists some proteins and their electrochemically active groups that have been studied with the mercury electrode. The active groups are thiol, bisulphite, prosthetic groups (acceptors of Fe^{3+}, Cu^{2+}, FAD, etc.) and aromatic groups,

Table 17.1. Proteins and their groups for electron exchange (adapted from Ref. 8)

Protein	MW/10^3	Electron acceptor	$E_{1/2}$ (pH 7) (V vs. SCE)
Insulin (dimer)	6(12)	S–S	−0.6
Ribonuclease	13.6	S–S	−0.8
Lysozyme	14.5	S–S	−0.8
Trypsin	23.8	S–S	−0.49
Chymotrypsin	25	S–S	−0.45
Pepsin	35	S–S	
β-Lactoglobulin	38	S–S	
Ovalbumin	40	S–S	−0.8
Human serum albumin	69	S–S	≈ −0.6
Bovine serum albumin	69	S–S	−0.63
Cytochrome c	13	Fe^{3+}	−0.13
Cytochrome c_3	14	Fe^{3+}	−0.53
Cytochrome b_5	≈14	Fe^{3+}	−0.58
Methaemoglobin	64	Fe^{3+}	−0.60
Metmyoglobin	16	Fe^{3+}	−1.05
Bacteriorhodopsin		R=CH—CH=R	−0.97
Cytochrome oxidase	200	Fe^{3+}, Cu^{2+}	−0.2
Tryptophan oxygenase	67	Fe^{3+}, Cu^{2+}	≈ −0.2
Glycogen phosphorylase	200	R—CH=N—R	−0.82[a]
Xanthine oxidase	200	FAD, SH	−0.59
Cholesterol oxidase		FAD	−0.33
Glucose oxidase	186	FAD, S–S	−0.36
Ferredoxin (spinach)	13.5	Fe^{3+}	−0.6
Ferritin	700	Fe^{3+}	−0.38[b]

[a] pH 4.9; [b] pH 2.

double bonds etc., all of them being molecular structures that can effect electron transfer. Proteins (because of their amphoteric properties) and nucleic acids (because of their phosphate groups) are both polyions, exhibiting the behaviour of a polyelectrolyte in solution.

Cellular membranes are usually made up of approximately 40 per cent lipids and 60 per cent proteins. These percentages can vary in certain cases: for example, the internal membrane of mitochondria has 20 per cent lipids and 80 per cent proteins, and myelin has 80 per cent lipids and 20 per cent proteins.

Figure 17.1a shows how the lipids are organized in a cellular membrane, forming a bilayer with the hydrophilic part on the outside and the hydrophobic part on the inside of the membrane. The proteins can be found on either side of the bilayer or across the bilayer from one side to the other. Thus the bilayer is not, in reality, a perfect bidimensional arrangement because the proteins can extend a considerable distance into the intra- or extracellular phase of the lipid wall.

An electric potential difference is created between the intra- and extracellular layers of the membrane (Fig. 17.1c), called the *surface potential*, E_s. This is due to the ionization of the amine and carboxyl groups, or others such as thio groups of proteins, on the surface of the membrane, which are oriented according to an asymmetric distribution. An electrical double layer is thus formed.

The variation of the surface potential is given by

$$E_s = E_D + E_{DL} \tag{17.1}$$

in which E_D is the potential difference due to the dipoles and E_{DL} is the potential difference due to the double layer, obtained from the Poisson–Boltzmann equation of a diffuse layer (Chapter 3).

Experimentally it is the *transmembrane potential difference* that is observed by the use of potential-sensitive fluorescent dyes. Two of the components of the transmembrane potential difference are the intra- and extracellular potential differences (Fig. 17.2). If ϕ_i is the potential in the interior of the cell and ϕ_o the exterior potential, then the transmembrane potential, E_m, is given, in the case of a symmetric membrane where E_D for the two sides cancels, by

$$E_m = \phi_i - \phi_o = E_{i,DL} - E_{o,DL} + E_{Diff} \tag{17.2}$$

in which $E_{i,DL}$ and $E_{o,DL}$ are the internal and external surface potentials, respectively, and E_{Diff} expresses the diffusion of ionic species through the symmetric membrane ($E_m = E_{Diff}$ when $E_i = E_o$). When there is a

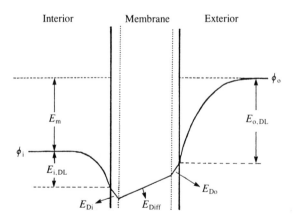

Fig. 17.2 Schematic representation of transmembrane potential profile. E_m, transmembrane potential difference; $E_{o,DL}$, exterior diffuse double layer potential difference; $E_{i,DL}$, interior diffuse double layer potential difference; E_D potential difference due to membrane molecular dipoles; $E_{Di} = E_{Do}$, symmetric membrane potential; E_{Diff} diffusion potential difference.

sufficiently large charge density on the membrane surfaces, the transmembrane potential is determined solely by the difference in surface
potentials, the diffusion potential being negligible. Supposing that the
intra- and extracellular electrolyte is the same, the only difference being
its concentration, then for a univalent ion

$$E_m = \frac{RT}{F} \ln \frac{c_o}{c_i} \tag{17.3}$$

which has the same form as the Nernst equation. c_o is the concentration
outside and c_i the concentration inside the cell.

At equilibrium, E_m is called the *resting potential,* and is affected by the
transport of all ions that can pass through the membrane, which is
permeable to almost all ions. In fact, the important ions are normally
limited to sodium and potassium and resting potentials vary in value from
tens to hundreds of millivolts.

It is therefore clear that the Donnan potential for equilibrium
potentials (Section 2.11) cannot be used in (17.3), except in special
circumstances, since there are various components in the total transmembrane potential. Donnan himself predicted that the phenomenon
would be more complicated for ion transfer processes between living cells
or tissue and the liquids that surround them[9]. This transmembrane
potential is not, however, the only one that can occur in the membrane[10].

The fact that the majority of *in vivo* processes occur on the surface of
or within the membrane and that electrical phenomena are very
important in membranes such as those found in the chloroplast, muscle
fibres, nerve fibres, mitochondria, etc., has recently led to intensive study
of the electrical properties of bilayer lipid membranes (BLM) in an
attempt to reproduce a model of the cell membrane. Membranes of
5–10 nm thickness have been studied, the membranes consisting of two
parallel sheets of lipids with a hydrophobic environment in the interior of
the membrane and the hydrophilic groups directed to the exterior
aqueous medium.

Artificial bilayer lipid membranes (BLM) have an electrical conduc-
tivity of $\kappa = 10^{-14}$–10^{-12} S cm^{-1}, less than cell membranes by a factor of
10^6. The conductivity is increased to physiological levels with the
introduction of electron acceptors or proteins in the artificial membranes
that form charge transfer complexes with the lipids, as for solid state
lipids[11].

The similarities between experimental results from lipids in the solid
state and in the membrane can be considered as experimental evidence
for a mechanism for the two states identical to that of semiconductors[12].

Natural membranes have two interfaces with aqueous solutions, and
interfacial properties, such as surface potential, concentration, and

capacitance should affect ion transfer processes. The interaction of charged groups on the surface with counter-ions in solution forms electrical double layers[13].

The surface compartment model (SCM)[14,15], which is a theory of ion transport focused on ionic process in electrical double layers at membrane protein surfaces, can explain these phenomena. The steady state physical properties of the discrete surface compartments are calculated from electrical double layer theory.

Artificial transmembrane material release and uptake can be achieved by *electroporation*[16]. The electroporation technique, dating from 1982, is a very efficient way of manipulating biological cells and cell tissue, through transient permeabilization of membranes by electrical pulses. Pulse characteristics are field strength of $0.1-30 \, kV \, cm^{-1}$ and $0.01-30 \, ms$ pulse duration. Another consequence of electroporation is that electroporated membranes are more likely to undergo fusion if two cells are brought into contact (cell *electrofusion*) as well as to undergo insertion of foreign glycoproteins and DNA (*electroinsertion* or *electrotransfection*).

Although the mechanisms of electroporation, electrofusion, and electroinsertion are not known, biophysical data suggest that the primary field pulse effect is interfacial polarization by ion accumulation at the membrane surfaces. The resulting transmembrane electric field causes rearrangements of the lipids such that pores are formed[17,18]. Electropores anneal slowly (over a period of minutes) when the pulse is switched off.

The direct electroporative transfer of genes, i.e. DNA (electrotransfection), to transform cells is of great interest for molecular biology, genetic engineering, therapy, and biotechnology. Other nucleic acids and proteins can also be efficiently transferred to recipient cells, microorganisms, and tissue. The main advantage of electroporative gene transfer is that intact, chemically untreated cell material can be transferred with high efficiency. In particular, the stable electrotransformation of intact bacteria, yeast, and plant cells is an important biological and biotechnological challenge.

17.3 Nerve impulse and cardiovascular electrochemistry

The importance of transmembrane potentials in cells has been demonstrated. Since the cells are totally enclosed by a membrane they naturally form an electrochemical cell. The cellular fluids contain sufficient concentrations of sodium, potassium, and chloride ions to be a good electrolyte, and potential differences originate in the intra- and extracellular membrane surfaces. We now discuss what happens when

there is an external depolarizing or hyperpolarizing stimulus in the cases
of the nerve impulse and cardiovascular problems. The action potential is
the response to the stimulus which puts the biological electrochemical cell
outside equilibrium.

The nerve impulse

The nerve cell membrane separates the external from the internal cell
fluid, as does any cell membrane. As is true of virtually all cells, the
intra- and extracellular fluids are electrolytic solutions of almost equal
conductivity, but their chemical composition is very different. The ions
present in largest quantities are sodium and potassium. The species in the
external fluid are made up of more than 90 per cent sodium and chloride
ions: in the cell interior there are principally potassium and organic ions
that cannot pass through the membrane, only 10 per cent of the ions
being sodium and chloride.

The nerve impulse is called the *action potential*, and consists principally
of two events that occur consecutively: an influx of positive charge

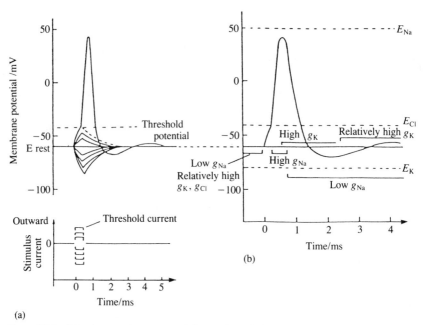

Fig. 17.3. (a) An action potential produced by a nerve-cell membrane in
response to a depolarization (above the threshold) stimulus. (b) Sketch of
time-dependent conductivity change of a nerve axon membrane (from Ref. 22
with permission).

through the membrane due mostly to the movement of sodium ions into the cell, followed by the efflux of positive charge, mainly due to potassium ions. The interpretation of this phenomenon by Hodgkin and Huxley[19] is that these movements reflect a transient opening of the membrane to sodium ions and then to potassium ions. Propagation of this impulse along the nerve is based on different kinds of transfer of the excitation: ion movement through the aqueous phase in and around the nerves, conduction by non-electrolytic carrier molecules, (for example acetylcholine at synapses), sometimes of a solid-state semiconductor type (Szent–Györgyi mechanism[20]), etc. Propagation of the impulse is the basis for communication through the nervous system.

The mathematical equations describing the ionic currents are empirical, but there is now proof for the existence of membrane-spanning proteins that act as channels for sodium and potassium ions and that open and close in response to changes in membrane polarization.

The nerve impulse, or action potential, may or may not be linked to an electric potential difference but has propagative characteristics[21]. The action potential is caused by the opening of the membrane to the entry of sodium ions. After the passage of the nerve impulse, the membrane tends to recover (Fig. 17.3), returning to its initial state. This occurs through an active transport mechanism of sodium from the interior to the exterior of the cell that involves a membrane-spanning enzyme, Na,K-ATPase. For each ATP molecule split by the enzyme, three sodium ions leave the nerve and two potassium ions enter, causing a net movement of charge through the membrane, which contributes to the transmembrane potential difference. This process is called the *electrogenic sodium pump* (Fig. 17.4). This pump is responsible for maintaining the transmembrane

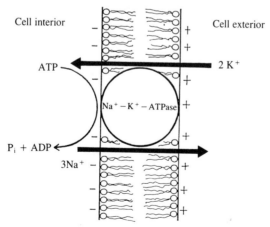

Fig. 17.4. Electrogenic sodium ion pump.

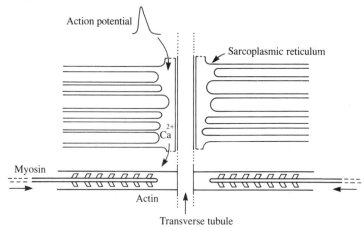

Fig. 17.5. Model for the release of calcium ions during muscular contraction.

resting potential, as it keeps the potassium concentration gradient constant.

Muscular contraction involves a similar process, since it is stimulated by the nerve impulse, and an action potential along the muscle membrane carries the impulse to the actin and myosin fibres. The muscle action potential causes the release of calcium ions from stores in the sarcoplasmic reticulum, and the interaction of actin and myosin provokes muscle contraction.

Figure 17.5 shows a scheme for excitation–contraction coupling of a skeletal muscle.

Cardiovascular problems

The mechanism leading to the formation of a blood clot has been much studied. It comprises a reaction sequence whose kinetics is not very clear, but there is no doubt that the initial step is an increase in the potential of the blood vessel walls, which are negatively charged in the normal state.

Investigation of coagulation is more easily carried out *in vitro.* Although the two processes are similar, it is necessary to be cautious in comparing *in vivo* and *in vitro* mechanisms. The clot formed in a test tube is different from that formed in the vascular system. Coagulation *in vitro* consists of a fibrin net in which there exist white and red cells and a small number of platelets. *In vivo,* the clots consist of large amorphous masses of platelets surrounded by white cells and very few red cells.

Blood coagulation factors are responsible for the coagulation phenom-

Table 17.2. Dependence of thrombus deposition at metal electrodes on position of metals in the electromotive series (adapted from Ref. 23 with permission)

Metal	M^{n+}/M standard electrode potential (vs. NHE)	Rest potential at metal-blood interface (vs. NHE)	Occurrence ($\checkmark$) or non-occurrence ($\times$) of thrombus deposition
Mg	−2.375	−1.360	×
Al	−1.670	−0.750	×
Cd	−0.402	−0.050	×
Cu	+0.346	+0.025	$\checkmark$
Ni	−0.230	+0.029	$\checkmark$
Au	+1.420	+0.120	$\checkmark$
Pt	+1.200	+0.125	$\checkmark$

enon that occurs in a reaction sequence, sometimes called the *cascade system*. Cyclic voltammograms of fibrinogen, thrombin, and prothrombin at platinum electrodes in ionic and pH conditions similar to blood showed the influence of oxygen and of adsorption phenomena occurring at the same time as electron transfer[23].

Experiments showed that coagulation increases for applied potential differences greater than +0.2 V vs. NHE; below this value, clot formation is very small. The rest potential of various materials used for vascular prostheses and cardiac valves was determined. In Table 17.2 some of the materials tested are mentioned. It was concluded that metallic electrodes with a negative potential vs. NHE in the blood are anticoagulant while those with positive potential are coagulant. Unfortunately, the metals most useful for prostheses are the most easily corroded: those of platinum and gold, not corroded, are unsuitable because of their positive rest potentials. Attempts to resolve the problem have utilized prostheses of plastic materials compatible in terms of their qualities of physical resistance, durability, etc. with their end use.

In the same way, the mechanism of action of drugs and coagulant and anticoagulant medication takes place through the variation of surface charge in the blood vessels.

The mechanism of blood coagulation is very complex, and has been the object of scientific and clinical investigation for more than a century[24,25]. The study of the mechanisms of the electrochemical reactions involved in cardiovascular processes and the selection of anticoagulant drugs and thrombogenic prosthesis materials, to be successful, needs to be carried out simultaneously through haematological and electrochemical research.

17.4   Oxidative phosphorylation

The electron transfer processes that occur within the membrane, such as
for example, phosphorylation (Fig. 17.6), are well known, but their
mechanisms remain unexplained. These electron transfer processes are of
primary importance in two types of membrane; chloroplasts in photosynthesis, and mitochondria in respiration.

   As an example we use mitochondria (Fig. 17.6). These are small
corpuscles that exist in large quantities within cells. They possess an
exterior and an interior membrane where the enzymes cytochrome b, c,
c_1, a and a_3, ATPase, and NADH are located. The interior membrane,
of non-repetitive structure, contains 80 per cent protein and 20 per cent
lipid. The Gibbs free energy variation of the conjugated redox pairs is
given by the formal potential, according to

$$\Delta G^{\ominus\prime} = -nF\,\Delta E^{\ominus\prime} \tag{17.4}$$

In the respiratory chain we start from the system $NAD/NADH_2$
$(E^{\ominus\prime} = -0.32\ V)$ and reach the system O_2/H_2O $(E^{\ominus\prime} = +0.82\ V)$. The
free energy change is thus $-220\ kJ\ mol^{-1}$, but this tells us nothing about
the mechanism of action of the mitochondria.

   Some of the steps of the electron transfer mechanism in biological
membranes are known, as they are for the associated proton transfer

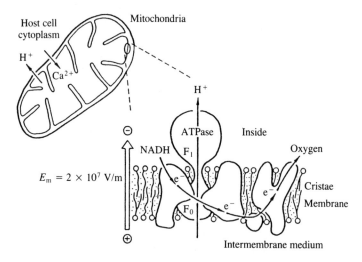

Fig. 17.6. The vectorial pumping of calcium ions and protons across the
mitochondrion membranes. A schematic enlargement of the inner (cristae)
membrane is shown to indicate the existence of protein-based electron (e^-) and
proton (H^+) conduction pathways (from Ref. 26 with permission).

mechanism. This proton flux, which accompanies the electron transfer, creates a proton gradient through the membrane and a potential difference given by

$$\Delta E_{H^+} = \Delta\phi - \frac{2.3RT}{F}\Delta pH \tag{17.5}$$

leading to the so-called *proton pump*.

The energy released by electron transfer can be used in the transport of protons through the membrane. One of the proton conduction mechanisms in proteins is through a chain of hydrogen bonds in the protein, i.e. a Grotthus mechanism (Section 2.9), similar to the mechanism of proton movement in ice. Protons are injected and removed by the various oxidation/reduction reactions which occur in the cell: there is no excess of protons or electrons in the final balance, and the reaction cycle is self-sustaining.

17.5 Bioenergetics

Bioenergetics is the study of energy flow in living organisms. In recent years new electrochemical techniques have been developed and existing ones perfected in order to allow the study of these processes[27]. Despite the importance of thermodynamic data in understanding the reactions, there is increasing interest in the investigation of the kinetics and mechanism, through the use of voltammetric techniques, especially using mediators.

The mechanisms of mediator action may be as follows:

1. The mediator is generated electrochemically and reacts with a biological molecule by homogeneous electron transfer;

2. The mediator is linked to the electrode surface, forming a surface-modified electrode, and the biological molecule links itself to the mediator layer by heterogeneous electron transfer.

In studying the kinetics of heterogeneous electron transfer in biomolecules, it is very important to take into account the solution conditions and electrode material. Mercury and platinum were the most used electrodes until 1970. After that, work done with glassy carbon, metallic oxide semiconductors and chemically modified electrodes began to appear. Proteins such as cytochromes c, c_3, c_7, b_3, and P450, myoglobin, haemoglobin, ferridoxin, peroxidase, and catalase are some of the most studied compounds[28,29] A cyclic voltammogram of ferricytochrome c at a gold electrode is shown in Fig. 17.7.

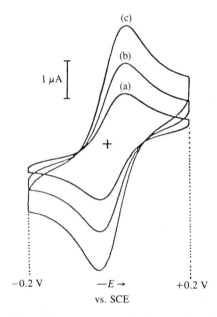

Fig. 17.7. D.c. cyclic voltammograms of horse heart ferricytochrome c (5 mg cm^{-3}) in NaClO$_4$ (0.1 M), phosphate buffer (0.02 M) at pH 7 in the presence of 4,4'-bipyridyl (0.01 M). Sweep rate (mV s^{-1}): (a) 20; (b) 50; (c) 100 (from Ref. 30 with permission).

There are groups of natural compounds that have a special position in the study of energy flow through living organisms, for example in the sequential exergonic oxidation of food. Lavoisier in the eighteenth century defined in a simplistic way the food metabolism of living organisms as being essentially combustion: food and oxygen are ingested, and carbon dioxide and water are ejected—the same total reaction as if the food was burnt. The correct elucidation of the mechanisms by which energy is stored in molecules such as carbohydrates, fats, and proteins, is a difficult challenge, as is how this energy is released to produce work in a controlled fashion.

In an intact cell, not all the energy release that occurs with the degradation of some compounds (catabolism) can be utilized in the synthesis (anabolism) of other cell components. This is a consequence of the *dissipation function*[31], which expresses the rate of loss of free energy in the thermodynamic coupling of two non-reversible processes, with one releasing and the other requiring energy. Since not all the energy released can be used, external energy from for example, photosynthesis, is necessary. There is also the possibility of energy retention in some chemical systems, which can be used in a controlled and efficient way when necessary. Indeed, some authors consider that organisms operate as fuel cells[32].

17.6 Bioelectrocatalysis

Enzymes are extremely important biomolecules because of their catalytic power and extraordinary specificity, superior to any synthetic catalyst.

The activity of redox enzymes as biological catalysts depends, in some cases, on their protein structure. In other situations the presence of non-proteic cofactors is necessary; the cofactors can be metals in the case of metalloenzymes, or organic molecules in the case of coenzymes (Table 17.3).

Bioelectrocatalysis can be defined as the group of phenomena associated with the acceleration of electrochemical reactions in the presence of biological catalysts—the enzymes.

For various reasons, it was only in the 1970s that enzymes began to be used as bioelectrochemical catalysts. Some of the reasons were difficulty in preparation of pure enzymes, their instability, and the lack of multiple applications. These problems have been largely overcome, and better purification methods and enzyme immobilization methods on electrode surfaces have been developed.

The principal applications of biocatalysts in electrochemical systems can be summarized as:

● development of biological catalysts for electrochemical applications, better than the existing inorganic catalysts

● development of bioelectrochemical systems that lead to substances of an organic nature with applications as fuels

● development of highly sensitive electrochemical sensors, applying the specific nature of enzymes' activity.

Table 17.3 Coenzymes in hydrogen atom and electron transfer reactions

NAD	nicotinamide adenine dinucleotide
NADP	nicotinamide adenine dinucleotide phosphate
FMN	flavin mononucleotide
FAD	flavin adenine dinucleotide
Coenzyme Q	ubiquinone

Despite their complexity, enzyme reactions follow kinetics of conventional chemical reactions[33,34]:

$$E + S \underset{k_{-1}}{\overset{k_1}{\rightleftharpoons}} ES$$

$$ES \overset{k_2}{\longrightarrow} E + product$$

where E is the enzyme and S the substrate (reagent) in the Michaelis–Menten formulation. The reaction rate is

$$-\frac{d[S]}{dt} = \frac{k_2[E][S]}{K_M} = \frac{[E_0][S]}{K_M + [S]} \tag{17.5}$$

with $[E_0]$ the initial concentration of free enzyme and K_M the Michaelis–Menten constant, given by $(k_{-1} + k_2)/k_1$.

It should be noted that there is, for enzymes:

• a proximity effect, a great affinity between enzyme and substrate

• a very strong orientation effect between the cofactors and the enzyme's active centre

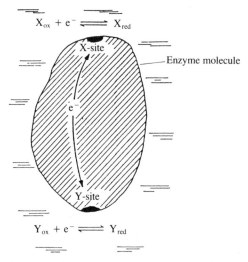

Fig. 17.8. Model of a particulate redox enzyme upon which the theory of electron conduction enzymes is based. Site X in the particle acts as electrode for the redox couple X and develops an equilibrium potential determined by the extent of the reduction of X. Site Y, on the opposite side, acts in the same way for Y. The potential difference causes an electronic current between sites X and Y and within the particle (from Ref. 35 with permission).

• an intramolecular catalysis effect owing to nucleophilic and electrophilic groups

• an effect from the polarization of the electroactive groups of the enzyme that causes a redistribution of electron density and activation of the substrate.

One of the models proposed for a possible enzyme redox reaction mechanistic pathway suggests (Fig. 17.8) that the enzyme contains simultaneously a part that acts as a solution cathode containing a so-called cathodic system where reduction occurs, and another that acts as a solution anode where there is oxidation[35]. The total charge transfer for the whole chemical reaction is therefore zero. This model is not completely correct, but the concept of a total chemical reaction without electron transfer to the exterior of the enzyme, although controlled by electron transfer, is interesting.

The explanation for biological enzyme conduction in enzyme oxidoreductases is not yet known, and could be an electron or proton transfer. However, there is no doubt that there is charge transfer from the cathodic to the anodic points.

Some methods for using enzymes in accelerating electrochemical processes have been developed and are shown in Fig. 17.9. In cases *a* and *b* there is an enzyme reaction whose intermediates, S*, undergo electrochemical transformation at a much lower overpotential than the initial substrate, S_{red}

$$S_{red} \rightarrow S^* \rightarrow S_{ox} + ne^-$$

In *c, d,* and *e* we have the typical case of a bioelectrocatalyst where, through a mediator, there is electron transfer between the electrode and the enzyme active centre where the substrate is in its turn activated and reacts. In *c* the components are in solution; in *d* and *e* the mediator or the enzyme are immobilized on the electrode surface, the electron transfer reaction occurring between mediator and electrode. In case *f* we have the ideal situation: direct electron exchange between the electrode and active centre of the enzyme, the mediator being eliminated. It is, nevertheless, very difficult to reconcile the enzyme characteristics and the electrochemical process, and it continues to be important to find adequate mediators and enzyme immobilization procedures.

As described above, the mediator is a compound that serves as intermediate between enzyme and electrode. Mediators need to have certain electrochemical properties, for example:

• the reaction between mediator and enzyme active centre must be

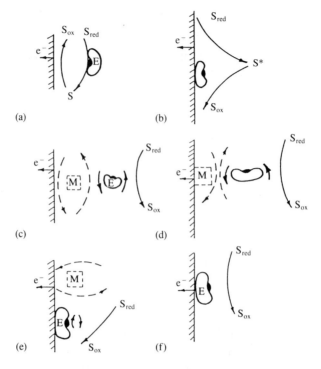

Fig. 17.9. Schemes showing the application of enzymes to promote electrochemical reactions (from Ref. 36 with permission).

rapid, that is the mediator itself must be a specific substrate for the
enzyme

• the redox potential of the mediator must be very close to that of the
reaction studied

• the oxidation–reduction reaction between mediator and electrode
material must be sufficiently fast to be in conditions close to reversibility

Work is being carried out with a view to the study and development of
optimum mediators.

The enzyme immobilization process is normally accompanied by a
reduction in activity. Nevertheless immobilized enzymes have several
advantages over soluble enzymes:

• a lesser total quantity of enzyme is required, and immobilized
enzymes can be reused

• the process can be operated continuously and can be readily
controlled

- the products are easily separated

- effluent problems and material handling are minimized

- in some cases enzyme properties, activity, and stability, can be altered favourably by immobilization

However, methods that lead to high-yield immobilization in water-insoluble matrices, that is, a small loss in enzyme activity, have been developed, Fig. 17.10.

Each method of immobilization has specific limitations[37], and it is necessary to find an appropriate procedure for any particular enzyme and an application that is simple, not too expensive, and leads to an immobilized enzyme with good retention of activity and good stability.

The principal immobilization methods for bioelectrocatalysts are:

1. *Physical immobilization methods*: there is no chemical modification of the enzyme.

- *Immobilization by adsorption.* This is the simplest method and consists in enzyme adsorption due to electrostatic, hydrophobic or dispersive forces on the electrode surface. The disadvantage is the high probability of enzyme desorption and denaturing.

- *Immobilization by inclusion* in gel is based on coupling enzymes to the lattice of a polymer matrix or enclosing them in semipermeable membranes. The enzyme does not bind to the matrix, so this method can be used for almost all enzymes, other biocatalysts, whole cells, or organelles, and the enzyme shows little deactivation compared with other

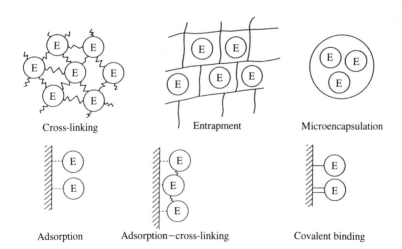

Cross-linking   Entrapment   Microencapsulation

Adsorption   Adsorption–cross-linking   Covalent binding

Fig. 17.10. Methods of enzyme immobilization (from Ref. 38 with permission).

Table 17.4. Classification of immobilization methods for insoluble enzymes

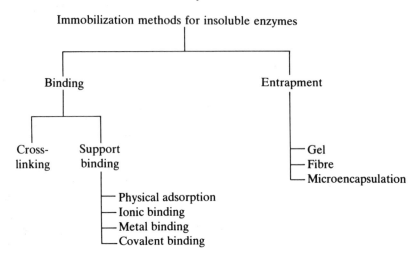

immobilization methods. Disadvantages are the possibility of enzyme leaving the gel net and the non-conducting properties of the gel.

2. Chemical immobilization methods. These can be accomplished via the formation of an array between the enzyme and surface active groups or via covalently linked transporters. The advantage is that the enzyme cannot escape to solution; the disadvantage is that partial or total enzyme deactivation can occur during immobilization owing to the formation of additional chemical links.

Research into new enzyme immobilization methods directly on the electrode surface is in progress.

Table 17.4 gives a classification of immobilization methods for insoluble enzymes that combines the nature of interaction responsible for immobilization and the nature of the support.

The mechanism and theory of bioelectrocatalysis is still under development. Electron transfer and variation of potential in the electrode–enzyme–electrolyte system has therefore to be investigated. Whether the enzyme is soluble and the electron transfer process occurs through a mediator, or whether there is direct enzyme immobilization on the electrode surface, the homogeneous process in the enzyme active centre has to be described by the laws of enzyme catalysis, and the heterogeneous processes on the electrode surface by the laws of electrochemical kinetics. Besides this there are other aspects outside electrochemistry or enzymology. Fundamental research in this dominion will elucidate the mechanisms of action of bioelectrocatalysts and develop the scientific basis for the appearance of optimized bioelectrochemical catalysts.

17.7 Bioelectroanalysis

Bioelectroanalysis is a new area in rapid development within electroanalysis[39]. The use of biological components: enzymes, antibodies, etc., to detect specific compounds has led to the development of *biosensors*[4,40].

A large part of analytical chemistry is linked to the qualitative and quantitative analysis of relatively simple species in complex matrices where various types of interferences occur. Bioelectroanalytical sensors permit the analysis of species with great specificity very rapidly, being sensitive, highly selective and, in principle, cheap. They can be used in clinical analysis, in on-line control processes for industry or environment, or even *in vivo*[41,42].

Bioelectrochemical sensors can be divided into two groups: potentiometric and amperometric. Since the description by Clark[43] of the first electrochemical biosensor based on the oxygen electrode, many potentiometric and amperometric sensors using enzymes[44] and other materials have been designed and applied.

In *potentiometric sensors* a membrane or surface sensitive to a species generates a potential proportional to the logarithm of the concentration of the active species, measured relative to a reference electrode (Chapter 13). The use of potentiometric electrodes in clinical analysis began at the beginning of the century with the pH glass electrode. Other electrodes for measurement in blood, such as ion-selective electrodes for sodium, potassium, lithium, and fluorine followed[45]. An important development was the appearance of the potassium electrode using valinomycin as neutral charge carrier.

The recently developed field-effect transistors (FETs)[41] have also been used as biosensors. The ion-selective field-effect transistor (ISFET) uses ion-selective membranes, identical to those used in ion-selective electrodes, over the gate.

Enzyme-selective electrodes (Fig. 17.11) have been made as a membrane containing immobilized enzymes placed over a pH electrode or over a gas electrode such as an ammonia electrode for potentiometric detection, or over an oxygen electrode for amperometric detection. The products of the reaction of enzyme with substrate are detected by the electrode.

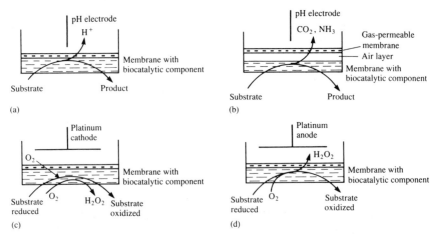

Fig. 17.11. (a) Potentiometric biosensor based on pH electrode; (b) Potentiometric biosensor based on gas electrode; (c) Oxygen electrode: determination of oxygen; (d) Oxygen electrode: determination of hydrogen peroxide (from Ref. 42 with permission).

Amperometric sensors measure the current produced by chemical reaction of an electroactive species at a constant applied voltage, which is related to the concentration of the species in solution (Chapter 14). Most biologically relevant compounds, such as glucose, urea, cholesterol, etc. are not electroactive and it is necessary to create an adequate combination of reactions to produce an electroactive species. Moreover, in many cases the selectivity given by a constant applied potential is not sufficient to discriminate between various electroactive species, and extra selectivity is required. Enzymes can provide this selectivity and sensitivity. Unfortunately, the ideal solution of direct immobilization of enzymes on electrodes (Section 17.6) tends to result in a significant loss of enzyme activity. Thus, the following two methods have been developed: one is placing membranes containing an immobilized enzyme over the electrode and the other is attaching mediators followed by the enzyme to the electrode surface (Fig. 17.12). Great care has to be taken in choosing mediators for a given enzyme reaction[46].

Conducting organic salt electrodes directly coupled to oxidases have been described such as, for example, N-methylphenazinium (NMP^+) cation and tetracyanoquinodimethane ($TCNQ^-$) anion as an electrode material for facilitating electron transfer of glucose oxidase[47]. Results with other salt cations, such as tetrathiafulvalene (TTF^+) and quinoline (Q^+), have been reported[48].

Electrochemical biosensors have the great advantage of being more

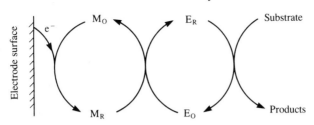

Fig. 17.12. General model for modified electrode, showing mediation of electron transfer. E_O, M_O and E_R, M_R are the oxidized and reduced forms of enzyme (E) and mediator (M) respectively.

economic than other methods and with rapid response and possibility of automation enabling a high sample throughput. The determination of about 80 different substances including substrates, cofactors, prosthetic groups, enzymes, antibodies, inhibitors, and activators using electrochemical biosensors has been proposed[42].

The use of biosensors for industrial and environmental analysis is very important for the control of food manufacturing processes, for the evaluation of food quality, for the control of fermentation processes and for monitoring of organic pollutants.

In environmental analysis biosensors have been developed for three major applications:

- the assay of environmental inorganic compounds such as nitrate, phosphate, and organic compounds such as methane, NTA, monomethylsulfate or dimethylformamide;

- monitoring of parameters such as biologically dissolved oxygen;

- the determination of toxic or mutagenic materials.

Biosensors have great advantages over spectrophotometric methods in the relatively little time consumed, ease of operation, and relatively low cost. Sensors using immobilized microorganisms instead of enzymes are being developed for environmental analysis.

Microbial sensors[49] are based on the contact between an electrode and immobilized living cells. The electrode converts the biochemical signal from the microorganism into an electric potential, for example the release of oxygen is registered as a current at constant potential at the Clark electrode. There are some advantages in the use of microbial sensors in relation to other enzymes for electroanalysis: there is no necessity to purify the microbial cells, the sensor is regenerated by immersion in a nutrient solution of the microbial culture (it is an *in vivo* sensor), there is no cofactor regeneration, and the cells fully catalyse metabolic transformations, which is not possible with just one enzyme. But there are also major disadvantages: the response time is long and the selectivity is low.

Microelectrodes[50] have been finding many possible applications due to their miniaturized geometry and possibility of *in vivo* applications, although biocompatibility of the materials used, the need for sterile conditions in implantation of the electrodes, and the risk of immune reactions or thrombosis are still major difficulties for any practical application of *in vivo* techniques.

The problems that occur with *in vivo* experiments are not completely solved. The points where the implanted electrodes cause tissue damage are rapidly regenerated and covered by conjunctive tissue or even by antibodies from electrode rejection. The formation and growth of conjunctive tissue is influenced by the form and nature of the electrode material. A material's biocompatibility is defined as its ability to perform with an appropriate host response in a specific application[51]. Therefore it is important to develop biomaterials for *in vivo* sensor applications, since neither the conjunctive tissue nor the antibody layer on the electrode is conducting, and a large decrease in electrode response after implantation is observed.

Electrochemical detectors have been used after high-performance liquid chromatography separation for the determination of catecholamines and similar metabolites. Catecholamines constitute a group of compounds of great biological interest, some being normally present in the chemical neurotransmission processes, others being neurotoxins[52] that can be responsible for schizophrenia, depression, and some mental perturbations, like Alzheimer's disease. There have been attempts to study their mechanism of action *in vivo*[53,54]. Nevertheless, the simultaneous appearance of ascorbic acid has made these investigations more difficult.

Microelectrodes are also used in electrophysiology as sensors of potential in intra- and extracellular measurements to study ion transport at the molecular level and to obtain measurements of the current passing through a single ion channel, as in the patch-clamp technique[55]. Analysis of electroencephalograms, electromyograms, and electrocardiograms, all measuring electrical signals generated in the human body, has been of great importance in the detection and treatment of the respective perturbations. The measurements are made through thin wires or electrodes of Ag | AgCl placed on the surface or in the interior of the organ that is to be analysed. For safety reasons the electrical signals from the body are converted into optical signals, then reconverted into electrical signals, using opto-isolators, to isolate the body electrically from the signal-processing instrument.

17.8 Future perspectives

Much remains to be done in the field of bioelectrochemistry, although it has already demonstrated great possibilities in elucidating biological reactions. The topics described can be consulted in greater depth in the recommended literature.

Our objective in this chapter has been to call attention to a fascinating and stimulating branch of electrochemistry in rapid development but still in its infancy. It will certainly contribute a great deal to the understanding of the kinetics and mechanisms of biological phenomena such as selectivity in ionic transport, excitability of membranes, nerve impulse conduction, muscle contraction, photosynthesis, energy conversion and storage, effects of hormones and drugs, clotting of blood, and many others.

References

1. S. Srinivasan, Yu. A. Chizmadzhev, J. O'M. Bockris, B. E. Conway, and E. Yeager (ed.), *Comprehensive treatise of electrochemistry*, Plenum, New York, Vol. 10, 1985.
2. G. Milazzo (ed.), *Topics in bioelectrochemistry and bioenergetics*, Wiley, 1978, 5 volumes.
3. F. Gutmann and H. Keyser (ed.), *Modern bioelectrochemistry*, Plenum, New York, 1986.
4. A. P. F. Turner, I. Karube, and G. S. Wilson (ed.), *Biosensors, fundamentals and applications*, Oxford University Press, 1987.
5. G. Milazzo *et al.*, *Experientia*, 1980, **36**, 1243.
6. A. Szent–Györgyi, *Science*, 1968, **161**, 988.
7. F. W. Cope, *Bull. Math. Biophysics*, 1969, **31**, 519.
8. H. Berg, in Ref. 1, p. 192.
9. F. G. Donnan, *Chem. Rev.*, 1924, **1**, 73.
10. H. Coster and J. R. Smith, in Ref. 2, Vol. 2, pp. 53–88.
11. Cr. Simionescu, Sv. Dumitrescu, and V. Percec in Ref. 2, Vol. 2, pp. 151–204.
12. B. Rosenberg, *Disc. Faraday Soc.*, 1971, **51**, 1.
13. M. J. Sparnaay, *The electrical double layer*, 1972, Pergamon Press, Oxford, pp. 1–19.
14. M. Blank, *Biochim. Biophys. Acta*, 1987, **906**, 277.
15. M. Blank, *Bioelectrochemistry* II, Vol. 32, Physical Series, Plenum, New York, 1988.
16. E. Neumann, A. E. Sowers, and C. A. Jordan (ed.), *Electroporation and electrofusion in cell biology*, Plenum, New York, 1989.
17. I. P. Sugar and E. Neumann, *Biophys. Chem.*, 1984, **19**, 211.
18. E. Neumann and K. Rosenheck, *J. Membrane Biol.*, 1972, **10**, 279.
19. A. L. Hodgkin and A. F. Huxley, *J. Physiol.*, 1952, **117**, 500.
20. A. Szent-Györgyi, *Nature*, 1941, **148**, 157.
21. P. K. J. Kinnunen and J. A. Virtanen, in Ref. 3, pp. 457–479.

22. S. Ohki, in Ref. 1, p. 94.
23. S. Srinivasan, in Ref. 1, p. 476.
24. R. C. Tolman and A. E. Stearn, *J. Am. Chem. Soc.*, 1918, **40**, 264.
25. E. J. Warburg, *Biochem. J.*, 1922, **16**, 153.
26. R. Pethig, in Ref. 3, p. 201.
27. E. F. Bowden, F. M. Hawkridge, and H. N. Blount, Ref. 1, pp. 297–346.
28. W. J. Albery, M. J. Eddowes, H. A. O. Hill, and A. R. Hillman, *J. Am. Chem. Soc.*, 1981, **103**, 3904.
29. K. Niki, Y. Kawasaki, N. Nishimura, Y. Higuchi, N. Yasuoka, and M. Kakudo, *J. Electroanal. Chem.*, 1984, **168**, 275.
30. M. J. Eddowes and H. A. O. Hill, *J. Am. Chem. Soc.*, 1979, **101**, 4461.
31. D. Walz, *Biochim. Biophys. Acta*, 1990, **1019**, 171.
32. J. O'M. Bockris and S. Srinivasan, *Nature*, 1967, **215**, 197.
33. G. Lehninger, *Biochemistry*, Worth Publishers, New York, 1975, Chapter 8.
34. L. Stryer, *Biochemistry*, Freeman, San Francisco, 1981.
35. F. W. Cope, *Bull. Math. Biophysics*, 1965, **27**, 237.
36. M. Tarasevich, in Ref. 1, p. 260.
37. J. F. Kennedy and J. M. S. Cabral in *Biotechnology*, VCH, Weinheim, 1987, Vol. 7a, Chapter 7.
38. S. A. Barker, in Ref. 4, p. 92.
39. J. P. Hart, *Electroanalysis of biologically important compounds*, Ellis Horwood, Chichester, 1990.
40. A. E. G. Cass (ed.), *Biosensors: a practical approach*, IRL Press, Oxford, 1990.
41. I. Karube, in *Biotechnology*, VCH, Weinheim, 1987, Vol. 7a, Chapter 13.
42. R. D. Schmid and I. Karube, in *Biotechnology*, VCH, Weinheim, 1987, Vol. 6b, Chapter 11.
43. L. C. Clark Jr. and C. Lyons, *Ann. N.Y. Acad. Sci.*, 1962, **102**, 29.
44. G. Nagy and E. Pungor, *Bioelectrochem. Bioenerget.*, 1988, **20**, 1.
45. J. Koryta, *Ions, electrodes and membranes*, 2nd edn, Wiley, Chichester, 1991.
46. P. N. Bartlett, P. Tebbutt, and R. G. Whitaker, *Prog. Reaction Kinetics*, 1991, **16**, 55.
47. J. J. Kulys, A. S. Samalius, and G. J. S. Svirmickas, *FEBS Letters*, 1980, **114**, 7.
48. W. J. Albery, P. N. Bartlett, and D. H. Craston, *J. Electroanal. Chem.*, 1985, **194**, 223.
49. K. Riedel, R. Renneberg, and P. Liebs, *Bioelectrochem. Bioenerget.*, 1988, **19**, 137.
50. R. M. Wightman in *Ultramicroelectrodes*, ed. M. Fleischmann, S. Pons, D. R. Rolison, and P. P. Schmidt, Datatech Systems, USA, 1987.
51. Biomaterials Society definition.
52. G. Dryhurst, *Chem. Rev.*, 1990, **90**, 758.
53. R. D. O'Neill, M. Fillenz, W. J. Albery, and N. J. Goddard, *Neuroscience*, 1983, **9**, 87.
54. R. M. Wightman, L. J. May, and A. C. Michael, *Anal. Chem.*, 1988, **60**, 769A.
55. E. Neher and B. Sakmann, *Nature*, 1976, **260**, 799.

Appendices

APPENDIX 1

USEFUL MATHEMATICAL RELATIONS

A1.1 The Laplace transform
A1.2 The Fourier transform
A1.3 Other useful functions and mathematical expressions

A1.1 The Laplace transform

Introduction

The Laplace transform is essential in order to transform a partial differential equation into a total differential equation. After solving the equation the transform is inverted in order to obtain the solution to the mathematical problem in real time and space.

The occurrence of partial differential equations in electrochemistry is due to the variation of concentration with distance and with time, which are two independent variables, and are expressed in Fick's second law or in the convective-diffusion equation, possibly with the addition of kinetic terms. As in the resolution of any differential equation, it is necessary to specify the conditions for its solution, otherwise there are many possible solutions. Examples of these boundary conditions and the utilization of the Laplace transform in resolving mass transport problems may be found in Chapter 5.

The transform

The definition of the Laplace transform $f(s)$, for a function $F(t)$, is

$$f(s) = \mathcal{L}\{F(t)\} = \int_0^\infty \exp(-st)F(t)\, dt \qquad (A1.1)$$

where s is a number sufficiently large for the integral to converge; $\mathcal{L}$ represents the transform. A simple example is

$$F(t) = \exp(\alpha t) \qquad (A1.2)$$

Then

$$f(s) = \int_0^\infty \exp\left(-(s - \alpha)\right)t \, dt \tag{A1.3}$$

$$= (s - \alpha)^{-1} \tag{A1.4}$$

Normally, instead of calculating the transform or its inverse, a table is consulted which gives the result directly. Table A1.1 shows useful examples, some of them easy to verify.

Table A1.1. Laplace transforms

$F(t)$	$f(s)$
1	$1/s$
t	$1/s^2$
$t^{n-1}/(n - 1)!$	$1/s^n$
$1/\sqrt{\pi t}$	$1/\sqrt{s}$
$2\sqrt{t/\pi}$	$s^{-3/2}$
t^{k-1}	$\Gamma(k)/s^k \quad (k > 0)$
$\exp(-at)$	$(s + a)^{-1}$
$t \exp(-at)$	$(s + a)^{-2}$
$a^{-1} \sin at$	$(s^2 + a^2)^{-1}$
$\cos at$	$s/(s^2 + a^2)$
$a^{-1} \sinh at$	$(s^2 - a^2)^{-1}$
$\cosh at$	$s/(s^2 - a^2)$
$\dfrac{k}{2\sqrt{\pi t^3}} \exp\left(\dfrac{-k^2}{4t}\right)$	$\exp(-k\sqrt{s}) \quad k > 0$
$\operatorname{erfc} \dfrac{k}{2\sqrt{t}}$	$\dfrac{1}{s} \exp(-k\sqrt{s}) \quad k \geq 0$
$\dfrac{1}{\sqrt{\pi t}} \exp\left(\dfrac{-k^2}{4t}\right)$	$\dfrac{1}{\sqrt{s}} \exp(-k\sqrt{s}) \quad k \geq 0$
$2\sqrt{\dfrac{t}{\pi}} \exp\left(\dfrac{-k^2}{4t}\right) - k \operatorname{erfc} \dfrac{k}{\sqrt{t}}$	$\dfrac{1}{s^{3/2}} \exp(-k\sqrt{s}) \quad k \geq 0$
$\dfrac{1}{\sqrt{\pi t}} - a \exp(a^2 t) \operatorname{erfc} a\sqrt{t}$	$(s^{1/2} + a)^{-1}$
$\dfrac{1}{\sqrt{\pi t}} + a \exp(a^2 t) \operatorname{erf} a\sqrt{t}$	$\dfrac{\sqrt{s}}{(s - a^2)}$
$\dfrac{1}{a} \exp(a^2 t) \operatorname{erf} a\sqrt{t}$	$[\sqrt{s}\,(s - a^2)]^{-1}$
$\exp(a^2 t) \operatorname{erfc} a\sqrt{t}$	$[\sqrt{s}\,(\sqrt{s} + a)]^{-1}$

Important properties

The Laplace transform has some properties which are extremely useful in aiding in the resolution of equations in electrochemistry and other branches of science.

1. The transform is linear:

$$\mathcal{L}\{aF(t) + bG(t)\} = af(s) + bg(s) \tag{A1.5}$$

2. The transform of a derivative is

$$\mathcal{L}\left\{\frac{dF(t)}{dt}\right\} = sf(s) - F(0) \tag{A1.6}$$

If $F(t)$ is redefined such that $F(0) = 0$, then $F(0)$ disappears. An example would be a dimensionless concentration variable

$$\gamma = \frac{c - c_\infty}{c_\infty} \tag{A1.7}$$

giving

$$\frac{\partial \bar{\gamma}}{\partial t} = \mathcal{L}\left\{\frac{\partial \gamma}{\partial t}\right\} = s\bar{\gamma} \tag{A1.8}$$

Note that the bar over the symbol γ represents the fact that the dimensionless concentration was transformed. All variables not subjected to direct transformation remain as they were before applying the transform.

3. The transform of an integral is

$$\mathcal{L}\left\{\int_0^t F(x)\, dx\right\} = \frac{1}{s} f(s) \tag{A1.9}$$

which involves, in the Laplace domain, simple division by the Laplace variable.

4. Inversion, when not corresponding to a tabulated expression, can often be made possible by separation of the solution in the Laplace domain into two parts which are tabulated, by means of the convolution integral

$$\mathcal{L}^{-1}\{f(s)g(s)\} = \int_0^t F(t - \tau)G(\tau)\, d\tau \tag{A1.10}$$

In other cases functions have to be written as their series expansions; this process can lead to errors, especially in expansions which contain terms alternately positive and negative.

A1.2 The Fourier transform

If, in the Laplace transform, we write

$$s = i\omega \tag{A1.11}$$

we obtain

$$f(s) = \int_0^\infty \exp(-i\omega t)F(t)\,dt \tag{A1.12}$$

which is the Fourier transform. This is, therefore, a special case of the Laplace transform, and corresponds to a transformation from real time to the frequency domain. Its importance is principally for the registering of frequency spectra, as in impedance studies (Section 11.12). It is particularly useful when low frequencies are being applied to a system and the response contains high-frequency noise—the noise is rejected by the transform, equivalent to filtering.

Any waveform can be described by a Fourier series $y(t)$, involving the summation of sinusoidal waves of different frequencies, phases and amplitudes:

$$y(t) = A_0 + \sum_{n=1}^{\infty} A_n \sin(\omega_0 nt + \phi_n) \tag{A1.13}$$

where A_n is the amplitude of component n with frequency $n\omega_0/2\pi$ Hz and phase angle ϕ_n. Numerical algorithms for synthesizing the waveform exist: generally the fast Fourier transform (FFT) is employed, which also performs the inversion, greater rapidity being given by computer control. An example of the approximate superposition of sinusoidal waves to give a square wave is shown schematically in Fig. A1.1.

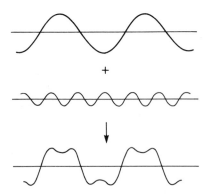

Fig. A1.1. Formation of a square wave by superposition of the components, showing the two most important components.

A1.3   Other useful functions and mathematical expressions

The Airy function

The Airy equation is very often encountered in the resolution of electrochemical mass transport problems in dimensionless variables, and has the form

$$\frac{d^2\omega}{dz^2} - \omega z = 0 \tag{A1.14}$$

The Airy functions, Ai(z) and Bi(z), are independent solutions of this equation, where

$$\text{Ai}(z) = \text{Ai}(0)f(z) + \text{Ai}'(0)g(z) \tag{A1.15}$$

$$\text{Bi}(z) = \sqrt{3}\,[\text{Ai}(0)f(z) - \text{Ai}'(0)g(z)] \tag{A1.16}$$

In these equations

$$f(z) = 1 + \frac{1}{3!}z^3 + \frac{1\,.\,4}{6!}z^6 + \frac{1\,.\,4\,.\,7}{9!}z^9 + \cdots \tag{A1.17}$$

$$g(z) = z + \frac{2}{4!}z^4 + \frac{2\,.\,5}{7!}z^7 + \frac{2\,.\,5\,.\,8}{10!}z^{10} + \cdots \tag{A1.18}$$

$$\text{Ai}(0) = 3^{-2/3}/\Gamma(2/3) = 0.35503 \tag{A1.19}$$

$$\text{Ai}'(0) = 3^{-1/3}/\Gamma(1/3) = 0.25882 \tag{A1.20}$$

where Γ is the gamma function (see below).
   An expression found in Chapter 5 is

$$\frac{\text{Ai}'(0)}{\text{Ai}(0)\Gamma(2/3)} = 0.53837 \tag{A1.21}$$

The gamma function

The gamma function appears in the inversion of Laplace transforms of the type $1/s^z \rightarrow t^{z-1}/\Gamma(z)$, in the numerical values of Ai(0), Bi(0) etc. Its definition is

$$\Gamma(z) = \int_0^\infty t^{z-1} \exp(-t)\,dt \tag{A1.22}$$

For integer values of z,

$$\Gamma(n+1) = n! \tag{A1.23}$$

Table A1.2. Values of the gamma function $\Gamma(z)$ for important fractions

z	$\Gamma(z)$
1/4	3.625610
1/3	2.678939
1/2	1.772454
2/3	1.354118
3/4	1.225416
1	1.000000

and, in general,

$$\Gamma(z+1) = z\Gamma(z) \tag{A1.24}$$

From tables listing values of Γ between 0 and 1 it is possible to calculate $\Gamma(z)$ for any positive value. Some values for fractional z are given in Table A1.2, and the variation of the gamma function with z is shown in Fig. A1.2.

The error function

The definition of the error function, erf (z), is

$$\text{erf}\,(z) = \frac{2}{\sqrt{\pi}} \int_0^z \exp\,(-t^2)\,\mathrm{d}t \tag{A1.25}$$

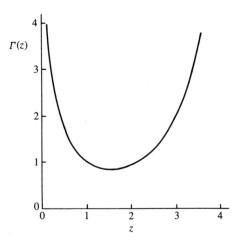

Fig. A1.2. Graphical representation of the gamma function, $\Gamma(z)$, for $z > 0$.

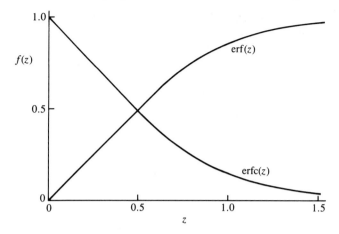

Fig. A1.3. Graphical representation of the functions erf (z) and erfc (z), showing the quasi-linearity for small z.

and of its complement, erfc (z)

$$\text{erfc}\,(z) = 1 - \text{erf}\,(z) \tag{A1.26}$$

As is easily seen by inspection, (A1.25) represents the integral of a curve of the same type as the Gaussian distribution—hence its name. It should be noted that erf $(0) = 0$ and that erf $(\infty) = 1$, as shown in Fig. A1.3. This function is frequently encountered in diffusion problems.

For $0 < z < 2$ the Maclaurin expansion can be used (see next subsection)

$$\text{erf}\,(z) = \frac{2}{\sqrt{\pi}} \left(z - \frac{z^3}{3!} + \frac{z^5}{5\,.\,2!} - \frac{z^7}{7\,.\,3!} + \frac{z^9}{9\,.\,4!} - \cdots \right) \tag{A1.27}$$

choosing the number of terms to give the accuracy needed. For $z < 0.1$ the first term is sufficient, that is

$$\text{erf}\,(z) \sim \frac{2z}{\pi^{1/2}} \tag{A1.28}$$

Values of the error function and its first derivative (the Gauss distribution) are tabulated[1].

Taylor and Maclaurin series

Taylor's theorem permits the expansion of certain functions, often in the form of a polynomial. Only the terms which contribute in a significant way to the response are utilized, in this way facilitating the mathematical resolution of problems. An example is the approximation that, close to zero, an exponential function is linear.

The theorem covers continuous functions which have continuous and unique value derivatives within the range considered. The expansion for $f(a+x)$ is

$$f(a+x)=f(a)+\frac{x}{1!}f'(a)+\frac{x^2}{2!}f''(a)+\frac{x^3}{3!}f'''(a)+\cdots+\frac{x^n}{n!}f^{(n)}(a)+E_n(x)$$

(A1.29)

where

$$E_n(x)=\frac{x^{n+1}}{(n+1)!}f^{(n+1)}(a+\theta x)$$

(A1.30)

and $0<\theta<1$, and represents the error on terminating the series at the nth term. If $\lim_{n\to\infty}E_n\to0$ then we may write

$$f(a+x)=\sum_{r=0}^{\infty}\frac{x^r}{r!}f^{(r)}(a)$$

(A1.31)

which is a Taylor series.

Table A1.3. Maclaurin series for some simple functions

Function	Expansion	Comments
$(1+x)^\alpha$	$1+\alpha x+\dfrac{\alpha(\alpha-1)}{2!}x^2+\dfrac{\alpha(\alpha-1)(\alpha-2)}{3!}x^3+\cdots$	$|x|<1$, all α
$\sin x$	$x-\dfrac{x^3}{3!}+\dfrac{x^5}{5!}-\dfrac{x^7}{7!}+\dfrac{x^9}{9!}-\cdots$	All x
$\cos x$	$1-\dfrac{x^2}{2!}+\dfrac{x^4}{4!}-\dfrac{x^6}{6!}+\dfrac{x^8}{8!}-\cdots$	All x
$\tan x$	$x+\dfrac{x^3}{3}+\dfrac{2x^5}{15}+\dfrac{17x^7}{315}+\cdots$	$-\frac{1}{2}\pi<x<\frac{1}{2}\pi$
$\log_e(1+x)$	$x-\dfrac{x^2}{2}+\dfrac{x^3}{3}-\dfrac{x^4}{4}+\dfrac{x^5}{5}-\cdots$	$-1\leq x\leq+1$
$e^{\alpha x}$	$1+\alpha x+\dfrac{(\alpha x)^2}{2!}+\dfrac{(\alpha x)^3}{3!}+\dfrac{(\alpha x)^4}{4!}+\cdots$	All x, all α
$\sinh x$	$x+\dfrac{x^3}{3!}+\dfrac{x^5}{5!}+\dfrac{x^7}{7!}+\dfrac{x^9}{9!}+\cdots$	All x
$\cosh x$	$1+\dfrac{x^2}{2!}+\dfrac{x^4}{4!}+\dfrac{x^6}{6!}+\dfrac{x^8}{8!}+\cdots$	All x

In the special case of $a = 0$, (A1.29) transforms into

$$f(x) = f(0) + xf'(0) + \frac{x^2}{2!}f''(0) + \cdots + \frac{x^n}{n!}f^{(n)}(0) + E_n(x) \quad \text{(A1.32)}$$

with

$$E_n(x) = \frac{x^{n+1}}{(n+1)!}f^{(n+1)}(\theta x) \quad \text{(A1.33)}$$

and for $\lim_{n \to \infty} E_n \to 0$

$$f(x) = \sum_{r=0}^{\infty} \frac{x^r}{r!}f^{(r)}(0) \quad \text{(A1.34)}$$

which is the Maclaurin series.

Exemplifying with the use of a Maclaurin series for $\exp(x)$ one obtains

$$\exp(x) = 1 + x + \frac{x^2}{2!} + \frac{x^3}{3!} + \cdots \quad \text{(A1.35)}$$

$$= \sum_{r=0}^{\infty} \frac{x^r}{r!} \quad \text{(A1.36)}$$

Other examples are to be found in Table A1.3.

Hyperbolic functions

Hyperbolic functions are combinations of exponentials. They are given in Table A1.4, and these functions are plotted in Fig. A1.4. Since they are continuous functions, with continuous derivatives obtained in the same way as normal trigonometric functions, that is

$$\frac{d}{dz}(\sinh z) = \cosh z \qquad \frac{d}{dz}(\cosh z) = \sinh z \quad \text{(A1.37)}$$

the appropriate Maclaurin series may be used (Table A1.3 gives examples).

Table A1.4. Hyperbolic functions

$\sinh x$	$\frac{1}{2}(e^x - e^{-x})$
$\cosh x$	$\frac{1}{2}(e^x + e^{-x})$
$\tanh x$	$\frac{\sinh x}{\cosh x} = \frac{e^x - e^{-x}}{e^x + e^{-x}}$
$\operatorname{csch} x$	$(\sinh x)^{-1}$
$\operatorname{sech} x$	$(\cosh x)^{-1}$
$\coth x$	$(\tanh x)^{-1}$

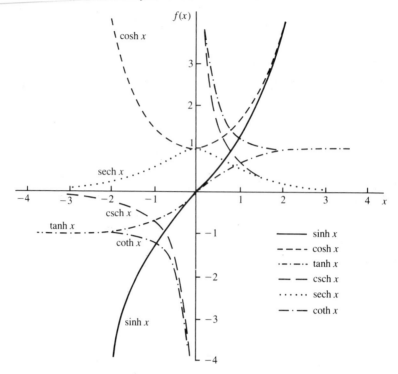

Fig. A1.4. Graphical representation of hyperbolic functions.

Hyperbolic functions frequently appear in electrochemical problems, for instance in the inversion of Laplace transforms. An important example of the use of the cosh function is in the expression for the differential capacity of the electrolyte double layer following the Gouy–Chapman model (Chapter 3) which has the minimum value and is symmetric around this minimum—compare Fig. 3.6 with the cosh function in Fig. A1.4.

Reference

1. M. Abramowitz and I. A. Stegun (ed.), *Handbook of mathematical functions*, Dover, New York, 1965.

APPENDIX 2

PRINCIPLES OF a.c. CIRCUITS

A2.1 Introduction

The electrochemical response to an a.c. perturbation is very important in impedance techniques (Chapter 11). This response cannot be understood without a knowledge of the fundamental principles of a.c. circuits[1], which is presented in this appendix.

We consider the application of a sinusoidal voltage

$$V = V_0 \sin \omega t \qquad (A2.1)$$

where V_0 is the maximum amplitude and ω the frequency (rad s^{-1}) to an electrical circuit that contains combinations of resistances and capacitances which, it is hoped, will adequately represent the electrochemical cell. The response is a current, given by

$$I = I_0 \sin (\omega t + \phi) \qquad (A2.2)$$

where ϕ is the phase angle between perturbation and response. The proportionality factor between V and I is the impedance, Z. Impedances consist of resistances, reactances (derived from capacitive elements) and inductances. Inductances will not be considered here, as for electrochemical cells, they only arise at very high frequencies (>1 MHz).

The problem of application of a sinusoidal current and response as a modulated potential obeys analogous mathematical considerations and will not be discussed.

A2.2 Resistance

In the case of a pure resistance, R, Ohm's law $V = IR$ leads to

$$I = \frac{V_0}{R} \sin \omega t \tag{A2.3}$$

and $\phi = 0$. There is no phase difference between potential and current.

A2.3 Capacitance

For a pure capacitor

$$I = C \frac{dV}{dt} \tag{A2.4}$$

Substituting for dV/dt, using (A2.1), one obtains

$$I = \omega C V_0 \sin \left(\omega t + \frac{\pi}{2} \right) \tag{A2.5}$$

$$= \frac{V_0}{X_C} \sin \left(\omega t + \frac{\pi}{2} \right) \tag{A2.6}$$

By comparing with (A2.2), we see that $\phi = \pi/2$, that is the current lags behind the potential by $\pi/2$. $X_C = (\omega C)^{-1}$ is known as the reactance (measured in ohms).

A2.4 Representation in the complex plane

Given the different phase angles of resistances and reactances described above, representation in two dimensions is useful. On the x-axis the

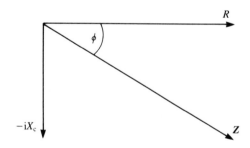

Fig. A2.1. Representation in the complex plane of an impedance containing resistive and capacitive components.

phase angle is zero; on rotating anticlockwise about the origin the phase angle increases; pure reactances are represented on the *y*-axis. The distance from the origin corresponds to the amplitude. This is precisely what is done with complex numbers as represented vectorially in the complex plane: here the real axis is for resistances and the imaginary axis for reactances. By convention, the current is always on the real axis. Thus it becomes necessary to multiply reactances by $-i$ ($\sqrt{-1}$) (Fig. A2.1).

A2.5   Resistance and capacitance in series

We exemplify the use of vectors in the complex plane with a resistance and capacitance in series (Fig. A2.2*a*). The total potential difference is the sum of the potential differences across the two elements. From Kirchhoff's law the currents have to be equal, that is

$$I = I_R = I_C \qquad (A2.7)$$

The differences in potential are proportional to R and X_C respectively. Their representation as vectors in the complex plane is shown in Fig. A2.1. The vectorial sum of $-iX_C$ and of R gives the impedance $\mathbf{Z}$. As a

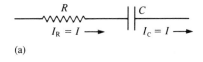

(a)

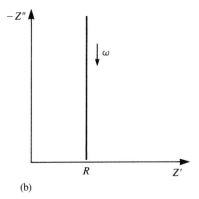

(b)

Fig. A2.2. Resistance and capacitance in series: (a) Electrical circuit; (b) Complex plane impedance plot.

vector, the impedance is

$$\mathbf{Z} = R - iX_C \tag{A2.8}$$

The magnitude of the impedance is

$$|\mathbf{Z}| = (R^2 + X_C^2)^{1/2} \tag{A2.9}$$

and the phase angle

$$\phi = \arctan \frac{|X_C|}{|R|} = \frac{1}{\omega RC} \tag{A2.10}$$

Often the in-phase component of the impedance is referred to as Z' and the out-of-phase component, i.e. at $\pi/2$, is called Z'', that is $\mathbf{Z} = Z' + iZ''$. Thus for this case

$$Z' = R, \qquad Z'' = -X_C \tag{A2.11}$$

This is a vertical line in the complex plane impedance plot, since Z' is constant but Z'' varies with frequency, as shown in Fig. A2.2b.

A2.6   Resistance and capacitance in parallel

The circuit is shown in Fig. A2.3a. The total current, I_{tot}, is the sum of the two parts, the potential difference across the two components being equal:

$$I_{tot} = \frac{V_0}{R} \sin \omega t + \frac{V_0}{X_C} \sin \left(\omega t + \frac{\pi}{2} \right) \tag{A2.12}$$

We need to calculate the vectorial sum of the currents shown in Fig. A2.3b. Thus

$$|I_{tot}| = (I_R^2 + I_C^2)^{1/2} \tag{A2.13}$$

$$= V \left(\frac{1}{R^2} + \frac{1}{X_C^2} \right)^{1/2} \tag{A2.14}$$

The magnitude of the impedance is

$$|\mathbf{Z}| = \left(\frac{1}{R^2} + \frac{1}{X_C^2} \right)^{-1/2} \tag{A2.15}$$

and the phase angle

$$\phi = \arctan \frac{I_C}{I_R} = \arctan \frac{1}{\omega RC} \tag{A2.16}$$

which is equal to the RC series combination (equation (A2.10)).

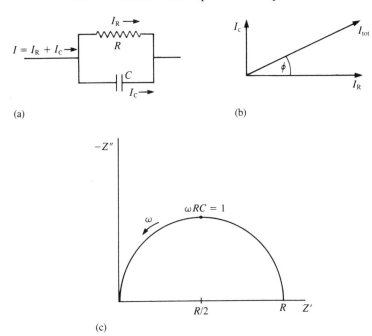

Fig. A2.3. Resistance and capacitance in parallel: (a) Electrical circuit; (b) Diagram showing the vectorial sum of the resistive and capacitive currents for a parallel RC combination; (c) Complex-plane impedance plot.

From (A2.12) we see that

$$\frac{1}{\mathbf{Z}} = \frac{1}{R} + i\omega C \tag{A2.17}$$

$$\mathbf{Z} = \frac{R}{1 + i\omega RC} \tag{A2.18}$$

This is easily separated into real and imaginary parts via multiplication by $(1 - i\omega RC)$. Thus

$$\mathbf{Z} = \frac{R(1 - i\omega RC)}{1 + (\omega RC)^2} \tag{A2.19}$$

and

$$Z' = \frac{R}{1 + (\omega RC)^2}, \qquad Z'' = \frac{-R^2 C}{1 + (\omega RC)^2} \tag{A2.20}$$

This is a semicircle in the complex plane (Fig. A2.3c), of radius $R/2$ and maximum value of $|Z''|$ defined by $\omega RC = 1$.

A2.7   Impedance in series and in parallel

Impedances can be combined in the same way as resistances:

$$\textit{in series}: \qquad \mathbf{Z} = \mathbf{Z}_1 + \mathbf{Z}_2 \tag{A2.21}$$

$$\textit{in parallel}: 1/\mathbf{Z} = 1/\mathbf{Z}_1 + 1/\mathbf{Z}_2 \tag{A2.22}$$

A2.8   Admittance

Admittance is the inverse of impedance, that is

$$\mathbf{Y} = 1/\mathbf{Z} \tag{A2.23}$$

It is represented by the symbol $\mathbf{Y}$, and can be especially useful in the analysis of parallel circuits, since admittances for elements in parallel are summed directly in the same way as one sums impedances for elements in series (and vice versa).

If the components of the impedance and admittance in phase are represented by Z' and Y' respectively and the components with phase angle of $\pi/2$ represented by Z'' and Y'', then

$$Y = \frac{1}{\mathbf{Z}} = \frac{1}{Z' + \mathrm{i}Z''} = \frac{Z' - \mathrm{i}Z''}{(Z')^2 + (Z'')^2} = Y' + \mathrm{i}Y'' \tag{A2.24}$$

Thus

$$Y' = \frac{Z'}{(Z)^2 + (Z'')^2} = G_\mathrm{p} \tag{A2.25}$$

and

$$Y'' = \frac{-Z''}{(Z)^2 + (Z'')^2} = B_\mathrm{p} = \omega C_\mathrm{p} \tag{A2.26}$$

where G_p is the conductance and B_p the susceptance.

A2.9   The Kramers–Kronig relations

An electrical system with linear properties does not generate harmonics in response to the perturbation signal, and the response to two or more superimposed excitation signals is equal to the sum of the two responses obtained by excitation independently. With electrochemical systems this linearity is possible to a good approximation for perturbations rather less than the thermal potential $(k_\mathrm{B}T/e) = 25$ mV at 298 K.

Since almost all equations used in impedance methods are derived assuming linearity, it is important to have some means of verifying this supposition. The Kramers–Kronig relations[2] link Z' with Z'' and allow the calculation of values for Z'' at any frequency from a knowledge of the full frequency spectrum of Z', and vice versa.

The relations are

$$Z'(\omega) - Z'(\infty) = \frac{2}{\pi} \int_0^\infty \frac{xZ''(x) - \omega Z''(\omega)}{x^2 - \omega^2} \, dx \qquad (A2.27)$$

$$Z'(\omega) - Z'(0) = \frac{2\omega}{\pi} \int_0^\infty \left[\frac{\omega}{x} Z''(x) - Z''(\omega) \right] \frac{1}{x^2 - \omega^2} \, dx \qquad (A2.28)$$

$$Z''(\omega) = -\left(\frac{2\omega}{\pi} \right) \int_0^\infty \frac{Z'(x) - Z'(\omega)}{x^2 - \omega^2} \, dx \qquad (A2.29)$$

$$\phi(\omega) = \frac{2\omega}{\pi} \int_0^\infty \frac{\ln |Z(x)|}{x^2 - \omega^2} \, dx \qquad (A2.30)$$

Although these have been applied to electrical systems for over 40 years, only recently have they been applied to electrochemical systems[3].

References

1. G. Lancaster, *Dc and ac circuits*, Oxford Physics Series, Oxford University Press, 1973.
2. H. W. Bode, *Network and feedback amplifier design*, van Nostrand, New York, 1945, Chapter 4.
3. D. D. Macdonald and M. Urquidi-Macdonald, *J. Electrochem. Soc.*, 1985, **132**, 2316.

APPENDIX 3

DIGITAL SIMULATION

A3.1 Introduction
A3.2 Simulation models
A3.3 Implicit methods

A3.1 Introduction

The theoretical solution to the equations for electrode processes nearly always has to involve approximations, not only for numerical but also for analytical solutions—such as, for example, the assumption that there is no convection within the diffusion layer of hydrodynamic electrodes. In other cases, of complex mechanism, it is not even possible to resolve the equations algebraically. There is another possibility for theoretical analysis, which is to simulate the electrode process digitally.

Digital simulation has two aims:

- to compare solutions obtained analytically and numerically and examine the agreement between the two methods

- to solve equations when algebraic solution is not possible.

The accuracy of the result obtained by simulation depends on the increments used in the variables (distance, time, concentration, etc.), and thence the computation time. Care must be taken not to introduce apparently small errors which can be propagated and grow along the simulation. Also, it should always be remembered that the experiment is correct; what may be wrong is our interpretation of the results, that is the simulation model.

Only an introduction to digital simulation will be given in this appendix. Extensive treatments exist, such as those in Refs. 1–4. In order to implement digital simulation FORTRAN has usually been employed.

A3.2 Simulation models

The method normally used in simulation is the finite difference method. The solution is divided into small volume elements within which concentrations are assumed to be uniform. Time is also incremented.

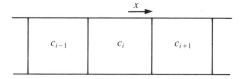

Fig. A3.1. Concentration elements for digital simulation.

The example to be given is semi-infinite linear diffusion to a planar electrode—concentration variation is only perpendicular to the electrode. Fick's second law is

$$\frac{\partial c}{\partial t} = D \frac{\partial^2 c}{\partial x^2} \tag{A3.1}$$

Three successive volume elements have concentrations c_{i-1}, c, and c_{i+1}, (Fig. A3.1), and the time increment is Δt. It is not difficult to show, by Taylor series expansion, that (A3.1) transforms into

$$\frac{c_i' - c_i}{\Delta t} = D \frac{(c_{i+1} - c_i)/\Delta x - (c_i - c_{i-1})/\Delta x}{\Delta x} \tag{A3.2}$$

$$= D \frac{c_{i-1} - 2c_i + c_{i+1}}{(\Delta x)^2} \tag{A3.3}$$

where c_i is the concentration in element i at time $(t + \Delta t)$. To do the computations boundary conditions have to be introduced, and particularly the initial conditions. It is also necessary to decide whether the starting point is the electrode surface or bulk solution.

The calculation can be done in two distinct ways:

1. Consider the elements as boxes—but should the electrode surface be in the middle of a box (these coordinates always have to be used in order to define the position of the box) or at the interface between two boxes? In the past the electrode surface has been put in the middle of a box, but this procedure has been criticized as not corresponding to physical reality.

2. Use as coordinates the points halfway between the interface of adjacent elements, calculating the variation of concentration with time at these points. In this case there is no problem with the positioning of the electrode surface, as it is at the edge of a box. The simulation formulae are the same. This method can be particularly advantageous for non-planar electrodes as it is not necessary to decide on the three-dimensional shape of the elements; only the position of the points matters.

This example is relatively simple. When there are convection, migration, or kinetic terms the simulation equations obviously become more complex.

Another method much used by engineers in solving heat transfer problems by digital simulation is the finite element method[5]; given the similarities with mass transfer in electrochemistry, it can also be used. After dividing the space into elements the variables for each element are defined by polynomials, the sum total for all space being known. An expression in matrix form permits the calculation of the values of the variables for each element. Up until now, applications have been few[6].

A3.3 Implicit methods

In the last section we considered explicit expressions which predict the concentrations in elements at $(t + \Delta t)$ from information at time t. An error is introduced due to asymmetry in relation to the simulation time. For this reason implicit methods, which predict what will be the next value and use this in the calculation, were developed. The version most used is the *Crank–Nicholson method*. Orthogonal collocation, which involves the resolution of a set of simultaneous differential equations, has also been employed. Accuracy is better, but computation time is greater, and the necessity of specifying the conditions can be difficult for a complex electrode mechanism. In this case the finite difference method is preferable[7].

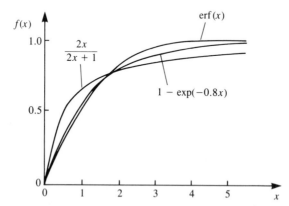

Fig. A3.2. Functions which reproduce variation of concentration with distance from an electrode. erf (x) is probably the most correct, but is more complicated to use in computation. In all cases, when $x \to \infty$, $f(x) \to 1$.

As implicit methods need much computation time it is important to minimize the number of elements as much as possible. Far from the electrode the variation in concentration is relatively small in comparison with the variation close to the electrode and, consequently, elements can be larger; this leads to elements of non-equal size or to non-equal intervals. It is useful to employ functions that reproduce the variation within the intervals to a good approximation. Some examples are shown in Fig. A3.2.

References

1. S. W. Feldberg, *Electroanalytical chemistry,* ed. A. J. Bard, Dekker, New York, Vol. 3, 1969, pp. 199–296.
2. D. Britz, *Digital simulation in electrochemistry,* Springer-Verlag, Berlin, 1981.
3. S. Pons, *Electroanalytical chemistry,* ed. A. J. Bard, Dekker, New York, Vol. 13, 1984, pp. 115–190.
4. J. Maloy, *Laboratory techniques in electroanalytical chemistry,* ed. P. T. Kissinger and W. R. Heinemann, Dekker, New York, 1984, Chapter 16.
5. S. S. Rao, *The finite element method in engineering,* Pergamon Press, New York, 1982.
6. J. Kwak and A. J. Bard, *Anal. Chem.,* 1989, **61,** 1221.
7. R. G. Compton, M. E. Laing, D. Mason, R. J. Northing, and P. R. Unwin, *Proc. R. Soc. Lond.,* 1988, **A418,** 113.

APPENDIX 4

STANDARD ELECTRODE POTENTIALS

The following is a list of standard electrode potentials of common half-reactions in aqueous solution, that is measured relative to the standard hydrogen electrode at 25°C (298.15 K) with all species at unit activity. Most of these values were taken from *Standard potentials in aqueous solution*, ed. A. J. Bard, R. Parsons, and J. Jordan, Dekker, New York, 1985, in which values for many other half-reactions may also be found.

	$E^{\ominus}/V$
$Ag^+ + e^- \rightarrow Ag$	+0.80
$Ag^{2+} + e^- \rightarrow Ag^+$	+1.98
$AgBr + e^- \rightarrow Ag + Br^-$	+0.07
$AgCl + e^- \rightarrow Ag + Cl^-$	+0.22
$AgI + e^- \rightarrow Ag + I^-$	−0.15
$Al^{3+} + 3e^- \rightarrow Al$	−1.68
$As + 3H^+ + 3e^- \rightarrow AsH_3$	−0.23
$As(OH)_3 + 3H^+ + 3e^- \rightarrow As + 3H_2O$	+0.24
$AsO(OH)_3 + 2H^+ + 2e^- \rightarrow As(OH)_3 + H_2O$	+0.56
$Au^+ + e^- \rightarrow Au$	+1.83
$Au^{3+} + 3e^- \rightarrow Au$	+1.52
$Ba^{2+} + 2e^- \rightarrow Ba$	−2.92
$Be^{2+} + 2e^- \rightarrow Be$	−1.97
$Br_2(l) + 2e^- \rightarrow 2Br^-$	+1.06
$Br_2(aq) + 2e^- \rightarrow 2Br^-$	+1.09
$BrO^- + H_2O + 2e^- \rightarrow Br^- + 2OH^-$	+0.76
$2HOBr + 2H^+ + 2e^- \rightarrow Br_2 + 2H_2O$	+1.60
$2BrO_3^- + 12H^+ + 10e^- \rightarrow Br_2 + 6H_2O$	+1.48
$BrO_4^- + 2H^+ + 2e^- \rightarrow BrO_3^- + H_2O$	+1.85
$CO_2 + 2H^+ + 2e^- \rightarrow CO + H_2O$	−0.11

$$E^{\ominus}/V$$

$$CO_2 + 2H^+ + 2e^- \rightarrow HCOOH \qquad -0.20$$
$$2CO_2 + 2H^+ + 2e^- \rightarrow H_2C_2O_4 \qquad -0.48$$
$$Ca^{2+} + 2e^- \rightarrow Ca \qquad -2.84$$
$$Cd(OH)_2 + 2e^- \rightarrow Cd + 2OH^- \qquad -0.82$$
$$Cd^{2+} + 2e^- \rightarrow Cd \qquad -0.40$$
$$Ce^{3+} + 3e^- \rightarrow Ce \qquad -2.34$$
$$Ce^{4+} + e^- \rightarrow Ce^{3+} \qquad +1.72$$
$$Cl_2 + 2e^- \rightarrow 2Cl^- \qquad +1.36$$
$$ClO^- + H_2O + 2e^- \rightarrow Cl^- + 2OH^- \qquad +0.89$$
$$2HOCl + 2H^+ + 2e^- \rightarrow Cl_2 + 2H_2O \qquad +1.63$$
$$HClO_2 + 2H^+ + 2e^- \rightarrow HOCl + H_2O \qquad +1.68$$
$$ClO_3^- + 3H^+ + 2e^- \rightarrow HClO_2 + H_2O \qquad +1.18$$
$$ClO_3^- + 2H^+ + e^- \rightarrow ClO_2 + H_2O \qquad +1.17$$
$$ClO_4^- + 2H^+ + 2e^- \rightarrow ClO_3^- + H_2O \qquad +1.20$$
$$Co^{2+} + 2e^- \rightarrow Co \qquad -0.28$$
$$Co^{3+} + e^- \rightarrow Co^{2+} \qquad +1.92$$
$$Co(NH_3)_6^{3+} + e^- \rightarrow Co(NH_3)_6^{2+} \qquad +0.06$$
$$Co(phen)_3^{3+} + e^- \rightarrow Co(phen)_3^{2+} \qquad +0.33$$
$$Co(C_2O_4)_3^{3-} + e^- \rightarrow Co(C_2O_4)_3^{4-} \qquad +0.57$$
$$Cr^{2+} + 2e^- \rightarrow Cr \qquad -0.90$$
$$Cr_2O_7^{2-} + 14H^+ + 6e^- \rightarrow 2Cr^{3+} + 7H_2O \qquad +1.38$$
$$Cr^{3+} + 3e^- \rightarrow Cr \qquad -0.74$$
$$Cs^+ + e^- \rightarrow Cs \qquad -2.92$$
$$Cu^+ + e^- \rightarrow Cu \qquad +0.52$$
$$Cu^{2+} + 2e^- \rightarrow Cu \qquad +0.34$$
$$Cu^{2+} + e^- \rightarrow Cu^+ \qquad +0.16$$
$$CuCl + e^- \rightarrow Cu + Cl^- \qquad +0.12$$
$$Cu(NH_3)_4^{2+} + 2e^- \rightarrow Cu + 4NH_3 \qquad -0.00$$
$$F_2 + 2e^- \rightarrow 2F^- \qquad +2.87$$
$$Fe^{2+} + 2e^- \rightarrow Fe \qquad -0.44$$
$$Fe^{3+} + 3e^- \rightarrow Fe \qquad -0.04$$
$$Fe^{3+} + e^- \rightarrow Fe^{2+} \qquad +0.77$$
$$Fe(phen)^{3+} + e^- \rightarrow Fe(phen)^{2+} \qquad +1.13$$
$$Fe(CN)_6^{3-} + e^- \rightarrow Fe(CN)_6^{4-} \qquad +0.36$$
$$Fe(CN)_6^{4-} + 2e^- \rightarrow Fe + 6CN^- \qquad -1.16$$
$$2H^+ + 2e^- \rightarrow H_2 \qquad 0 \text{ (by definition)}$$

	$E^{\ominus}/V$
$2H_2O + 2e^- \rightarrow H_2 + 2OH^-$	-0.83
$H_2O_2 + H^+ + e^- \rightarrow HO\cdot + H_2O$	$+0.71$
$H_2O_2 + 2H^+ + 2e^- \rightarrow 2H_2O$	$+1.76$
$Hg_2^{2+} + 2e^- \rightarrow 2Hg$	$+0.80$
$Hg_2Cl_2 + 2e^- \rightarrow 2Hg + 2Cl^-$	$+0.27$
$Hg^{2+} + 2e^- \rightarrow Hg$	$+0.86$
$2Hg^{2+} + 2e^- \rightarrow Hg_2^{2+}$	$+0.91$
$Hg_2SO_4 + 2e^- \rightarrow 2Hg + SO_4^{2-}$	$+0.62$
$I_2 + 2e^- \rightarrow 2I^-$	$+0.54$
$I_3^- + 2e^- \rightarrow 3I^-$	$+0.53$
$2HOI + 2H^+ + 2e^- \rightarrow I_2 + 2H_2O$	$+1.44$
$2IO_3^- + 12H^+ + 10e^- \rightarrow I_2 + 6H_2O$	$+1.20$
$IO(OH)_5 + H^+ + e^- \rightarrow IO_3^- + 3H_2O$	$+1.60$
$In^+ + e^- \rightarrow In$	-0.13
$In^{3+} + 2e^- \rightarrow In^+$	-0.44
$In^{3+} + 3e^- \rightarrow In$	-0.34
$K^+ + e^- \rightarrow K$	-2.93
$Li^+ + e^- \rightarrow Li$	-3.04
$Mg^{2+} + 2e^- \rightarrow Mg$	-2.36
$Mn^{2+} + 2e^- \rightarrow Mn$	-1.18
$Mn^{3+} + e^- \rightarrow Mn^{2+}$	$+1.51$
$MnO_2 + 4H^+ + 2e^- \rightarrow Mn^{2+} + 2H_2O$	$+1.23$
$MnO_4^- + 8H^+ + 5e^- \rightarrow Mn^{2+} + 4H_2O$	$+1.51$
$MnO_4^- + e^- \rightarrow MnO_4^{2-}$	$+0.56$
$MoO_4^{2-} + 4H_2O + 6e^- \rightarrow Mo + 8OH^-$	-0.91
$NO_3^- + 2H^+ + e^- \rightarrow NO_2 + H_2O$	$+0.80$
$NO_3^- + 4H^+ + 3e^- \rightarrow NO + 2H_2O$	$+0.96$
$NO_3^- + H_2O + 2e^- \rightarrow NO_2^- + 2OH^-$	$+0.01$
$Na^+ + e^- \rightarrow Na$	-2.71
$Ni^{2+} + 2e^- \rightarrow Ni$	-0.257
$Ni(OH)_2 + 2e^- \rightarrow Ni + 2OH^-$	-0.72
$NiO_2 + 2e^- \rightarrow Ni^{2+} + 2H_2O$	$+1.59$
$O_2 + 2H_2O + 4e^- \rightarrow 4OH^-$	$+0.40$

	$E^{\ominus}/V$
$O_2 + 4H^+ + 4e^- \rightarrow 2H_2O$	$+1.23$
$O_2 + e^- \rightarrow O_2^-$	-0.33
$O_2 + H_2O + 2e^- \rightarrow HO_2^- + OH$	-0.08
$O_2 + H^+ + e^- \rightarrow HO_2$	-0.13
$O_2 + 2H^+ + 2e^- \rightarrow H_2O_2$	$+0.70$
$P + 3H^+ + 3e^- \rightarrow PH_3$	-0.06
$HPO(OH)_2 + 3H^+ + 3e^- \rightarrow P + 3H_2O$	-0.50
$HPO(OH)_2 + 2H^+ + 2e^- \rightarrow H_2PO(OH) + H_2O$	-0.50
$PO(OH)_3 + 2H^+ + 2e^- \rightarrow HPO(OH)_2 + H_2O$	-0.28
$Pb^{2+} + 2e^- \rightarrow Pb$	-0.13
$PbO_2 + 4H^+ + 2e^- \rightarrow Pb^{2+} + 2H_2O$	$+1.70$
$PbSO_4 + 2e^- \rightarrow Pb + SO_4^{2-}$	-0.36
$Pt^{2+} + 2e^- \rightarrow Pt$	$+1.19$
$Rb^+ + e^- \rightarrow Rb$	-2.93
$S + 2e^- \rightarrow S^{2-}$	-0.48
$2SO_2(aq) + 2H^+ + 4e^- \rightarrow S_2O_3^{2-} + H_2O$	-0.40
$SO_2(aq) + 4H^+ + 4e^- \rightarrow S + 2H_2O$	$+0.50$
$S_4O_6^{2-} + 2e^- \rightarrow 2S_2O_3^{2-}$	$+0.08$
$SO_4^{2-} + H_2O + 2e^- \rightarrow SO_3^{2-} + 2OH^-$	-0.94
$2SO_4^{2-} + 4H^+ + 2e^- \rightarrow S_2O_6^{2-} + 2H_2O$	-0.25
$S_2O_8^{2-} + 2e^- \rightarrow 2SO_4^{2-}$	$+1.96$
$Sn^{2+} + 2e^- \rightarrow Sn$	-0.14
$Sn^{4+} + 2e^- \rightarrow Sn^{2+}$	$+0.15$
$Sr^{2+} + 2e^- \rightarrow Sr$	-2.89
$Ti^{2+} + 2e^- \rightarrow Ti$	-1.63
$Ti^{3+} + e^- \rightarrow Ti^{2+}$	-0.37
$TiO^{2+} + e^- \rightarrow Ti^{3+}$	$+0.10$
$Tl^+ + e^- \rightarrow Tl$	-0.34
$V^{2+} + 2e^- \rightarrow V$	-1.13
$V^{3+} + e^- \rightarrow V^{2+}$	-0.26
$VO^{2+} + 2H^+ + 2e^- \rightarrow V^{3+} + H_2O$	$+0.34$
$VO_2^+ + 2H^+ + e^- \rightarrow VO^{2+} + H_2O$	$+1.00$
$Zn^{2+} + 2e^- \rightarrow Zn$	-0.76
$Zn(OH)_4^{2-} + 2e^- \rightarrow Zn + 4OH^-$	-1.29

INDEX